LACTOGENIC HORMONES

*A Ciba Foundation Symposium
in memory of Professor S. J. Folley*

Edited by
G. E. W. WOLSTENHOLME
and
JULIE KNIGHT

CHURCHILL LIVINGSTONE
Edinburgh and London
1972

First published 1972

Containing 120 illustrations

I.S.B.N. 0 443 00871

© Longman Group Ltd. 1972

All rights reserved. No part of this publication may be reproduced, stored in a retrieval system, or transmitted, in any form or by any means, electronic, mechanical, photocopying, recording or otherwise, without the prior permission of the copyright owner.

Any correspondence in connection with this volume should be directed to the publishers at 104 Gloucester Place, London, W1H 4AE.

Printed in Great Britain

WOLSTENHOLME, GORDON ETHELB
LACTOGENIC HORMONES: A CIBA F
000077118

HCL   R35.C56.S98/V.102

THE UNIVERSITY OF LIVERPOOL
HAROLD COHEN LIBRARY

CONDITIONS OF BORROWING

Members of Council, members and retired members of the University staff, and students registered with the University for higher degrees — 20 volumes for one month. All other readers entitled to borrow — 6 volumes for 14 days in term or for the vacation.

Books may be recalled after one week for the use of another reader.

LACTOGENIC HORMONES

Photograph by Peter Grugeon

SYDNEY JOHN FOLLEY, F.R.S.

1906–1970

(*Frontispiece*)

Contents

J. C. Beck
    Introduction   1

C. H. Li
    Recent knowledge of the chemistry of lactogenic hormones   7
    Discussion   22

L. M. Sherwood, S. Handwerger and W. D. McLaurin
    The structure and function of human placental lactogen   27
    Discussion   45

R. M. MacLeod and J. E. Lehmeyer
    Regulation of the synthesis and release of prolactin   53
    Discussion   76

H. Friesen, C. Belanger, H. Guyda and P. Hwang
    The synthesis and secretion of placental lactogen and pituitary prolactin   83
    Discussion   103

R. W. Turkington
    Molecular biological aspects of prolactin   111
    Discussion   127

A. G. Frantz, D. L. Kleinberg and G. L. Noel
    Physiological and pathological secretion of human prolactin studied by *in vitro* bioassay   137

I. A. Forsyth
    Use of a rabbit mammary gland organ culture system to detect lactogenic activity in blood   151

R. W. Turkington
    Measurement of prolactin activity in human serum by the induction of specific milk proteins *in vitro*: results in various clinical disorders   169
    Discussion   184

G. D. BRYANT and F. C. GREENWOOD
    The concentrations of human prolactin in plasma measured by radioimmunoassay: experimental and physiological modifications   197

T. M. SILER, L. L. MORGENSTERN and F. C. GREENWOOD
    The release of prolactin and other peptide hormones from human anterior pituitary tissue cultures   207
    Discussion   217

W. N. SPELLACY
    Immunoassay of human placental lactogen: physiological studies in normal and abnormal pregnancy   223
    Discussion   235

J. L. PASTEELS
    Morphology of prolactin secretion   241

C. S. NICOLL
    Secretion of prolactin and growth hormone by adenohypophyses of rhesus monkeys *in vitro*   257

J. L. PASTEELS
    Tissue culture of human hypophyses: evidence of a specific prolactin in man   269
    Discussion   277

J. SWANSON BECK
    Immunofluorescence studies on the adenohypophysis in pregnancy   287
    Discussion   294

C. A. NICOLL and H. A. BERN
    On the actions of prolactin among the vertebrates: is there a common denominator?   299
    Discussion   317

J. MEITES
    Hypothalamic control of prolactin secretion   325
    Discussion   338

M. APOSTOLAKIS, S. KAPETANAKIS, G. LAZOS and A. MADENA-PYRGAKI
    Plasma prolactin activity in patients with galactorrhoea after treatment with psychotropic drugs   349
    Discussion   354

E. E. McGarry and J. C. Beck
    Biological effects of non-primate prolactin and human placental lactogen   361
    Discussion   383

    General Discussion   391

J. C. Beck
    Conclusions   403

Index of Contributors   405

Subject Index   406

Contributors

Symposium on Lactogenic Hormones,
held 11th–13th May 1971

J. C. Beck (Chairman)	McGill University Clinic, Royal Victoria Hospital, 687 Pine Avenue West, Montreal 112, P.Q., Canada
M. Apostolakis*	Institute of Experimental Physiology, Faculty of Medicine, Aristotelion University of Thessaloniki, Thessaloniki, Greece
J. Swanson Beck	Department of Pathology, The University, Dundee DD1 4HN, Scotland
G. D. Bryant	Department of Biochemistry and Biophysics, University of Hawaii, Snyder Hall, 2538 The Mall, Honolulu, Hawaii
P. M. Cotes	MRC Division of Biological Standards, National Institute for Medical Research, Mill Hill, London N.W.7, England
A. T. Cowie	Department of Physiology, National Institute for Research in Dairying, Shinfield, Reading, RG2 9AT, England
R. Denamur	Laboratoire de Physiologie de la Lactation, Institut National de la Recherche Agronomique, 78-Jouey-en-Josas, France
I. A. Forsyth	National Institute for Research in Dairying, Shinfield, Reading RG2 9AT, England
A. G. Frantz	Department of Medicine, College of Physicians and Surgeons of Columbia University, 630 West 168th Street, New York, N.Y. 10032, U.S.A.
H. Friesen	Royal Victoria Hospital, 687 Pine Avenue West, Montreal 112, P.Q., Canada

* *In absentia*

F. C. Greenwood	Department of Biochemistry and Biophysics, University of Hawaii, Snyder Hall, 2538 The Mall, Honolulu, Hawaii
M. Herlant	Laboratoire d'histologie, Faculté de Médicine et de Pharmacie, Université Libre de Bruxelles, 97 rue aux Laines, Bruxelles, Belgium
C. H. Li	Hormone Research Laboratory, University of California, San Francisco, California 94122, U.S.A.
R. M. MacLeod	Division of Cancer Studies, University of Virginia Medical Center, Charlottesville, Virginia 22901, U.S.A.
J. Meites	Department of Physiology, Giltner Hall, Michigan State University, East Lansing, Michigan 48823, U.S.A.
C. S. Nicoll	Department of Physiology-Anatomy, University of California, Berkeley, California 94720, U.S.A.
J. L. Pasteels	Laboratoires d'histologie et de microscopie électronique, Université Libre de Bruxelles, 97 rue aux Laines, Bruxelles, Belgium
F. J. A. Prop	Pathologisch-Anatomisch Laboratorium, Academisch Ziekenhuis, Eerste Helmersstraat 104, Amsterdam-Oud west, Holland
L. M. Sherwood	Endocrine Unit, Beth Israel Hospital, 330 Brookline Avenue, Boston, Massachusetts 02215, U.S.A.
R. V. Short	Department of Veterinary Clinical Studies, School of Veterinary Medicine, University of Cambridge, Madingley Road, Cambridge CB3 0ES, England
W. N. Spellacy	Department of Obstetrics-Gynecology, School of Medicine, University of Miami, Florida 33152, U.S.A.
R. W. Turkington	Department of Medicine, University of Wisconsin, Madison, Wisconsin 53706, U.S.A.
A. E. Wilhelmi	Department of Biochemistry, Woodruff Medical Center of Emory University, Atlanta, Georgia 30322, U.S.A.

The Ciba Foundation

The Ciba Foundation was opened in 1949 to promote international cooperation in medical and chemical research. It owes its existence to the generosity of CIBA Ltd, Basle (now CIBA-GEIGY Ltd), who, recognizing the obstacles to scientific communication created by war, man's natural secretiveness, disciplinary divisions, academic prejudices, distance, and differences of language, decided to set up a philanthropic institution whose aim would be to overcome such barriers. London was chosen as its site for reasons dictated by the special advantages of English charitable trust law (ensuring the independence of its actions), as well as those of language and geography.

The Foundation's house at 41 Portland Place, London, has become well known to workers in many fields of science. Every year the Foundation organizes six to ten three-day symposia and three or four shorter study groups, all of which are published in book form. Many other scientific meetings are held, organized either by the Foundation or by other groups in need of a meeting place. Accommodation is also provided for scientists visiting London, whether or not they are attending a meeting in the house.

The Foundation's many activities are controlled by a small group of distinguished trustees. Within the general framework of biological science, interpreted in its broadest sense, these activities are well summed up by the motto of the Ciba Foundation: *Consocient Gentes*—let the peoples come together.

INTRODUCTION

J. C. BECK

It is my very real pleasure, as Chairman of this meeting, to pay special tribute to Professor S. J. Folley, the man to whom this symposium is dedicated. His untimely death was a great loss and it is our profound regret that he did not live to join us in these deliberations.

This opportunity to pay tribute to one of the giants of endocrinology is an honour and privilege. The work of Sydney John Folley and his associates has laid much of the groundwork that brings this family of acquaintances and admirers together at this symposium. Inevitably, advances in technology and a new generation of creative younger scientists 'standing upon the shoulders of past giants' move the frontiers of our knowledge forward by a major stride, and I believe that in the course of the symposium we see this phenomenon demonstrated to an unusual degree. Observations made on the lactogenic hormones during the last quarter of a century are suddenly being brought into fresh focus. These observations are simultaneously generating new knowledge on the role of lactogenic hormones in health and disease and contributing unique information on the relationships between the structure and function of a group of polypeptide hormones at the molecular level.

At this point I would like to comment on 'relevant' research. It is almost irreverent today not to use this progressively more hackneyed term, but if there is one thing that the history of science has taught us, it is that no generation in its time is necessarily competent to judge which science of its day will ultimately be highly relevant or what datum may eventually prove 'the missing link' in the solution of some vital problem. It is upon the premises that were established 10, 20, 50 and 100 years or more ago that a so-called breakthrough or giant 'stride' suddenly takes place. It is difficult at a given time to know what is relevant to what.

At best we can only judge whether a worker is competent to do the job for which he seeks support, whether he is an honest individual, whether he works with care and meticulous detail, and whether he has built all the appropriate controls into his experimental design so that his work can be interpreted. If all these qualities are present, it is my belief that sooner or later the data obtained not only will be relevant, but will be highly useful. They will represent one more contribution to our culture's intellectual 'bank' and may one day provide an answer to what otherwise might remain a puzzle. I believe that in John Folley such a scientific personality is exemplified, and that in the course of this symposium we see an example of how some things suddenly fit together. After long years of toil in many places with a variety of tools and the accumulation of many data, the stage has finally been set.

Sydney John Folley obtained the degree of B.Sc. with first class honours in chemistry at the University of Manchester in 1927, and was awarded the Mercer Scholarship in chemistry. He then took an M.Sc. degree for research in colloid chemistry and joined the Department of Physiology as Research Assistant to H. S. Raper, who had recently succeeded Professor A. V. Hill as the Head of the Department. In 1931, he moved to Liverpool as an Assistant Lecturer in Biochemistry.

In 1932 he had to relinquish his teaching appointment at the University of Liverpool because of ill health, and through the influence of H. S. Raper went to the National Institute for Research in Dairying at Shinfield. At that time the Physiology Department occupied a converted back bedroom in the Manor House. The staff consisted of a biochemist from Oxford, G. L. Peskett, and a technical assistant, S. C. Watson. This is a far cry from the department which existed at John Folley's death—it is a modern facility with excellent equipment, and a staff which, together with visiting workers, sometimes exceeded fifty.

Initially in his new task, John Folley assisted Peskett in studies concerned with the relationship between blood electrolytes and the secretion of the lipid constituents of milk. When Peskett left the Institute in 1934, Folley wanted to carry on the physiological work, and with the new Director, Dr H. D. Kay, who was appointed in 1933, Folley collaborated in studies on the properties of the alkaline phosphatase of the mammary gland— his first venture into a new field which was later to become his life's work. Kay had recently returned from Canada and it was under his influence that Folley became interested in the relationship between the thyroid gland and lactation. This was his initiation into endocrinology; he rapidly became aware of the changes that were occurring in this field, particularly in the relationship between the knowledge of the anterior pituitary and the gonadal hormones and their influence on reproduction. His interest in the role and function of the mammary gland and in reproductive physiology was first aroused by A. S. Parkes' book entitled *The Internal Secretions of the Ovary*, and shortly thereafter Folley came into contact with the extremely active group headed by Parkes at the National Institute for Medical Research, then at Hampstead. This led to Folley's first experiments on the effects of the oestrogenic hormones on the composition of the milk of cattle, and brought about the first of many contributions Folley made over his lifetime on the hormonal control of lactation.

Through his acquaintance with F. G. Young, one of A. S. Parkes' colleagues at Hampstead, Folley became interested in the anterior pituitary hormone, prolactin. At the time, Frank Young was primarily concerned with the effect of anterior pituitary hormones on carbohydrate metabolism, but their combined interest in the effects of anterior pituitary hormones on the mammary glands resulted in productive collaboration for a number of years. They were the first to suggest the concept of co-lactogens—that is, a complex of anterior pituitary hormones responsible for lactogenesis as well as for the maintenance of lactation. The role of anterior pituitary growth hormone in lactation led to the discovery of the galactopoietic action of highly purified bovine growth hormone. It is of interest that some years later a group of workers, headed by Dr A. T. Cowie in Folley's laboratory, first made the critical observation that after pituitary stalk section the pituitary continues to secrete prolactin but not the other anterior pituitary tropic hormones. This latter work also emphasized that prolactin and growth hormone were the most critical anterior pituitary hormones for the maintenance of lactation in the goat.

In the middle 1930's, F. H. A. Marshall invited Folley to contribute a chapter on lactation for a treatise on the physiology of reproduction. At a meeting of the contributors, Folley came into contact with other distinguished reproductive physiologists, among whom was Solly Zuckerman. Folley and Zuckerman joined forces in a study of the control of mammary growth in the rhesus monkey, and correlated this with the careful observations which Zuckerman had made on the menstrual cycle of primates. This began a long series of studies by members of Folley's department on various aspects of mammary growth and development in a wide variety of animal species, with a long-term objective of describing the conditions under which mammary glands might be grown by artificial means. During this time Folley repeatedly emphasized the importance of the traditional internal glandular morphology of the mammary gland.

During the Second World War, the major interests of the department concerned projects in the field of physiology of lactation and reproduction which might have practical applications to the efficiency of dairy farming. One of these projects was concerned with the hormonal stimulation of udder growth and milk secretion, whereby it was hoped that sterile cows and heifers might be brought into milk production. With a gift of stilboestrol from Charles E. Dodds, the Folley group were the first to show that its effects in the cow with respect to milk composition were comparable to those of natural oestrogens.

In the last decade the interests of Folley and his associates were in the neuroendocrine control of those hormones known to play an important role in lactation. Although they concentrated on two main themes—the factors concerned with the secretion of prolactin, and those involved in the mechanisms responsible for the removal of milk from the mammary gland during suckling—the interests of the group were mainly concerned with the latter. In the late 1950's the group developed a sensitive and specific method for estimating oxytocin, the milk ejection hormone, in biological fluids. The method depended upon measuring the milk ejection pressure in the cannulated mammary gland of the anaesthetized lactating guinea pig after retrograde arterial injection of the test solution; this enabled the group to study the release of oxytocin in the jugular blood of cows during machine milking, of sows during suckling and of goats during suckling and hand milking, and showed that the release of the hormone in the cow could be conditioned by auditory and visual stimuli.

The concept of species specificity of anterior pituitary protein hormones led Folley and his associates, together with Carl Gemzell, to study the prolactin-like activity of various preparations of human growth hormone. They found, as did others, that human growth hormone possessed weak crop gland-stimulating properties, but they went on to show that, by the mammary intraductal test for lactogenic activity in pseudopregnant rabbits, human growth hormone was as active as purified sheep prolactin. This series of observations together with some made by others caused uncertainty as to whether prolactin existed as a separate lactogenic hormone in primates as it does in animals such as the sheep and ox. The whole question of the possible existence of a separate human prolactin and the role of human placental lactogen in lactogenesis was raised as a fruitful area for future research by Forsyth and Folley in a recent paper (1970).

Folley was awarded the degree of D.Sc. by his university in 1940 and he was elected a Fellow of the Royal Society in 1951. In 1964 he was given the title of Research Professor in the University of Reading and in the same year he was awarded an honorary doctorate by the University of Ghent. In 1969 he was awarded the Dale Medal by the Society for Endocrinology. The medal was presented by Professor H. Heller at the Meeting House of the Zoological Society of London after Folley had delivered the annual Dale Lecture. His international contacts were many and varied and he maintained a keen interest in organizations fostering the study of endocrinology. He was a distinguished member of the Society for Endocrinology and one of its founder members; he was the Society's first secretary from 1946 to 1951 and was Chairman from 1951 to 1956. He attended to the affairs of the *Journal of Endocrinology* with great energy and devotion and was Chairman of the Editorial Board from 1959 until his death.

John Folley's contacts with the Ciba Foundation have been many. He served as a member of the Ciba Foundation Scientific Advisory Panel from 1953 until his death. He participated in the first eight colloquia on endocrinology and in three symposia (*Isotopes in Biochemistry*, 1951; *Toxaemias of Pregnancy*, 1950; and *Mammalian Germ Cells*, 1952). It was through the aid of the Foundation that his classic monograph *Recherches récentes sur la physiologie et la biochimie de la sécrétion lactée* was translated into English (*The Physiology and Biochemistry of Lactation*, 1956).

In the time that I knew Folley, I was always reminded of the parallelism between his extra-scientific interests and those of J. S. L. Browne. Both had been beset through much of their careers with serious health problems and in later years with near blindness. Both had a deep and abiding appreciation of music and painting, and both had the same favourite composers. They both prided themselves on the capabilities of their high-fidelity equipment and would set the volume control at a level appropriate for a major concert hall. Folley's interests in art included the modern and he was a devotee of Picasso.

John Folley died on the 29th June 1970 after a brief illness and his untimely demise has saddened all who knew him. His career was a distinguished one and the dedication of this symposium to his honour is a fitting tribute to his major contributions to our understanding of the biochemistry and physiology of the lactogenic hormones.

Before closing, I want to comment briefly on the nomenclature of one of the lactogenic hormones discussed in the symposium. The new protein hormone isolated from human placenta was named human placental lactogen (HPL) by Josimovich and MacLaren (1962) and chorionic growth hormone-prolactin (CGP) by Kaplan and Grumbach (1964a, b). Subsequently it was designated as purified placental protein (human) (PPP(H)) by Bell and his associates (Florini *et al.* 1966) and placental protein, the most non-commital description of all, by Friesen (1965).

As a result of a discussion during a Round Table Conference on Human Placental Lactogen held at the University of Siena in September 1967, Li, Grumbach, Kaplan, Josimovich, Friesen and Catt proposed the name 'human chorionic somatomammotropin' (HCS) (Li *et al.* 1968). This was an attempt to prevent further confusion by the use of different terms for the same hormonal agent. Since the hormone was located in the syncytio-trophoblastic layer of the human placenta, according to immunofluorescence studies, and since it has both growth hormone (somatotropin) and lactogenic hormone (mammotropin) activities, the term was in line with the established name for the other gonadotropin produced by the human placenta, human chorionic gonadotropin. 'HCS' thus indicated the origin of the hormone as well as the known biological properties. However, this name has not been entirely satisfactory or acceptable in practice, and readers will find on pp. 400-402 a discussion of the advantages of retaining, at least until we know more about its functions, the original name of human placental lactogen (HPL). In accordance with this view, HPL has been adopted as consistently as is feasible in this volume.

REFERENCES

FLORINI, J. R., TONELLI, G., BREUER, C. B., COPPOLA, J., RINGLER, I. and BELL, P. H. (1966) *Endocrinology* **79,** 692–708.

FOLLEY, S. J. (1956) *The Physiology and Biochemistry of Lactation*. Springfield, Ill.: Thomas.

FORSYTH, I. A. and FOLLEY, S. J. (1970) In *Ovo-implantation: Human Gonadotropins and Prolactin*, pp. 266–278, ed. Hubinont, P. O., Leroy, F., Robyn, C. and Leleux, P. Basel: Karger.

FRIESEN, H. (1965) *Endocrinology* **76,** 369–381.

JOSIMOVICH, J. B. and MACLAREN, J. A. (1962) *Endocrinology* **71,** 209–220.

KAPLAN, S. L. and GRUMBACH, M. M. (1964a) *J. Clin. Endocrinol. & Metab.* **24,** 80–100.

KAPLAN, S. L. and GRUMBACH, M. M. (1964b) *Trans. N.Y. Acad. Sci.* **27,** 167–188.

LI, C. H., GRUMBACH, M. M., KAPLAN, S. L., JOSIMOVICH, J. B., FRIESEN, H. and CATT, K. J. (1968) *Experientia* **24,** 1288.

RECENT KNOWLEDGE OF THE CHEMISTRY OF LACTOGENIC HORMONES

Choh Hao Li

Hormone Research Laboratory, University of California, San Francisco

THERE are three lactogenic hormones which have been isolated and chemically characterized from three different sources, namely the human pituitary, ovine pituitary and human placenta. These are: human growth hormone (HGH), ovine prolactin and human chorionic somatomammotropin (HCS; human placental lactogen, HPL). This paper summarizes briefly some aspects of the chemistry of these three lactogenic hormones.

HUMAN PITUITARY GROWTH HORMONE

HGH was isolated and characterized (Li and Papkoff 1956; Li 1957a) fifteen years ago and is a protein of molecular weight 21 500 (Li and Starman 1964) and isoelectric point pH 4·9 (Li 1957a). Some physicochemical properties of HGH are presented in Table I. It consists of a single polypeptide

TABLE I

SOME PHYSICOCHEMICAL PROPERTIES OF HGH, OVINE PROLACTIN AND HCS

Properties	HGH	Ovine prolactin	HCS
Molecular weight	21 500	23 300	21 600
Isoelectric point, pH	4·9	5·7	
Sedimentation coefficient, $s_{20,w}$	2·18	2·19	
Diffusion coefficient, $D_{20} \times 10^7$	8·88	8·44	
$[\alpha]_D^{25}$ (0·1M-acetic acid)	$-39°$	$-41°$	
Ellipticity $[\theta]$, at 221 nm	$-19\,700$	$-21\,400$	$-16\,700$
α-Helix content, percentage	55	55	45
pK_a of tyrosine residues	10·8	11·2	10·9
$E_{1cm}^{0.1}$ at 277 nm	0·931	0·894	0·822

chain with one tryptophan residue and two disulphide bridges (Li and Papkoff 1956; Li 1957a). The complete amino acid sequence (Li, Liu and Dixon 1966; Li, Dixon and Liu 1969) of HGH was first proposed in 1966 to be a single chain protein of 188 amino acids with the tryptophan residue at position 25 and the two disulphide bridges formed by residues 68–162 and 179–186, as shown in Fig. 1. This proposed structure has

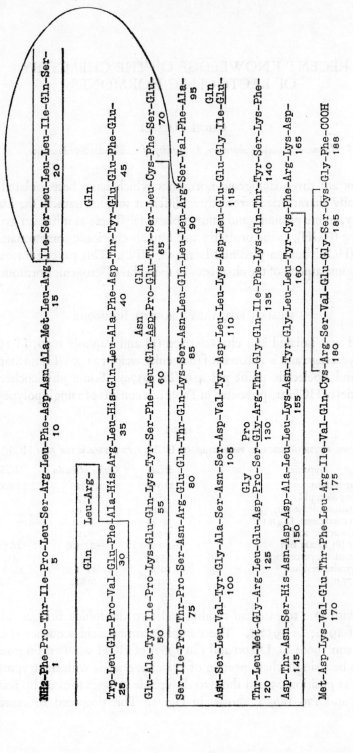

Fig. 1. The amino acid sequence of HGH (1966).

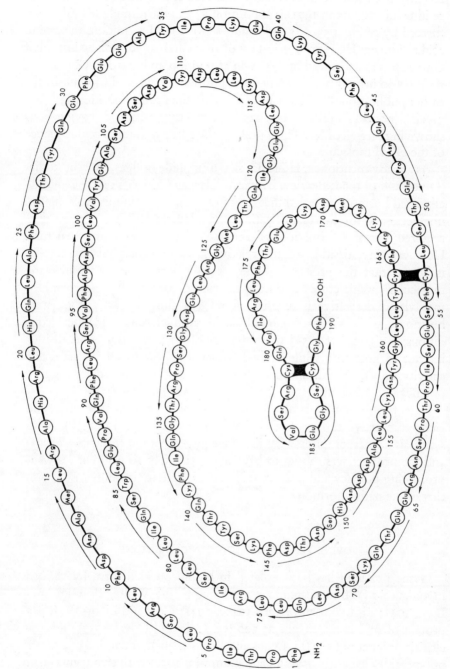

FIG. 2. The amino acid sequence of HGH (1971).

recently been re-investigated and it has been found that the tryptophan*
residue (Li and Dixon 1971) is at position 85 and the disulphide bridges are
formed by residues 53–164 and residues 181–188. In addition, the amino
acid residues in HGH are 190 instead of 188; Gln is at position 29 instead of
Glu; Asn is at 47 instead of Asp, Gln at position 49 instead of Glu, Gln at
position 90 instead of Glu and Gln at position 121 instead of Glu. In the
earlier publication (Li, Dixon and Liu 1969), the proline (position 130) and
glycine (position 132) residues were incorrectly positioned, having been
shown in reversed order. Fig. 2 presents the corrected amino acid sequence
of the HGH molecule.

As already mentioned, HGH has two disulphide bridges. Recent studies
(Dixon and Li 1966; Bewley, Brovetto-Cruz and Li 1969) on the reduced-
alkylated hormone revealed that the disulphide bonds in this molecule are
not necessary for the manifestation of biological activity, nor are they
required for the formation of the secondary and tertiary structure.
However, investigation of the relative rates of proteolysis by trypsin
indicated that the presence of these bonds does serve to stabilize the
molecular architecture against perturbing forces. The carbamidomethylated derivative is digested about 1·5 times as fast as the native
hormone, while the carboxymethylated product, under identical conditions, is digested at 2·5–3 times the rate of the native hormone. It is of
interest to note that while both derivatives appear to retain lactogenic
activity, only the carbamidomethylated product retains growth-promoting
potency (see Table II).

Recently, the synthesis of a protein possessing growth-promoting and
lactogenic activities has been achieved (Li and Yamashiro 1970). This
synthesis was accomplished by the solid-phase method (Merrifield 1963)
and the product was shown to have approximately 10 per cent growth-
promoting potency and 5 per cent lactogenic activity in comparison with
that of the native hormone.

OVINE PITUITARY LACTOGENIC HORMONE

Ovine prolactin has been isolated in highly purified form by various
investigators (White, Catchpole and Long 1937; White, Bonsuess and
Long 1942; Li, Lyons and Evans 1940a, 1941; Cole and Li 1955). It is a
protein of molecular weight 23 300 (Li, Cole and Coval 1957) and isoelectric point at pH 5·7 (Li, Lyons and Evans 1940b). It consists of a single
polypeptide chain (Li 1957b; Li and Cummins 1958) with two tryptophan

---

* Niall (1971) also reported that the tryptophan residue is at position 85.

TABLE II

BIOLOGICAL ACTIVITIES OF HGH AND ITS DERIVATIVES

Preparation*	Rat tibia test** Tibia width (μm) for total dose (μg)			Pigeon local crop assay** Dry mucosal weight (mg) for dose (μg)	
	20	60	80	2	8
Native	211 ± 4 (12)	269 ± 5 (11)		13.5 ± 3.1 (6)	18.2 ± 4.7 (5)
RCAM	197 ± 6 (12)	257 ± 4 (12)		12.6 ± 1.8 (5)	20.1 ± 5.8 (5)
RCOM		194 ± 14 (5)	199 ± 22 (4)	11.9 ± 2.8 (6)	15.1 ± 3.9 (4)

* RCAM, reduced-tetra-S-carbamidomethylated derivative; RCOM, reduced-tetra-S-carboxymethylated derivative.
** Expressed as mean ± standard error followed by the number of test animals in parentheses.

residues and three disulphide bridges (Li 1949). Table I (p. 7) summarizes some physicochemical properties of ovine prolactin.

There are seven tyrosine residues in prolactin. One of these tyrosine residues is 'buried', as revealed by spectrophotometric titrations of the hormone in KCl solution (Ma, Brovetto-Cruz and Li 1970). This 'buried' tyrosine becomes ionized only after extensive alkali denaturation. However, all seven tyrosyl residues were found to react with tetranitromethane (Ma, Brovetto-Cruz and Li 1970). When the nitro-prolactin was assayed by the pigeon local crop test, no loss of lactogenic activity was observed, as shown in Table III. Nitro-prolactin cross-reacts completely with the

TABLE III
LACTOGENIC ACTIVITY OF NITRO-PROLACTIN

Preparation	Dry mucosal weight (mg) for dose (μg)	
	1	4
Prolactin	$18 \cdot 36 \pm 4 \cdot 77$	$36 \cdot 43 \pm 8.53$
Nitro-prolactin	$14 \cdot 76 \pm 3 \cdot 58$	$37 \cdot 26 \pm 9 \cdot 92$

Assayed by the pigeon local crop test; expressed as mean± standard error; six birds were used for each assay.

rabbit antiserum to the native hormone, as revealed by the Ouchterlony agar diffusion test (Fig. 3). However, spectrophotometric titrations and rates of tryptic digestion indicated that nitration produced significant conformational changes in the protein molecule. Apparently, the biological and immunological properties of prolactin do not depend upon the integrity of tyrosine residues as well as the molecular conformation.

The complete amino acid sequence of ovine prolactin has recently been elucidated, as shown in Fig. 4 (Li et al. 1969, 1970). It is a protein consisting of 198 amino acids with threonine at the amino terminus and half-cystine at the carboxyl end. The three disulphide bridges are formed between residues 4 and 11, between residues 190 and 198 and between 58 and 173. The two tryptophan residues are in positions 90 and 149 and the seven methionine residues are in positions 24, 36, 53, 81, 104, 129 and 131.

HUMAN CHORIONIC SOMATOMAMMOTROPIN

The presence of lactogenic activity in the human placenta was first described by Ehrhardt in 1936. Later studies of Josimovich and MacLaren (1962) and others (Kaplan and Grumbach 1964; Friesen 1965; Florini et al. 1966) showed that the human placenta contains a protein hormone possessing biological properties in common with HGH. The hormone has since been designated human chorionic somatomammotropin (HCS) (Li et al. 1968).

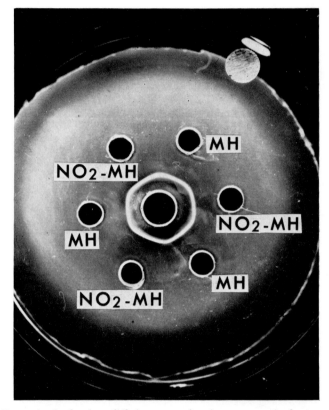

FIG. 3. An Ouchterlony diffusion pattern showing cross-reaction between nitro-prolactin (NO$_2$-MH) and rabbit antiserum to ovine prolactin (MH).

(facing page 12)

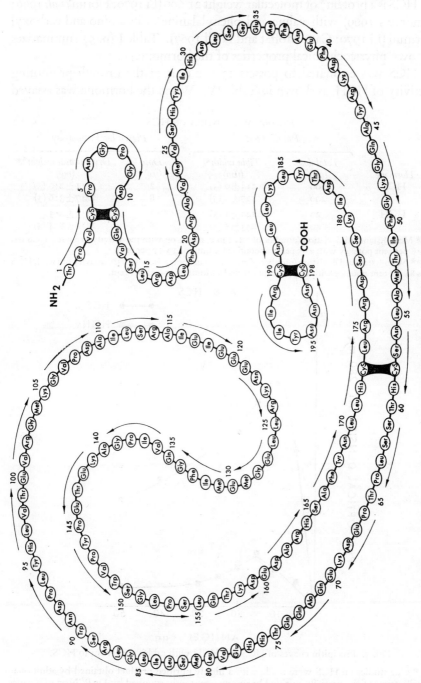

FIG. 4. The amino acid sequence of ovine prolactin.

HCS is a protein* of molecular weight 21 600 (Li 1970; Florini et al. 1966; Andrews 1969) with valine and phenylalanine at its amino and carboxyl termini (Li 1970; Catt, Moffat and Niall 1967). Table I (p. 7) summarizes known physicochemical properties of the hormone.

HCS was estimated to possess 13 per cent of the growth-promoting activity of HGH, as shown in Table IV. When the hormone was assayed

TABLE IV
BIOLOGICAL ACTIVITIES OF HCS

Hormone	Rat tibia test		Pigeon local crop assay	
	Total dose (μg)	Tibia width* (μm)	Dose (μg)	Dry mucosal weight** (mg)
HCS	100	213±9 (5)	2	11·1±2·0 (5)
	200	227±6 (5)	8	22·7±3·6 (5)
HGH	20	220±4 (5)	2	11·3±2·3 (5)
	40	243±5 (5)	8	23·1±8·4 (5)

* Mean± standard deviation (number of rats); in comparison with the activity of HGH, HCS has 12·8 per cent potency with a 95 per cent confidence limit of 10–17 per cent.

** Mean± standard deviation (number of crop sacs); in comparison with the activity of HGH, HCS has 84 per cent potency with a 95 per cent confidence limit of 57–122 per cent.

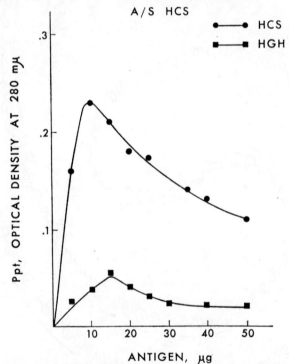

FIG. 5. Precipitin curves of HCS and HGH with rabbit antiserum to HCS.

* Our studies on HCS were made with a highly purified product obtained by additional purification of the partially purified hormone generously supplied by Dr P. Neri of Sclavo Institute, Siena, Italy.

for prolactin activity in the pigeon local crop assay, it gave 84 per cent of the potency of HGH (Table IV). As shown by various investigators (Josimovich and MacLaren 1962; Kaplan and Grumbach 1964; Friesen 1965; Florini *et al*. 1966), the rabbit antiserum to HCS cross-reacts with HGH in Ouchterlony diffusion experiments but does not give a continuous and identical line. This difference in antigenic sites between HCS and HGH is clearly shown by the precipitin curves, as illustrated in Fig. 5.

The complete amino acid sequence of HCS has very recently* been established (Li, Dixon and Chung 1971), as shown in Fig. 6. It may be noted that the single tryptophan residue is located at position 85 and that the six methionine residues are at positions 14, 64, 95, 124, 169 and 178. The two disulphide bridges form two loops: one between residues 53 and 164 and the other between residues 181 and 188.

When the primary structure of HCS is compared with that of HGH (see Fig. 2), it is remarkable that both of them contain 190 amino acid residues and that 160 out of 190 residues are located in identical positions (see Table V). The 30 positions showing differences contain 19 highly

TABLE V

DIFFERENCES OF RESIDUE POSITIONS IN THE HGH AND HCS STRUCTURES

Residue position*	HGH	HCS
1–2	Phe-Pro	Val-Gln
4	Ile	Val
12	Asn	His
16	Arg	Gln
20	Leu	Ala
25	Phe	Ile
34	Ala	Thr
39	Glu	Asp
46–49	Gln-Asn-Pro-Gln	His-Asp-Ser-Glu
52	Leu	Phe
56	Glu	Asp
64	Arg	Met
73	Gln	Glu
83	Gln	Glu
90	Gln	Arg
95	Val	Met
99	Ser	Asn
103–104	Gly-Ala	Asp-Thr
106	Asn	Asp
109	Val	Asp
111	Asp	His
132	Pro	Arg
138	Phe	Leu
152	Asp	His
178	Ile	Met

* For HGH, see Fig. 2; for HCS, see Fig. 6.

* The complete amino acid sequence of HCS was first reported at the Prolactin Workshop at the National Institutes of Health, Bethesda, January 11, 1971.

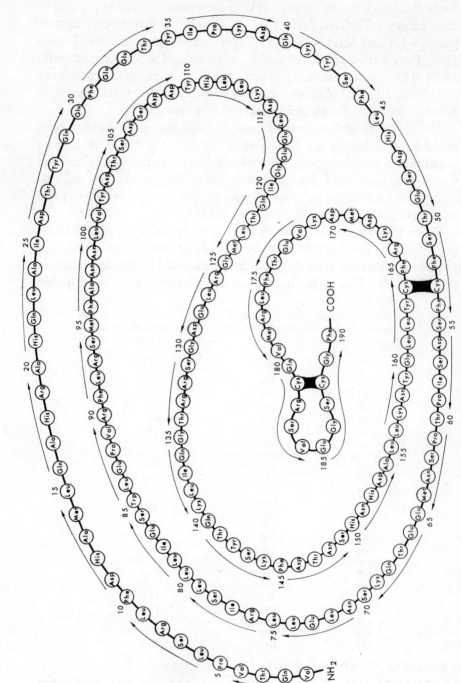

Fig. 6. The amino acid sequence of HCS.

acceptable replacements and four acceptable replacements, with only seven positions being occupied by non-homologous amino acids. Thus, there are 183 homologous positions in HGH and HCS, amounting to 96 per cent of either chain. The very high degree of homology between these two hormonal proteins may also be seen from the minimum number of nucleotide base differences in their genetic codes. All but one of the 30 replacements can be explained by the change of a single base in the codon triplet for that position. The replacement of Leu_{20} in HGH for Ala_{20} in HCS is the only one that requires a minimum of two base changes. This gives a total of only 31 base differences in the 570 nucleotides which code for these two hormonal proteins.

CIRCULAR DICHROISM SPECTRA OF HGH, HCS AND OVINE PROLACTIN

Fig. 7 gives the circular dichroism (CD) spectra of HGH, HCS and ovine prolactin in the region of side-chain absorption (Bewley and Li 1971a, b). Although the spectra of HGH and prolactin show considerable

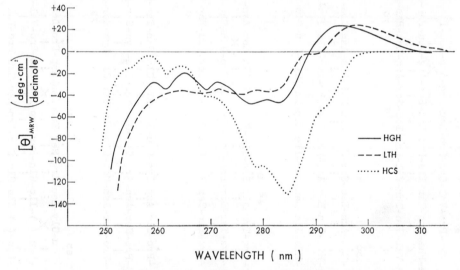

FIG. 7. Side-chain circular dichroism spectra of HGH, HCS and ovine prolactin in buffers of pH 8·4.

similarity in the conformations of these two molecules, they are not completely identical. Both proteins show an asymmetrical positive peak above 290 nm. This asymmetry is probably due to overlapping with the negative bands below 286 nm. The shape of the spectrum of ovine prolactin suggests that the positive band in this protein overlaps with an additional

18   CHOH HAO LI

```
                         2                  5        15                 20
Prolactin: 		            -Pro-Val-Cys-Pro-Leu-Arg-Asp-Phe-Arg-Ala-Val-Met-Val-( )-Ser-His-Tyr-Ile-    29
HGH:       NH2-Phe-Pro-Thr-Ile-Pro-Leu-Ser-Arg-Leu-Phe-Asp-Asn-Ala- X -Met-Leu-Arg-Ala-His-Arg-Leu-
HCS:       NH2-Val-Gln-Thr-Val-Pro-Leu-Ser-Arg-Leu-Phe-Asp-His-Ala-( )-Met-Leu-Gln-Ala-His-Arg-Ala-     20
                                        1                         10

                         30                                      40                                 41
Prolactin:   -His-Asn-Leu-Ser-Ser-Glu-Met-Phe-Asn-Glu-Phe-                                          -Asp-Lys-
HGH:         -His-Gln-Leu-Ala-Phe-Asp-Thr-Tyr-Gln-Glu-Phe- X -Glu-Ala-Tyr-Ile-Pro-Lys-Glu-Gln-
HCS:         -His-Gln-Leu-Ala-Ile-Asp-Thr-Tyr-Gln-Glu-Phe- X -Glu-Thr-Tyr-Ile-Pro-Asp-Gln-            40
                                          30                              65

                   45    50                                60                              84
Prolactin:   -Arg-Tyr-Ala-Phe-Ile-Thr-Met-Ala-Leu-Asn-Ser-( )-Cys-His-Thr-Ser-Ser-Leu-Pro-Thr-
HGH:         -Lys-Tyr-Ser-Phe-Leu-Gln-Asn-Pro-Gln-Thr-Ser-Leu- X -Phe-Ser-Glu-Ser-Ile-Pro-Thr-
HCS:         -Lys-Tyr-Ser-Phe-Leu-His-Asp-Ser-Glu-Thr-Ser-Phe-Cys-Phe-Ser-Asp-Ser-Ile-Pro-Thr-
                                                 50                              60

                                         70                            80
Prolactin:   -Pro-Glu-Asp-Lys-Glu-Gln-Ala-Gln-Gln-Thr-His-His-Glu-Val-Leu-( )-Met-Ser-Leu-Ile-
HGH:         -Pro-Ser-Asn-Arg-Glu-Glu-Thr-Gln-Gln-Lys-Ser-Asn-Leu-Gln-Leu-Leu-Arg-Ile-Ser-Leu-Leu-
HCS:         -Pro-Ser-Asn-Met-Glu-Glu-Thr-Gln-Gln-Lys-Ser-Asn-Leu-Glu-Leu-Leu-Arg-Ile-Ser-Leu-Leu-
                                                                 70                              80

                 85              90
Prolactin:   -Leu-Gly-Leu-Arg-Ser-Trp-Asn-Asp-Pro-Leu-
HGH:         -Leu-( )-Ile-Gln-Ser-Trp-Leu-Glu-Pro-Val-Gln-Phe-Leu-Arg-Ser-Val-Phe-Ala-Asn-Ser-Leu-
HCS:         -Leu-( )-Ile-Glu-Ser-Trp-Leu-Glu-Pro-Val-Arg-Phe-Leu-Arg-Ser-Met-Phe-Ala-Asn-Asn-Leu-    100
                                                                 90
```

CHEMISTRY OF LACTOGENIC HORMONES

```
                                       95                    100
Prolactin:                         -Tyr-His-Leu-Val-Thr-Glu-Val-
                                     |  :  X  |  |  |  .
     HGH: -Val-Tyr-Gly-Ala-Ser-Asp-Val-Tyr-Asp-Leu-Lys-Asp-Leu-Glu-Gly-Ile-
             |  |  :  :  X  |  :  :  X  |  |  |  |  |  |  |
     HCS: -Val-Tyr-Asp-Thr-Ser-Asp-Ser-Asp-Asp-Tyr-His-Leu-Leu-Asp-Leu-Gly-Ile-
                                                          110                120

                                   125                             135
Prolactin:                    -Arg-Leu-Glu-Gly-Met-           -Gly-Gln-Val-Ile-
                                |  X  |  |  .                   |  :  :  :
     HGH: -Gln-Thr-Leu-Met-Gly-Arg-Leu-Glu-Asp-Gly-Ser-Pro-Arg-Thr-Gly-Gln-Ile-Phe-Lys-Gln-
             |  |  |  |  |  |  |  |  :  |  |  |  |  |  |  |  |  |  |  |
     HCS: -Gln-Thr-Leu-Met-Gly-Arg-Leu-Glu-Asp-Gly-Ser-Arg-Arg-Thr-Gly-Gln-Ile-Leu-Lys-Gln-
                                          130                                140

                                              160                                        169
Prolactin:              -Gln-Thr-Lys-Asp-Glu-Asp-Ala-Arg-His-Ser-Ala-Phe-Tyr-Asn-
                          :  :  :  |  |  |  |  X  |  |  :  X  |  |
     HGH: -Thr-Tyr-Ser-Lys-Phe-Asp-Thr-Asn-Ser-His-Asn-Asp-Asp-Ala-Leu-Leu-Lys-Asn-( )-Tyr-Gly-
             |  |  |  |  |  |  |  |  |  |  |  X  |  |  |  |  |  |      |  |
     HCS: -Thr-Tyr-Ser-Lys-Phe-Asp-Thr-Asn-Ser-His-Asn-His-Asp-Ala-Leu-Leu-Lys-Asn-( )-Tyr-Gly-
                                     150                                              160

                                                         180                                 189
Prolactin: -Leu-Leu-His-Cys-Leu-Arg-Arg-Asp-Ser-Ser-Lys-Ile-Asp-Thr-Tyr-Leu-Lys-Leu-Leu-Asn-
              :  |  :  |  |  :  :  |  |  :  :  :  |  |  :  :  |  :  |  :
     HGH: -Leu-Leu-Tyr-Cys-Phe-Arg-Lys-Asp-Met-Asp-Lys-Val-Glu-Thr-Phe-Leu-Arg-Ile-Val-Gln-
             |  |  |  |  |  |  |  |  |  |  |  |  |  |  :  |  :  |  |  |
     HCS: -Leu-Leu-Tyr-Cys-Phe-Arg-Lys-Asp-Met-Asp-Lys-Val-Glu-Thr-Phe-Leu-Arg-Met-Val-Gln-
                              170                                      180

               190                   198
Prolactin: -Cys-Arg-Ile-Ile-Tyr-Asn-Asn-Asn-Cys-COOH
              |  :  X  X  |  :  :  :  |
     HGH: -Cys-Arg-Ser-Val-( )-Gly-Ser-Cys-Gly-Phe-COOH
             |  |  |  |      |  |  |  |  |  |
     HCS: -Cys-Arg-Ser-Val-( )-Gly-Ser-Cys-Gly-Phe-COOH
                                                    190
```

FIG. 8. Comparison of the structures of HGH, HCS and ovine prolactin. The residue position numbers for HGH and HCS are the same, and appear below the HCS sequence. The residue position numbers for ovine prolactin appear above the prolactin sequence. Homology is indicated by; identical pairs, vertical bar; highly acceptable replacements, three dots, and acceptable replacements, single dot. Unacceptable replacements are indicated by x. An asterisk has been placed over those residues which are common to all three proteins.

negative band, also above 290 nm, which is not found in HGH. Simple graphical subtraction of the HGH spectrum from that of ovine prolactin between 288 and 310 nm produces a weak negative band with a maximum at 291–292 nm. Since this 'hidden' band in prolactin lies above 290 nm, it may tentatively be assigned to tryptophan, along with the positive band. This would be consistent with the fact that ovine prolactin contains two tryptophan residues, one showing a positive CD band and the other a negative one, while the single tryptophan in HGH gives rise only to a positive band. The spectrum of HCS in the same region of side-chain absorption is quite different from those of HGH and ovine prolactin. The two negative peaks shown by HCS at 269 and 261·5 nm are almost identical to corresponding peaks in the HGH spectrum which have been assigned to phenylalanine residues. These peaks are almost entirely absent from the spectrum of ovine prolactin. At the present time we cannot make definite chromophore assignments to the negative dichroism in the HCS spectrum between 270 and 290 nm. The two peaks at 279 and 284·5 nm are probably due largely to tyrosine residues although both tryptophan and the disulphide bonds undoubtedly contribute to this region also. It may be pointed out that it is in this region of the spectrum that HCS shows the greatest differences from the other two hormones. The negative shoulder above 290 nm in the spectrum of HCS is almost certainly due to the protein's single tryptophan residue. It corresponds very closely to the 'hidden' tryptophan band in ovine prolactin, while differing from the positive tryptophan in HGH.

STRUCTURAL COMPARISON OF OVINE PROLACTIN, HCS AND HGH

To examine the structures for areas of homology, we have aligned the three sequences (Bewley and Li 1971c) according to the best possible fit of certain reference residues. Half-cystine, tryptophan, tyrosine, histidine and proline were used as references because of their limited content in these proteins and their low relative mutability. Occasionally, the introduction of a gap into one or the other sequence was required to obtain the best alignment.

In Fig. 8, an asterisk has been placed above those residue positions which contain the same amino acid in all three structures. There are 50 such positions. These common positions would seem to be more or less randomly distributed, there being no particular area(s) in which they are clearly concentrated. They are also about equally distributed between hydrophilic and hydrophobic residues. However, in terms of the total content in all three proteins, some residue types appear in the common positions much more than others. For example, with the exception of the small disulphide ring near the amino terminal of ovine prolactin, each half-cystine residue is homologous with a corresponding half-cystine in the other two molecules. Thus, 12 of the 14 cystine residues, or 86 per cent of the total cystine content, appear in one or another homologous position in the three hormones. Similarly, 75 per cent of the tryptophan residues (one in each sequence, or three out of a total content of four) appear in a single homologous position.

Other residues whose positions appear to be conserved by identity are: proline, 50 per cent; leucine, 45 per cent; tyrosine, 39 per cent; histidine, 33 per cent; phenylalanine, 30 per cent; alanine and arginine, 27 per cent; serine, aspartic acid and glycine, 24 per cent each. Although glutamic acid, methionine, threonine, glutamine and lysine also occur in common positions in all three sequences, the percentages of their total contents, conserved in this fashion, are all well below 25 per cent.

The fact that the amino acid sequence of HGH, HCS and ovine prolactin is in close similarity is especially of interest in view of the fact that these three molecules are active as growth-promoting and lactogenic hormones in spite of their differences in origin. It may be that these three hormones are derived through evolution from a common ancestral molecule.

SUMMARY

There are three hormones which are presently known to possess lactogenic activities, namely, non-primate pituitary prolactin, human pituitary growth hormone (HGH) and human chorionic somatomammotropin (HCS). The primary structures of the first two hormones are already known. The complete amino acid sequence of HCS is presented; it is a protein consisting of 190 amino acids with valine and phenylalanine as amino-terminal and carboxyl-terminal residues respectively. The single tryptophan residue in the HCS molecule is located at position 85 and the six methionine residues are at positions 14, 64, 95, 124, 169 and 178. The two disulphide bridges in HCS form two loops: one between residues 53 and 164 and the other between residues 181 and 188.

Some physicochemical properties and data on structure–activity relationships of HGH, ovine prolactin and HCS are discussed, and the proposed structure of HCS is compared with that of HGH and of ovine prolactin.

Acknowledgements

I take this opportunity to express my gratitude to the following colleagues for their participation in these investigations: T. A. Bewley, J. Brovetto-Cruz, D. Chung, J. S. Dixon, T.-B. Lo, L. Ma, Y. A. Pankov, K. D. Schmidt and D. Yamashiro. The technical assistance of D. Gordon, J. Knorr and J. D. Nelson is gratefully acknowledged. I also thank the American Cancer Society, the National Institutes of Health, the Allen Foundation and the Geffen Foundation for research grants.

REFERENCES

ANDREWS, P. (1969) *Biochem. J.* **111**, 799.
BEWLEY, T. A., BROVETTO-CRUZ, J. and LI, C. H. (1969) *Biochemistry* **8**, 4701.
BEWLEY, T. A. and LI, C. H. (1971a) *Biochemistry* In press.
BEWLEY, T. A. and LI, C. H. (1971b) *Arch. Biochem. & Biophys.* **144**, 589.
BEWLEY, T. A. and LI, C. H. (1971c) *Experientia* In press.
CATT, K. J., MOFFAT, B. and NIALL, H. D. (1967) *Science* **157**, 321.
COLE, R. D. and LI, C. H. (1955) *J. Biol. Chem.* **213**, 197.
DIXON, J. S. and LI, C. H. (1966) *Science* **154**, 785.
EHRHARDT, K. (1936) *Muench. Med. Wochenschr.* **83**, 1163.
FLORINI, J. R., TONELLI, G., BREUER, C. B., COPPOLA, J., RINGLER, I. and BELL, P. H. (1966) *Endocrinology* **79**, 692.
FRIESEN, H. (1965) *Endocrinology* **76**, 692.
JOSIMOVICH, J. B. and MACLAREN, J. A. (1962) *Endocrinology* **71**, 209.
KAPLAN, S. L. and GRUMBACH, M. M. (1964) *J. Clin. Endocrinol. & Metab.* **24**, 80.
LI, C. H. (1949) *J. Biol. Chem.* **178**, 459.
LI, C. H. (1957a) *Fed. Proc. Fed. Am. Soc. Exp. Biol.* **16**, 775.
LI, C. H. (1957b) *J. Biol. Chem.* **229**, 157.
LI, C. H. (1970) *Ann. Sclavo (Siena)* **12**, 651.
LI, C. H., COLE, R. D. and COVAL, M. J. (1957) *J. Biol. Chem.* **229**, 153.
LI, C. H. and CUMMINS, J. T. (1958) *J. Biol. Chem.* **233**, 73.
LI, C. H. and DIXON, J. S. (1971) *Arch. Biochem. & Biophys.* **146**, in press (Sept.).
LI, C. H., DIXON, J. S. and CHUNG, D. (1971) *Science* **173**, 56.
LI, C. H., DIXON, J. S. and LIU, W.-K. (1969) *Arch. Biochem. & Biophys.* **133**, 70.
LI, C. H., DIXON, J. S., LO, T.-B., PANKOV, Y. A. and SCHMIDT, K. D. (1969) *Nature (Lond.)* **224**, 695.
LI, C. H., DIXON, J. S., LO, T.-B., SCHMIDT, K. D. and PANKOV, Y. A. (1970) *Arch. Biochem. & Biophys.* **141**, 705.
LI, C. H., GRUMBACH, M. M., KAPLAN, S. L., JOSIMOVICH, J. B., FRIESEN, H. and CATT, K. J. (1968) *Experientia* **24**, 1288.
LI, C. H., LIU, W.-K. and DIXON, J. S. (1966) *J. Am. Chem. Soc.* **88**, 2050.
LI, C. H., LYONS, W. R. and EVANS, H. M. (1940a) *J. Gen. Physiol.* **23**, 433.
LI, C. H., LYONS, W. R. and EVANS, H. M. (1940b) *J. Am. Chem. Soc.* **62**, 2925.
LI, C. H., LYONS, W. R. and EVANS, H. M. (1941) *J. Biol. Chem.* **140**, 43.
LI, C. H. and PAPKOFF, H. (1956) *Science* **124**, 1293.
LI, C. H. and STARMAN, B. (1964) *Biochim. & Biophys. Acta* **86**, 175.
LI, C. H. and YAMASHIRO, D. (1970) *J. Am. Chem. Soc.* **92**, 7608.
MA, L., BROVETTO-CRUZ, J. and LI, C. H. (1970) *Biochemistry* **9**, 2302.
MERRIFIELD, R. B. (1963) *J. Am. Chem. Soc.* **85**, 2149.
NIALL, H. D. (1971) *Nature New Biol.* **230**, 90.
WHITE, A., BONSUESS, R. W. and LONG, C. N. H. (1942) *J. Biol. Chem.* **143**, 447.
WHITE, A., CATCHPOLE, H. R. and LONG, C. N. H. (1937) *Science* **86**, 82.

DISCUSSION

Sherwood: The most intriguing biochemical difference between ovine prolactin and the other two hormones (HGH and HPL) is the presence of an extra disulphide bond at the amino-terminal end of the ovine hormone.

Is there any evidence that prolactins from other lower forms have an additional disulphide bond?

Li: Dr R. E. Fellows reported at the Prolactin Workshop at the National Institutes of Health in January 1971 that the structure of bovine prolactin is very close to that of ovine prolactin, but I do not know any data on other species.

Wilhelmi: The bovine hormone has six half-cystine residues just like the ovine. Dr Fellows has told me that the sequence at the amino-terminal end is very similar to that of the ovine hormone. Also, of course, they cross-react nearly identically, immunologically, so the probability is that bovine and ovine prolactin are very nearly the same. But that is the limit of our knowledge.

Nicoll: Dr S. Eppstein (1964) reported that there are a large number of disulphide groups in porcine prolactin (14 half-cystine residues). Has that been confirmed?

Li: Dr Eppstein did the initial work in our laboratory. Porcine prolactin was a rather difficult molecule to isolate at that time and the whole problem is open again, but his data clearly showed more S-S bridges in porcine than ovine or bovine prolactin. The data have not been confirmed but no one else has yet obtained a preparation pure enough to show differences from Eppstein's data.

Wilhelmi: One of my students, Mr Sam Hagan, is working on the purification of porcine prolactin now. Our data suggest that it's very different from Dr Eppstein's preparation. We rather think that his preparation had a contaminant, perhaps a small peptide containing a lot of cystine, possibly a posterior lobe hormone. We can't be sure yet, because the porcine molecule is, as you say, difficult to purify. Its specific activity, as one purifies it, never gets as high as that of highly purified ovine and bovine prolactin, and this worries us. Until we are certain, we don't want to start any chemistry on an insufficiently purified molecule.

Nicoll: Dr Li, you report that HPL has 13 per cent of the growth-promoting activity of HGH in the rat tibia test. Have you tested ovine or bovine prolactin using highly purified material, to see how they compare with HGH in this assay?

Li: We have done this repeatedly. We have not been able to show by the tibia test any growth-promoting activity of ovine or bovine prolactin. Some other test might reveal such activity. The Merck group showed that in the normal rat, ovine prolactin promoted growth, but that might be due to something else (such as contamination with growth hormone) (Reisfeld *et al.* 1961). I think the rat tibia test is the most specific test for growth-promoting activity.

Beck: Many people would challenge that, Dr Li! This is really very crucial; how specific is the rat tibia as a bioassay for growth-promoting activity?

Frantz: Dr Li, aren't your figures for the growth-promoting activity of HPL (13 per cent of HGH activity) somewhat higher than those reported by others?

Li: We found that the dimer of HCS (HPL) has very little growth-promoting activity (unlike the monomer), so my feeling is that the monomer is the one with high growth-promoting activity by the tibia test.

Frantz: Is the form in which the hormone has been supplied by Lederle chiefly that of the dimer?

Li: I can only judge from the literature. I think the Lederle material is not all monomer.

Friesen: It looked to me as if your ratio of elution volume to void volume was two, which is identical with that of the Lederle preparation and of some of ours.

Li: Have you determined the actual molecular weight of the monomer by an osmotic pressure measurement?

Friesen: No.

Li: We used gel filtration for estimating the molecular weight, and we are now investigating the biophysical behaviour of the dimer, including the tertiary structure, which might be very similar to that of the dimer of HGH.

Nicoll: I have been working with Dr Paul Licht in Berkeley on growth hormone and prolactin in different vertebrate species. We are using a growth test in the Californian toad (*Bufo boreas*) which measures the increase in body length of the animals. We found that ovine and bovine prolactin are about 5 per cent as active as ovine or bovine growth hormones. Porcine prolactin and HPL from Lederle were completely inactive in this test. We found that some prolactins of non-mammalian species are highly active (Nicoll and Licht 1971).

Li: There must be some overlap in growth-promoting and lactogenic activity because of the structural similarity, with approximately 50 per cent of the amino acids identical.

Sherwood: We have used the Lederle preparation of HPL exclusively; on gel filtration at neutral pH it tends to behave as a dimer. Monomer was produced by spinning it in the ultracentrifuge in 5M-guanidine.

Li: Did you assay the growth-promoting activity of the monomer?

Sherwood: We have not attempted to bioassay the growth-promoting activity of the monomer. Dr Friesen (1965) has bioassayed a similar Lederle preparation very extensively, and found that it had much less than 13 per cent of the growth-promoting activity of growth hormone.

Friesen: The Lederle group (Breuer 1969) has reported that the monomer in their hands was no more active in growth promotion than the dimer. Dr Li, I wonder whether the slope of the dose–response curve of HPL is identical with that of growth hormone? If not, expressing it as a percentage of activity is not perfectly valid. It looks as if HPL has a flatter slope.

Li: The slope is not completely identical as we expected, but one can roughly estimate the amount of growth-promoting activity in comparison with that of HGH.

Turkington: The biological activities of these molecules will ultimately be resolved in terms of their interaction with a specific receptor molecule. In addition to analysing these hormones in terms of their amino acid sequences, Dr Li, have you compared them by model building, examining their tertiary structures, to determine what groups may be available to bond at the biologically active site?

Li: Not yet. This is why we hope to prepare biologically active and radioactive molecules, to study this type of binding site. The nitro-lactogenic hormone (nitro-prolactin), for example, is fully active; when we produce a radioactive nitro-lactogenic hormone, we shall be able to do such experiments.

Beck: It would be interesting to know what you intend to use as your receptor site!

Greenwood: Surely, the translocation of a sequence in HGH and the fact that the molecule you synthesized is nevertheless biologically active answers Dr Turkington's question. The tertiary structure is not involved in biological activity.

Turkington: It must be involved.

Li: The tertiary structure might reside in a small segment; even the decapeptides sometimes have a tertiary structure. That is what Dr Turkington is referring to. Certainly the synthesis of a protein using the 1966 proposed structure which has activity indicates that the whole molecule is not necessary for hormonal activity.

Wilhelmi: Does nitrophenylation of prolactin have any effect on its tertiary structure, as seen in the circular dichroism, for instance?

Li: We have not done this. The circular dichroism spectrum is somewhat difficult to determine because of the nitro group. This is why we employed immunochemical methods.

Bryant: We find that immunological activity is very easily reduced by iodinating ovine prolactin, unlike human growth hormone. This is contrary to what you showed.

Li: This is why we are interested in the nitro-prolactin.

Greenwood: Two points on nomenclature. Firstly, on 'human placental lactogen' or 'human chorionic somatomammotropin'. This latter term (conceived in a smoke-filled room in Siena!) will presumably have to be modified, since Dr Friesen has shown that human prolactin increases in pregnancy. Secondly, it might be useful to reach agreement on human prolactin; I don't like the term 'luteotropic hormone', or mammotropin.

REFERENCES

BREUER, C. B. (1969) *Endocrinology* **85**, 989–999.
EPPSTEIN, S. H. (1964) *Nature (Lond.)* **202**, 899–900.
FRIESEN, H. G. (1965) *Endocrinology* **76**, 369–381.
NICOLL, C. S. and LICHT, P. (1971) *Gen. & Comp. Endocrinol.* in press.
REISFELD, R. A., TONG, G. L., RICKES, E. L. and BRINK, N. G. (1961) *J. Am. Chem. Soc.* **83**, 3717–3719.

[For nomenclature, see General Discussion, p. 391 and p. 400. For further discussion of HGH and HPL structure, see pp. 45–51.]

THE STRUCTURE AND FUNCTION OF HUMAN PLACENTAL LACTOGEN

Louis M. Sherwood, Stuart Handwerger and William D. McLaurin

Endocrine Unit, Department of Medicine, Beth Israel Hospital and Harvard Medical School, Boston

In earlier reports from this laboratory (Sherwood 1967, 1969) striking homologies were noted in the amino acid compositions of the tryptic peptides of human placental lactogen (HPL) and human growth hormone (HGH). Alignment of the tryptic peptides of HPL in homologous positions with those of HGH further suggested a remarkable similarity in the amino acid sequences of the two hormones. At the carboxyl-terminal end of HPL, fourteen residues were identical with HGH except for a single substitution in HPL of methionine for isoleucine (Sherwood 1969); at the amino terminus, eleven of the first seventeen residues of the two hormones were homologous (Catt, Moffat and Niall 1967). We now report the complete amino acid sequence of HPL (Sherwood *et al.* 1971a) and a detailed comparison of its structure with that of HGH (Li, Liu and Dixon 1966; Niall 1971). Since the two hormones are identical in 86 per cent of their amino acid sequence, there is a firm basis for their similarities in biological, immunological and physicochemical properties.

MATERIALS AND METHODS

Partially purified HPL was kindly supplied by Drs Paul H. Bell and Charles Breuer of Lederle Laboratories. This material was further purified by gel filtration as described previously (Sherwood 1967) and was shown to be chemically homogeneous before we began the sequence studies. The purified protein was a single peak on gel filtration and in the ultracentrifuge (in 5M-guanidine) and behaved as a single band in immunodiffusion studies with antiserum against the crude hormone. A single amino-terminal valine was determined by the dansyl technique except in occasional preparations where an amino-terminal threonine was also noted. The latter finding has been reported (Catt, Moffat and Niall 1967) and we have observed amino-terminal peptides without the two first residues in the sequence studies. Whether this difference is due to aminopeptidase activity in placental extracts or represents an intraspecies variant is not yet known. On polyacrylamide gel electrophoresis, minor microheterogeneity, probably due to partial deamidation (Sherwood 1967), was noted.

The purified hormone was biologically active in causing weight gain in hypophysectomized mice (Florini et al. 1966), in stimulating lactation in mid-pregnant rabbits (Friesen 1966) and in inducing mammary gland differentiation and casein synthesis by mouse mammary tissue in organ culture (Turkington and Topper 1966).

Enzyme digestion

Trypsin. One gramme of native HPL was reduced and alkylated with iodoacetic acid in order to disrupt the disulphide bonds (Crestfield, Moore and Stein 1963; Sherwood 1967). The modified protein was analysed for complete conversion and then cleaved with trypsin (2 per cent ratio of enzyme to substrate) at 37°C in 0·2M-ammonium bicarbonate (pH 7·8) with twice recrystallized trypsin (Worthington). After lyophilization, the enzyme digest was fractionated by ion exchange chromatography as previously described (Sherwood 1967). The peptide fractions were analysed after alkaline hydrolysis (Hirs, Moore and Stein 1956) and subjected, after lyophilization, to high voltage electrophoresis and/or paper chromatography in butanol–glacial acetic acid–water (4–1–5) by the method of Katz, Dreyer and Anfinsen (1959). Homogeneous peptides were eluted from paper with 10 per cent acetic acid followed by 2 per cent ammonium hydroxide and then hydrolysed with 6N-hydrochloric acid for 22 hours at 110°C in a sealed, evacuated dessicator flushed with nitrogen (Hirs, Stein and Moore 1954). The hydrolysates were analysed on a Beckman 121 automatic amino acid analyser equipped for high sensitivity analysis (0·01 μM with an accuracy of 3 per cent).

Chymotrypsin. Purified HPL (860 mg) was oxidized with performic acid (in the ratio 90 per cent formic acid/10 per cent hydrogen peroxide) for 2 hours at −10°C in an alcohol-ice mixture by the method of Hirs (1967). After dilution with distilled water, the preparation was lyophilized carefully to dry powder. An aliquot was tested for complete conversion by amino acid analysis. Performic-oxidized HPL was then incubated in 0·2M-ammonium bicarbonate (pH 8.4) with α-chymotrypsin (Worthington, twice recrystallized) at 37°C for two hours (Canfield 1963; Canfield and Anfinsen 1963a). After lyophilization, the mixture of peptides was separated on a column of Dowex 50-X2 (160 × 2·5 cm) at 55°C with a four-bottle gradient elution system of pyridine acetate as follows: 500 ml each of pyridine acetate pH 2·8 (0·2M), pH 3·4 (0·4M), pH 4·5 (0·6M) and pH 5·50 (1·0M). The column was then eluted with 3M-pyridine acetate buffer (pH 5·5) followed by 3M-pyridine. The effluent fractions were analysed after alkaline hydrolysis and purified as above.

Pepsin. Three hundred mg of native HPL were dissolved in 50 ml of 0·01N-HCl adjusted to pH 2·0. One mg of pepsin (Worthington) was added and the digestion performed at room temperature for 30 minutes with constant stirring. The reaction was stopped by the addition of pyridine to pH 5 and the digest was then lyophilized. The digest was subjected to ion exchange chromatography on Dowex 50-X2 (60 × 1·5 cm column) with a four-bottle system as above.

Since the pepsin digestion was performed on native HPL, the disulphide bonds were left intact. One disulphide bond at the carboxyl-terminal end was intact in a single peptic peptide (P 6D). The remaining disulphide bond joined two peptic peptides, P14-1 and P14-2, which were then separated by performic acid oxidation and high voltage electrophoresis on Whatman 3MM paper (see **Results**).

Cyanogen bromide cleavage

Three hundred mg of native HPL were cleaved with cyanogen bromide by the method of Gross and Witkop (1962). The hormone was dissolved in 100 ml of 70 per cent formic acid which was bubbled with nitrogen to prevent oxidation of methionine. One gramme of solid cyanogen bromide (Eastman Chemical) was added and incubated at 25°C in a sealed container with gentle stirring for six hours. The reaction was stopped by diluting the sample with distilled water and freeze-drying. To measure the extent of reaction, an aliquot of the reaction mixture was oxidized with performic acid and subjected to amino acid analysis following acid hydrolysis, to calculate the content of unreacted methionine as methionine sulphone. The peptides were separated on a column of Sephadex G-50 (200 × 2·5 cm) using 0·15M-acetic acid and the eluted peptides were analysed by optical density at 280 nm and at 570 nm after alkaline hydrolysis and reaction with ninhydrin. The fractions were divided and analysed further by high voltage electrophoresis and chromatography (Fig. 1). The largest peptide (C5), which contained two half-cystine residues, was rechromatographed on Sephadex G-50 and then oxidized with performic acid. The two peptide fragments C5A and C5B were separated on a smaller column of Sephadex G-50 (65 × 2 cm) in 0·2M-ammonium formate buffer pH 2·8 (Fig. 2). The remaining large peptides (C7A and C7B) were separated by ion exchange chromatography on a column of carboxymethyl cellulose using a two-bottle gradient of ammonium acetate from pH 5·0 (0·01M) to pH 7·0 (0·33M).

2*

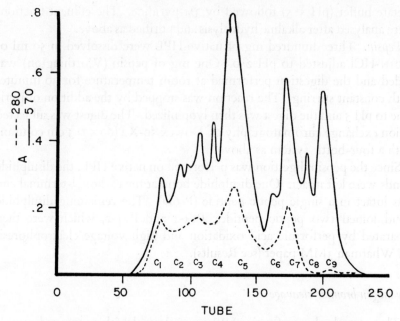

FIG. 1. Elution pattern of peptides from cyanogen bromide cleavage of HPL on Sephadex G-50 in 0·15M-acetic acid. Fractions were analysed for optical density at 280 nm and at 570 nm after alkaline hydrolysis and reaction with ninhydrin.

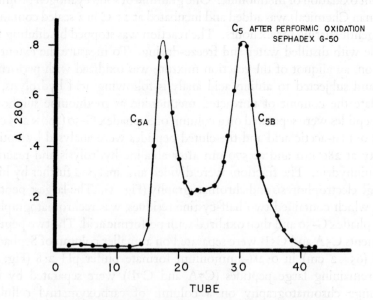

FIG. 2. Separation of C5A and C5B on Sephadex G-50 after cleavage of the disulphide bond of C5 with performic acid. The column was developed with 0·2M-ammonium formate buffer (pH 2·8).

Edman degradation

Sequential degradation of individual peptides was done by a modification of the method of Edman (1950) as published by Elzinga (1970). Peptides ranging in size from three to twelve residues were generally subjected to an average of six to eight cycles, although in some instances it was possible to remove up to ten amino acids. Sequenator grade reagents (pyridine, phenylisothiocyanate and trifluoroacetic acid) were obtained from the Pierce Chemical Company. The peptide (0·025 to 0·05 μmoles) was dissolved in 0·2 ml of pyridine and 15μl of phenylisothiocyanate. After blowing with nitrogen, the reaction mixture was incubated at 37°C for one hour and dried down. Trifluoroacetic acid (0·2 ml) was then added and the mixture was blown with nitrogen; after incubation for 15 minutes at 37°C it was dried down. This procedure was repeated until the desired number of Edman steps was completed. After the last step, the remaining material was dissolved in 1N-HCl and N-butyl acetate and extracted three times with N-butyl acetate. After evaporation of the aqueous phase, the peptide was hydrolysed for 22 hours at 110°C in constant boiling 6N-HCl.

For some peptides, sequential degradation was performed by the dansyl-Edman technique as described by Gray (1967). After cleavage and extraction of the phenylthiohydantoin derivative, the remaining amino-terminal amino acid was reacted with dimethylnaphthylene sulphonyl chloride to convert it to the dansyl derivative. The reaction was performed at pH 9·7 in sodium bicarbonate buffer. The peptide was then hydrolysed in 6N-hydrochloric acid for 16 hours; after drying, the hydrolysate was analysed by thin layer chromatography on polyamide paper (Cheng Chin Trading Company) using the two-solvent system described by Woods and Wang (1967). The location of the dansyl-amino acid was determined with an ultraviolet source.

RESULTS

A series of unique peptides was obtained from the cleavage of HPL by each of the four methods described earlier. The peptides obtained from trypsin cleavage have previously been described (Sherwood 1967) and included a total of 19 peptides which were obtained from cleavage at 18 of the 20 lysine and arginine residues (see Table I). The only peptide not identified in the original study was T14 (a tripeptide of sequence Phe–Leu–Arg). Peptide T6B identified in very small yield in the original digest proved to be non-homogeneous and not significant. Otherwise, the alignment of homologous peptides with HGH predicted in the earlier study proved to be correct. In parallel with recent studies (Niall 1971) showing that the

TABLE I
TRYPTIC PEPTIDES OF HUMAN PLACENTAL LACTOGEN

Peptide	Sequence
T18C	Val, Gln, Thr, Val, Pro, Leu, Ser, *Arg*
T26	Leu, Phe, Asp, His, Ala, Met, Leu, Gln, Ala, His, *Arg*
T18	Ala, His, Gln, Leu, Ala, Ile, Asp, Thr, Tyr, Gln, Glu, Phe, Glu, Glu, Thr, Tyr, Ile, Pro, *Lys*
T7B	Asp, Glx, *Lys*
T12	Tyr, Ser, Phe, Leu, His, Asp, Ser, Glx, Thr, Ser, Phe, Cys, Phe, Ser, Asx, Ser, Ile, Pro, Thr, Pro, Ser, Asx, Met, Glx, Glx, Thr, Glx, *Lys*
T17A	Ser, Asx, Leu, Glx, Leu, Leu, *Arg*
T24	Ile, Ser, Leu, Leu, Leu, Ile, Glx, Ser, Trp, Leu, Glx, Pro, Val, *Arg*
T14	Phe, Leu, *Arg*
T13B	Ser, Met, Phe, Ala, Asx, Asx, Leu, Val, Tyr, Asx, Thr, Ser, Asx, Asx, Asx, Ser, Tyr, His, Leu, Leu, *Lys*
T8A	Asx, Leu, Glx, Glx, Gly, Ile, Glx, Thr, Leu, Met, Gly, *Arg*
T9B	Leu, Glx, Asx, Gly, Ser, Arg, *Arg*
T16C	Thr, Gly, Glx, Ile, Leu, *Lys*
T7A	Glx, Thr, Tyr, Ser, *Lys*
T20D	Phe, Asx, Thr, Asx, Ser, His, Asx, His, Asx, Ala, Leu, Leu, *Lys*
T16A	Asx, Tyr, Gly, Leu, Leu, Tyr, Cys, Phe, *Arg*
T12D	*Lys*
T14A	Asx, Met, Asx, *Lys*, Val, Glx, Thr, Phe, Leu, *Arg*
T13C	Met, Val, Gln, Cys, *Arg*
T5A	Ser, Val, Glu, Gly, Ser, Cys, Gly, Phe

tryptophan-containing peptide of HGH was located incorrectly in the original report by Li, Liu and Dixon (1966), the homologous tryptophan peptide of HPL (T24) has been reassigned to a position between T17A and T13B.

A total of 19 chymotryptic peptides were obtained and their amino acid compositions used to align the sequence of the tryptic peptides in the molecule. As shown in detail in Table II, 15 chymotryptic peptides overlapped areas containing lysine and arginine residues and confirmed the alignment of the tryptic peptides. Chymotryptic peptides were in general smaller than the tryptic peptides and frequently ended in carboxyl-terminal leucine, phenylalanine and tyrosine. The four peptides containing cysteine were identified by their content of cysteic acid, since performic-oxidized HPL had been used for the chymotryptic digest. In similar fashion, the 21 peptic peptides permitted 15 of the tryptic peptides to be overlapped (Table II), thus completing the definitive alignment of all the tryptic peptides in the molecule.

Peptic peptide P14 failed to move from the origin and proved to be a peptide containing two half-cystine residues. After its elution from paper and purification on Sephadex G-25, P14 was oxidized with performic acid and separated into two peptides on high voltage electrophoresis: P14-1 (composition Asp_2, Thr_2, Ser_3, Pro_2, Cys, Phe) and P14-2 (composition Cys, Tyr, Phe).

TABLE II
DETERMINATION OF THE AMINO ACID SEQUENCE OF PLACENTAL LACTOGEN

```
         VAL-GLN-THR-VAL-PRO-LEU-SER-ARG-LEU-PHE-ASP-HIS-ALA-MET-LEU-GLN-ALA-HIS-ARG-ALA-HIS-GLN-LEU-ALA-ILE-
          1   2   3   4   5   6   7   8   9  10  11  12  13  14  15  16  17  18  19  20  21  22  23  24  25
T18C     Val,Gln,Thr,Val,Pro,Leu,Ser,Arg
CT17D    Val,Gln,Thr,Val,Pro,Leu
P5C      Val,Gln,Thr,Val,Pro,Leu,Ser,Arg,Leu
C8                       Val,Pro,Leu,Ser,Arg,Leu,Phe,Asp,His,Ala,Met
T26                                      Leu,Phe,Asp,His,Ala,Met,Leu,Gln,Ala,His,Arg
CT31D                                            Asp,His,Ala,Met,Leu,Gln
P10C                                                 His,Ala,Met,Leu,Gln
C5A                                                          Leu,Gln,Ala,His,Arg,Ala,His,Gln,Leu,Ala,Ile,
T18                                                                      Ala,His,Gln,Leu,Ala,Ile,
CT29A                                                                                        Ala,Ile,

         ASP-THR-TYR-GLN-GLU-PHE-GLU-GLU-THR-ILE-PRO-LYS-ASP-GLX-LYS-TYR-SER-PHE-LEU-HIS-ASP-SER-GLX-THR-
          26  27  28  29  30  31  32  33  34  35  36  37  38  39  40  41  42  43  44  45  46  47  48  49  50
T18 cont.    Asp,Thr,Tyr,Gln,Glu,Phe,Glu,Glu,Thr,Tyr,Ile,Pro,Lys
CT29A cont.  Asp,Thr
C5A cont.    Asp,Thr,Tyr,Gln,Glu,Phe,Glu,Glu,Thr,Tyr,Ile,Pro,Lys,Asp,Glx,Lys,Tyr,Ser,Phe,Leu,His,Asp,Ser,Glx,Thr
P4C                          Tyr,Gln,Glu,Phe,Glu
P11B                                     Glu,Glu,Thr,Tyr,Ile,Pro,Lys,Asp,Glx,Lys,Tyr,Ser,Phe,Leu
CT49B                                                        Ile,Pro,Lys,Asp,Glx,Lys,Tyr
T7A                                                                      Asp,Glx,Lys
T12                                                                                  Tyr,Ser,Phe,Leu,His,Asp,Ser,Glx,Thr,
CT37C                                                                                    Ser,Phe,Leu,His,Asp,Ser,Glx,Thr,

         SER-PHE-CYS-PHE-SER-SER-ASX-SER-ILE-PRO-THR-PRO-SER-ASX-MET-GLX-THR-GLX-LYS-SER-ASX-LEU-GLX-LEU-LEU
          51  52  53  54  55  56  57  58  59  60  61  62  63  64  65  66  67  68  69  70  71  72  73  74  75
T12 cont.    Ser,Phe,Cys,Phe,Ser,Asx,Ser,Ile,Pro,Thr,Pro,Ser,Asx,Met,Glx,Thr,Glx,Lys
CT37C cont.  Ser,Phe,Cys,Phe
C5A cont.    Ser,Phe,Cys,Phe,Ser,Asx,Ser,Ile,Pro,Thr,Pro,Ser,Asx,Met
P14-1                        Cys,Phe,Ser,Asx,Ser,Ile,Pro,Thr,Pro,Ser,Asx,Met
3P1H                                                         Ser,Asx,Met,Glx,Thr,Glx,Lys,Ser,Asx,Leu
CT13D                                                                    Glx,Thr,Glx,Lys,Ser,Asx,Leu,Glx,Leu,Leu
P3B                                                                      Glx,Thr,Glx,Lys,Ser,Asx,Leu,Glx,Leu,
C7A                                                                      Glx,Thr,Glx,Lys,Ser,Asx,Leu,Glx,Leu,Leu,
T17A                                                                             Glx,Lys,Ser,Asx,Leu,Glx,Leu,Leu,
```

Table II continued

	76	77	78	79	80	81	82	83	84	85	86	87	88	89	90	91	92	93	94	95	96	97	98	99	100
	ARG	ILE	SER	LEU	LEU	ILE	GLX	SER	TRP	LEU	ARG	PHE	LEU	ARG	SER	MET	PHE	ALA	ASX	LEU
T17A cont.	Arg
CT47C	Arg, Ile, Ser, Leu
P11C	Arg, Ile, Ser, Leu, Leu
P10D	Ile, Ser, Leu, Leu
T24	Ile, Ser, Leu, Leu, Ile, Glx, Ser, Trp, Leu, Glx, Pro, Val, Arg
P3C	Leu, Leu, Ile, Glx, Ser
P9B	Leu, Ile, Glx, Ser, Trp, Leu, Glx, Pro, Val
CT37E	Glx, Pro, Val, Arg, Phe, Leu
T14	Phe, Leu, Arg
CT52B	Phe, Leu, Arg, Ser
CT47B	Arg, Ser, Met, Phe
C7A cont.	Arg, Ile, Ser, Leu, Leu, Ile, Glx, Ser, Trp, Leu, Glx, Pro, Val, Arg, Phe, Leu, Arg, Ser, Met
T13B	Ser, Met, Phe, Ala, Asx, Leu
C7B	Phe, Ala, Asx, Asx, Leu
CT51C	Ala, Asx, Asx, Leu
P5A	Leu

	101	102	103	104	105	106	107	108	109	110	111	112	113	114	115	116	117	118	119	120	121	122	123	124	125
	VAL	TYR	ASX	THR	SER	ASX	ASX	ASX	SER	TYR	HIS	LEU	LEU	LYS	ASX	LEU	GLX	GLX	GLY	ILE	GLX	THR	LEU	MET	GLY
CT51C cont.	Val
T13B cont.	Val, Tyr, Asx, Thr, Ser, Asx, Asx, Ser, Tyr, His, Leu, Leu, Lys
CT37F	Asx, Thr, Ser, Asx, Asx, Asx, Ser, Tyr, His, Leu, Leu, Lys
P5A cont.	Val, Tyr, Asx, Thr, Ser, Asx, Asx, Asx, Ser, Tyr, His, Leu, Leu, Lys, Asx, Leu, Glx, Glx, Gly
C7B cont.	Val, Tyr, Asx, Thr, Ser, Asx, Asx, Asx, Ser, Tyr, His, Leu, Leu, Lys, Asx, Leu, Glx, Glx, Gly, Ile, Glx, Thr, Leu, Met
CT30D	Lys, Asx, Leu, Glx, Glx, Gly, Ile, Glx, Thr, Leu, Met, Gly
3P1E	Leu, Glx, Glx, Gly, Ile, Glx, Thr, Leu
3P1K	Leu, Met, Gly
T8A	Leu, Met, Gly
CT24B	Asx, Leu, Glx, Glx, Gly, Ile, Glx, Thr, Leu, Met, Gly
P5B	Leu, Glx, Glx																							Gly,
C5B																									Gly,

TABLE II continued

```
          ARG-LEU-GLX-ASX-GLY-SER-ARG-ARG-THR-GLY-GLX-ILE-LEU-LYS-THR-TYR-SER-LYS-PHE-ASX-THR-ASX-SER-HIS-
          126 127 128 129 130 131 132 133 134 135 136 137 138 139 140 141 142 143 144 145 146 147 148 149 150
T8A  cont.  Arg
P5B  cont.  Arg,Leu,Glx,Asx,Gly,Ser
P11A        Leu,Glx,Asx,Gly,Ser
3P1K cont.  Arg,Leu,Glx,Asx,Gly,Ser,Arg,Arg,Thr
T9B         Leu,Glx,Asx,Gly,Ser,Arg,Arg
T16C                                    Thr,Gly,Glx,Ile,Leu,Lys
3P1M                                        Ile,Leu,Lys,Glx,Thr,Tyr,Ser,Lys,Phe
T7A                                                 Lys,Glx,Thr,Tyr,Ser
T20D                                                                   Lys,Phe,Asx,Thr,Asx,Ser,His
CT32C                                                                          Asx,Thr,Asx,Ser,His
C5B  cont.                                                                         Asx,Thr,Asx,Ser,His,

          ASX-HIS-ASX-ALA-LEU-LEU-LYS-ASX-TYR-GLY-LEU-LEU-TYR-CYS-PHE-ARG-LYS-ASX-MET-ASX-LYS-VAL-GLX-THR-PHE-
          151 152 153 154 155 156 157 158 159 160 161 162 163 164 165 166 167 168 169 170 171 172 173 174 175
T20D cont.  Asx,His,Asx,Ala,Leu,Leu,Lys
CT32C cont. Asx,His,Asx,Ala,Leu,Leu
C5B  cont.  Asx,His,Asx,Ala,Leu,Leu,Lys,Asx,Tyr,Gly,Leu,Leu,Tyr,Cys,Phe,Arg,Lys,Asx,Met
CT45C                                    Lys,Asx,Tyr
CT37B                                                Gly,Leu,Leu
P11B                                                             Gly
P14-2                Ala,Leu,Leu,Lys,Asx,Tyr,Gly
T16A                                                                 Cys,Phe,Arg
CT37D                                                Asx,Tyr,Gly,Leu,Leu,Tyr,Cys,Phe,Arg
T12D                                                                            Cys,Phe,Arg,Lys,Asx,Met,Asx,Lys,Val,Glx,Thr,Phe,
T14A                                                                                                 Lys
C9A4                                                                                            Asx,Met,Asx,Lys,Val,Glx,Thr,Phe,
                                                                                                 Asx,Lys,Val,Glx,Thr,Phe,

           LEU-ARG-MET-VAL-GLN-CYS-ARG-SER-VAL-GLU-GLY-SER-CYS-GLY-PHE
           176 177 178 179 180 181 182 183 184 185 186 187 188 189 190
T14A cont.  Leu,Arg
C9A4 cont.  Leu,Arg,Met
P11D        Leu,Arg,Met
T13C            Met,Val,Gln,Cys,Arg
CT3A            Arg,Met,Val,Gln,Cys,Arg,Ser,Val,Glu,Gly,Ser,Cys,Gly,Phe
P6D                 Gln,Cys,Arg,Ser,Val,Glu,Gly,Ser,Cys,Gly,Phe
C9B2                    Val,Gln,Cys,Arg,Ser,Val,Glu,Gly,Ser,Cys,Gly,Phe
T5A                             Ser,Val,Glu,Gly,Ser,Cys,Gly,Phe
```

The reaction of HPL with cyanogen bromide resulted in 95 per cent cleavage of the native molecule. After separation of the fragments on Sephadex G-50 (Fig. 1), a small amount of unreacted hormone appeared at the void volume, followed by two major fractions absorbing at 280 nm (C5 and C7) and two fractions (C8 and C9) detectable only after alkaline hydrolysis. The first two peaks were rechromatographed on Sephadex G-50. After performic oxidation and rechromatography on Sephadex G-50, C5, which contained two half-cystine residues, separated into two peaks (C5A and C5B) (Fig. 2). C7 was fractionated further on carboxymethyl cellulose and contained two peptides C7A and C7B (see Table III).

TABLE III

CYANOGEN BROMIDE PEPTIDES OF HPL

Amino acid	C8	C5A	C7A	C7B	C5B	C9A4	C9B2	Total
Lysine	0	2	1	1	4	1	0	9
Histidine	1	3	0	1	2	0	0	7
Arginine	1	1	3	0	4	1	1	11
Tryptophan	0	0	1	0	0	0	0	1
Aspartic acid	1	5	1	7	7	1	0	22
Threonine	1	4	1	2	3	1	0	12
Serine	1	6	4	2	3	0	2	18
Glutamic acid	1	8	6	3	3	1	2	24
Proline	1	3	1	0	0	0	0	5
Half-cystine	0	1	0	0	1	0	2	4
Glycine	0	0	0	1	4	0	2	7
Alanine	1	3	0	1	1	0	0	6
Valine	2	0	1	1	0	1	2	7
*Methionine**	1	1	1	1	1	1	0	6
Isoleucine	0	3	2	1	1	0	0	7
Leucine	2	3	8	5	6	1	0	25
Tyrosine	0	3	0	2	3	0	0	8
Phenylalanine	1	4	1	1	2	1	1	11
Total	14	50	31	29	45	9	12	190

* As homoserine lactone.

Three smaller peptides were identified by high voltage electrophoresis of peaks C8 and 9 (C8, C9A4 and C9B2). After being shown to be homogeneous, these were eluted from paper and subjected to amino acid analysis. Amino acid analyses of the cyanogen bromide fragments (which constitute the entire molecule) are shown in Table III. Since there are six methionine residues, the anticipated seven peptides were obtained once it was appreciated that C5 actually represented two peptides joined by a disulphide bond. The sequence of cyanogen bromide peptides in the molecule based on the tryptic peptide alignment was C8, C5A, C7A, C7B, C5B, C9A4 and C9B2, beginning at the amino terminus (Fig. 3, Table II).

From the homologous assignment of the tryptic peptides of HPL and HGH and the overlapping information for all peptides provided by the results of chymotrypsin, pepsin and cyanogen bromide digestion, it was possible to align all the tryptic peptides in their ordered sequence (see Table II). The individual details of all peptides and their overlaps are shown.

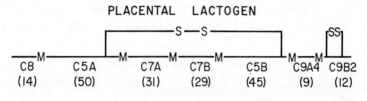

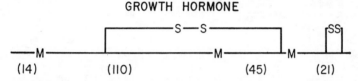

FIG. 3. Diagrammatic model indicating the sites of cyanogen bromide cleavage at methionine residues (M) in HPL and HGH. Cleavage of HPL produced 7 fragments; cleavage of HGH, 4 fragments. The number of amino acid residues in each fragment is indicated.

The amino acid sequence of individual peptides was determined by Edman degradation, permitting the determination of the unique sequence of amino acids in the HPL molecule. Smaller peptides were generally used for sequencing and an average of six to eight cycles were performed. For some of the larger tryptic peptides, additional cleavage with chymotrypsin or pepsin produced fragments small enough to degrade by the Edman method directly. The limiting factor in Edman degradation was incomplete reaction of the insoluble residue after solvent extraction for six or more cycles. The majority of the sequence information was provided by the subtractive method in which the amino-terminal phenylthiohydantoin derivative was removed and the remaining peptide analysed after acid hydrolysis by high sensitivity amino acid analysis. In some instances, direct identification of the amino-terminal amino acid remaining after removal of the Edman derivative was performed by the dansyl technique (Gray 1967).

The tryptophan-containing tryptic peptide T24 was identified by Ehrlich staining (Canfield and Anfinsen 1963b) and by appropriate spectrophotometric techniques. Peptic peptide P9B which contained tryptophan was cleaved at the tryptophan residue with N-bromosuccinimide (Ramachandran and Witkop 1967) producing two fragments P9A1 (Leu, Ile, Glx, Ser) and P9A2 (Leu, Glx, Pro, Val) whose sequence was then determined by Edman degradation.

The assignment of the disulphide bonds was made from the cyanogen bromide and peptic fragments. C5 and P14 each consisted of two peptides joined by a disulphide bond and were easily separated by performic oxidation. The disulphide bond at the carboxyl terminus remained intact in cyanogen bromide fragment C9B2, establishing the disulphide bond from

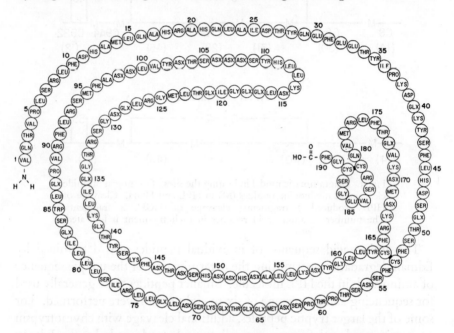

FIG. 4. The amino acid sequence of HPL based on the overlapping peptides from tryptic, chymotryptic, peptic and cyanogen bromide cleavage and Edman degradation at each position. Amide groups not yet determined are indicated as Glx and Asx respectively.

residue 181 and 188. The content of amides has not yet been determined completely. For those already identified, the peptide was digested with leucine aminopeptidase (Sherwood 1969) and then analysed on the amino acid analyser under modified conditions which separate asparagine and glutamine from aspartic and glutamic acids respectively.

The complete sequence of HPL is indicated in Fig. 4 and shows a small disulphide bond at the carboxyl terminus between residues 181 and 188 and a larger one joining residues 53 and 164. A detailed comparison of the sequence of the placental hormone is made with HGH in Table IV which indicates corresponding residues and the substitutions that are present.

TABLE IV

A COMPARISON OF THE AMINO ACID SEQUENCES OF PLACENTAL LACTOGEN AND GROWTH HORMONE

```
HPL: Val-Gln-Thr-Val-Pro-Leu-Ser-Arg-Leu-Phe-Asp-His-Ala-Met-Leu-Gln-Ala-His-Arg-Ala-His-Gln-Leu-Ala-Ile
        x   x   x   x       x   x           x   x   x   x   x       x   x   x       x   x   x   x   x   x
HGH: Phe-Pro-Thr-Ile-Pro-Leu-Ser-Arg-Leu-Phe-Asp-Asn-Ala-Met-Leu-Arg-Ala-His-Arg-Leu-His-Gln-Leu-Ala-Phe
      1                                                                                              25

HPL: Asp-Thr-Tyr-Gln-Glu-Phe-Glu-Glu-Thr-Tyr-Ile-Pro-Lys-Asp-Gln-Lys-Tyr-Ser-Phe-Leu-His-Asp-Ser-Glx-Thr
        x   x   x   x   x       x       x   x   x   x   x       x       x   x   x   x   x   x   x   x
HGH: Asp-Thr-Tyr-Gln-Glu-Phe-Glu-Glu-Ala-Tyr-Ile-Pro-Lys-Glu-Gln-Lys-Tyr-Ser-Phe-Leu-Gln-Asn-Pro-Glu-Thr
      26                                                                                             50

HPL: Ser-Phe-Cys-Phe-Ser-Asp-Ser-Ile-Pro-Thr-Pro-Ser-Asx-Met-Glx-Glx-Thr-Glx-Lys-Ser-Asx-Leu-Glx-Leu-Leu
        x   x   x   x           x   x   x   x   x       x       x   x   x   x   x       x   x       x   x
HGH: Ser-Leu-Cys-Phe-Ser-Glu-Ser-Ile-Pro-Thr-Pro-Ser-Asn-Arg-Glu-Glu-Thr-Gln-Gln-Lys-Ser-Asn-Leu-Gln-Leu
      51                                                                                             75

HPL: Arg-Ile-Ser-Leu-Leu-Leu-Ile-Glx-Ser-Trp-Leu-Glx-Pro-Val-Arg-Phe-Leu-Arg-Ser-Met-Phe-Ala-Asx-Asx-Leu
        x   x   x   x   x   x   x       x   x   x       x   x   x   x   x   x   x   x   x   x       x   x
HGH: Arg-Ile-Ser-Leu-Leu-Leu-Ile-Gln-Ser-Trp-Leu-Glu-Pro-Val-Gln-Phe-Leu-Arg-Ser-Val-Phe-Ala-Asn-Ser-Leu
      76                                                                                            100

HPL: Val-Tyr-Asx-Thr-Ser-Asx-Asx-Ser-Tyr-His-Leu-Leu-Lys-Asx-Leu-Glx-Glx-Gly-Ile-Glx-Thr-Leu-Met-Gly
        x   x   x   x   x   x   x   x   x           x   x       x   x   x   x   x   x   x   x   x   x
HGH: Val-Tyr-Gly-Ala-Ser-Asp-Ser-Asn-Val-Tyr-Asp-Leu-Leu-Lys-Asp-Leu-Glu-Glu-Gly-Ile-Gln-Thr-Leu-Met-Gly
     101                                                                                             125

HPL: Arg-Leu-Glx-Asx-Gly-Ser-Arg-Arg-Thr-Gly-Glx-Ile-Leu-Lys-Gln-Thr-Tyr-Ser-Lys-Phe-Asx-Thr-Asx-Ser-His
        x   x   x   x   x   x   x   x   x       x   x   x   x       x   x   x   x   x   x   x   x   x   x
HGH: Arg-Leu-Asp-Gly-Ser-Pro-Arg-Thr-Gly-Gln-Ile-Phe-Lys-Gln-Thr-Tyr-Ser-Lys-Phe-Asp-Thr-Asn-Ser-His
     126                                                                                             150

HPL: Asx-His-Asx-Ala-Leu-Leu-Lys-Asx-Tyr-Gly-Leu-Leu-Tyr-Cys-Phe-Arg-Lys-Asx-Met-Asx-Lys-Val-Glx-Thr-Phe
        x   x   x   x   x   x   x   x   x   x   x   x   x   x   x   x   x   x   x   x   x   x   x   x   x
HGH: Asn-Asp-Asp-Ala-Leu-Leu-Lys-Asn-Tyr-Gly-Leu-Leu-Tyr-Cys-Phe-Arg-Lys-Asp-Met-Asp-Lys-Val-Glu-Thr-Phe
     151                                                                                             175

HPL: Leu-Arg-Met-Val-Gln-Cys-Arg-Ser-Val-Glu-Gly-Ser-Cys-Gly-Phe-OH
        x   x   x   x   x   x   x   x   x   x   x   x   x   x   x
HGH: Leu-Arg-Ile-Val-Gln-Cys-Arg-Ser-Val-Glu-Gly-Ser-Cys-Gly-Phe-OH
     176                                                         190
```

DISCUSSION

This paper describes the chemical findings that led to the determination of the complete amino acid sequence of HPL. A communication describing this structure has recently appeared (Sherwood *et al.* 1971*a*). HPL is a single chain polypeptide of 190 amino acids with two intrachain disulphide bonds. It contains six methionine residues which have been assigned to positions 14, 64, 95, 124, 169 and 178 and a single tryptophan residue in position 85. The disulphide bonds connect half-cystine residues at positions 53 and 164 and 181 and 188 respectively.

A large number of peptides obtained from selective cleavage of native or modified HPL with trypsin, chymotrypsin, pepsin and cyanogen bromide plus sequential degradation of individual peptides by the Edman (1950) technique provided the unique sequence of amino acids outlined in detail in Table II and shown in Fig. 4.

A detailed comparison of the structure of HPL with that of HGH has been facilitated by recent observations (Niall 1971; Niall *et al.* 1971) showing that the original HGH sequence published by Li, Liu and Dixon (1966) was not completely correct. In his studies with the automatic sequenator, Niall failed to find tryptophan at position 25 in HGH, the position in which Li had suggested it was present. Instead, the tryptophan region comprising 14 amino acids was shifted toward the carboxyl terminus so that tryptophan occupies position 85 in the revised sequence. Likewise, an additional leucine and arginine have been added in positions 92 and 93. These revisions make the homologies between HPL and HGH even more striking than they originally seemed. Earlier, it had appeared that there might be significant differences in the sequences of the two hormones at the amino terminus, but the revised sequence of HGH indicates that the differences are only minor.

A detailed comparison of HPL and HGH is made in Table IV. A total of 163 of 190 residues in the two hormones are identical, accounting for 86 per cent homology in their amino acid sequences. Although the amides have not yet been completely determined, most are identical with those in HGH. Where amino acid substitutions occur, they are usually relatively minor and consist of residues closely related in chemical properties. The four half-cystines and the two disulphide bonds are in identical positions in the two peptides. One major difference is the presence of six methionine residues in HPL compared with three in HGH. This accounts for the seven cyanogen bromide fragments of HPL compared with four of HGH (see Fig. 3). Likewise, there is a significant difference in histidine content, with HPL containing seven and HGH containing three residues (see Table V).

TABLE V
AMINO ACID COMPOSITIONS OF HPL AND HGH

Amino acid	HPL	HGH
Lysine	9	9
Histidine	7	3
Arginine	11	11
Tryptophan	1	1
Aspartic acid	22	20
Threonine	12	10
Serine	18	18
Glutamic acid	24	26
Proline	5	8
Half-cystine	4	4
Glycine	7	8
Alanine	6	7
Valine	7	7
Methionine	6	3
Isoleucine	7	8
Leucine	25	26
Tyrosine	8	8
Phenylalanine	11	13
Total	190	190

The amino acid substitutions in HPL are scattered throughout the molecule, although more are concentrated near the amino terminus, as follows: positions 1–25 (7 substitutions); 26–50 (5), 51–75 (3); 76–100 (3); 101–125 (5); 126–150 (2); 151–175 (1), and 176–190 (1). Of the 27 substitutions, all but three are highly favoured or favoured substitutions (Dayhoff 1969) (Table VI). The three non-favoured substitutions are at

TABLE VI
AMINO ACID SUBSTITUTIONS IN HPL

	Number of substitutions	Position
Non-favoured	3	1, 20*, 132
Acceptable (7–20 × chance)	3	103, 109*, 111, 152
Favoured (21–39 × chance)	5	2, 48, 52, 90, 138
Highly favoured (40 × chance)	15	4, 12, 16, 25, 29, 34, 39, 46, 56, 64, 95, 99, 104, 107*, 178

* Substitution requiring more than a single base change.

positions 1, 20 and 132 and one requires more than a single base change. Of the 24 favoured substitutions, only two require more than a single base change in the triplet codon.

As might be predicted from their marked homologies in primary structure, HPL and HGH share striking similarities in physicochemical properties, secondary and tertiary structure and structure–function relationships. Both hormones have a tendency to aggregate in solution (Sherwood 1967), and in starch, polyacrylamide and agar gel electrophoresis, the two hormones exhibit similar mobilities, although HPL is slightly more acidic (Florini *et al.* 1966; Sherwood 1967). After incubation at alkaline pH, each undergoes deamidation, with the production of electrophoretic heterogeneity (Lewis and Cheever 1965; Sherwood and Handwerger 1969). By optical rotatory dispersion, about half the amino acid residues of each hormone are in α-helical conformation, and the helical content of each is remarkably independent of pH. In urea at pH 8·0, each hormone shows little change in helical content until the urea concentration exceeds 5M (Bewley and Li 1967; Aloj and Edelhoch 1971).

The tertiary structure contributed by the disulphide bonds of HPL and HGH is not essential for lactogenic activity. Dixon and Li (1966) and Bewley, Brovetto-Cruz and Li (1969) showed that reduction and alkylation of HGH with either iodoacetic acid or iodoacetamide caused no decrease in lactogenic potency in the pigeon crop sac assay. Using the mouse mammary gland assay described by Turkington and Topper (1966), we have shown in recent studies that reduction and alkylation or performic acid oxidation of HPL and HGH caused no loss of lactogenic activity (Handwerger, Pang and Sherwood 1971; Sherwood *et al.* 1971*b*). Performic acid oxidation, in addition to converting the half-cystine residues to cysteic acid, converts the methionine residues to methionine sulphone and modifies the single tryptophan residue in each molecule. The two disulphide bonds of HGH are also unnecessary for somatotropic activity as measured by the rat tibia test (Dixon and Li 1966; Bewley, Brovetto-Cruz and Li 1969). Breuer (1969), however, suggested that complete reduction and alkylation of HPL abolished its ability to stimulate protein synthesis in costal cartilage.

Other studies of the relation between structure and function in our laboratory indicated (1) that complete cleavage of HPL and HGH at their methionine residues by cyanogen bromide caused no loss of lactogenic activity and (2) that deamidation of HPL by incubation at alkaline pH resulted in a small but statistically significant enhancement of lactogenic activity (Handwerger, Pang and Sherwood 1971; Sherwood and Handwerger 1969). Similar studies on the somatotropic activity of the modified hormones are now in progress.

With the complete determination of the amino acid sequence of HPL and its homologous alignment with HGH, an earlier prediction of a common ancestral polypeptide for the two hormones has been considerably strengthened (Sherwood 1967). It is also clear that the two molecules are chemically related, although less closely, to ovine prolactin

(Li et al. 1969), but their chemical relationship to human pituitary prolactin remains to be defined (Sherwood 1971). HPL and HGH are so strikingly similar in structure that it is not surprising that there are marked overlaps in their biological and immunological effects. Although the lactogenic activity of the two hormones is identical (Chadwick, Folley and Gemzell 1961; Ferguson and Wall 1961; Josimovich and MacLaren 1962; Handwerger, Pang and Sherwood 1971) somatotropic activity is markedly different, in some cases as much as 100 to 1000-fold (Friesen 1965). Since the amino acid differences between HPL and HGH are relatively few and are concentrated primarily at the amino-terminal end of each hormone, it is possible that the active site for somatotropic activity is located in that portion of the polypeptide chain. It is likely that the lactogenic activity will be found in an identical or nearly identical portion of the polypeptide backbone common to both molecules, and that the somatotropic site is separate or at least only partially overlapping. These studies hold the exciting promise that a small fragment of HGH which might be readily synthesized may have sufficient growth-promoting activity to make it a source of urgently needed HGH. Synthesis of the 190 amino acid peptide recently performed by Li and Yamashiro (1970) is not currently a commercially feasible venture, and synthesis of HGH in tissue culture has so far had only limited success.

SUMMARY

The complete amino acid sequence of human placental lactogen was determined by overlapping the peptides obtained from cleavage of the hormone with trypsin, chymotrypsin, pepsin and cyanogen bromide and by sequential degradation of individual peptides by the method of Edman. Placental lactogen, like human growth hormone, is an 190 amino acid polypeptide containing two intrachain disulphide bonds in identical locations. Eighty-six per cent of the residues in the two molecules are identical, and most of the substitutions are highly favoured, being explained by a single base change in the triplet codon. The striking similarities in chemical structure amplify further the possibility that potent somatotropic activity might be eventually identified in a peptide fragment less than 190 amino acids in length.

Acknowledgements

We are indebted to Mr Michael Lanner for his expert technical assistance; to Dr M. Elzinga for many helpful discussions in the course of this work; to Drs Charles Breuer and Paul Bell of Lederle Laboratories for their generous gifts of HPL; to Dr Alfred Wilhelmi and the National Pituitary Agency for generous gifts of HGH; to Dr H. Hiatt for his support and encouragement; and to Mrs Anita Heber for excellent secretarial assistance. This work was supported by grants HD 03388, CA 10736, and T01 AM 05116 from the United States

Public Health Service and a grant from the John A. Hartford Foundation, Inc. Dr Sherwood is the recipient of a Research Career Development Award K04 AM 46, 426 and Dr Handwerger of a Special Fellowship F03 GM40345 from the U.S.P.H.S.

REFERENCES

ALOJ, S. M. and EDELHOCH, H. (1971) *J. Biol. Chem.* **246,** 5047–5052.
BEWLEY, T. A. and LI, C. H. (1967) *Biochim. & Biophys. Acta* **140,** 201–207.
BEWLEY, T. A., BROVETTO-CRUZ, J. and LI, C. H. (1969) *Biochemistry* **8,** 4701–4708.
BREUER, C. B. (1969) *Endocrinology* **85,** 989–999.
CANFIELD, R. E. (1963) *J. Biol. Chem.* **238,** 2691–2697.
CANFIELD, R. E. and ANFINSEN, C. B. (1963a) *J. Biol. Chem.* **238,** 2684–2690.
CANFIELD, R. E. and ANFINSEN, C. B. (1963b) In *The Proteins*, vol. I, pp. 311–378, ed. Neurath, H. New York: Academic Press.
CATT, K. J., MOFFAT, B. and NIALL, H. D. (1967) *Science* **157,** 321.
CHADWICK, A., FOLLEY, S. J. and GEMZELL, C. A. (1961) *Lancet* **2,** 241–243.
CRESTFIELD, A. M., MOORE, S. and STEIN, W. H. (1963) *J. Biol. Chem.* **238,** 622–627.
DAYHOFF, M. O. (1969) In *Atlas of Protein Sequence and Structure*, vol. 4, pp. 85–87. Silver Spring, Md.: National Biomedical Research Foundation.
DIXON, J. S. and LI, C. H. (1966) *Science* **154,** 785–786.
EDMAN, P. (1950) *Acta Chem. Scand.* **4,** 283–293.
ELZINGA, M. (1970) *Biochemistry* **9,** 1365–1374.
FERGUSON, K. A. and WALL, A. L. C. (1961) *Nature (Lond.)* **190,** 632–633.
FLORINI, J. R., TONELLI, G., BREUER, C. B., COPPOLA, J., RINGLER, I. and BELL, P. H. (1966) *Endocrinology* **79,** 692–708.
FRIESEN, H. G. (1965) *Endocrinology* **76,** 369–381.
FRIESEN, H. G. (1966) *Endocrinology* **79,** 212–213.
GRAY, W. R. (1967) In *Methods in Enzymology*, vol. 11, pp. 469–475, ed. Hirs, C. H. W. New York: Academic Press.
GROSS, E. and WITKOP, B. (1962) *J. Biol. Chem.* **237,** 1856–1860.
HANDWERGER, S., PANG, E. C. and SHERWOOD, L. M. (1971) Submitted.
HIRS, C. H. W. (1967) In *Methods in Enzymology*, vol. 11, pp. 197–199, ed. Hirs, C. H. W. New York: Academic Press.
HIRS, C. H. W., STEIN, W. H. and MOORE, S. (1954) *J. Biol. Chem.* **211,** 941–950.
HIRS, C. H. W., MOORE, S. and STEIN, W. H. (1956) *J. Biol. Chem.* **219,** 623–642.
JOSIMOVICH, J. B. and MACLAREN, J. A. (1962) *Endocrinology* **71,** 209–220.
KATZ, A. M., DREYER, W. J. and ANFINSEN, C. B. (1959) *J. Biol. Chem.* **234,** 2897–2900.
LEWIS, U. J. and CHEEVER, E. V. (1965) *J. Biol. Chem.* **240,** 247–252.
LI, C. H., LIU, W.-K. and DIXON, J. S. (1966) *J. Am. Chem. Soc.* **88,** 2050–2051.
LI, C. H., DIXON, J. S., LO, T.-B., SCHMIDT, K. D. and PANKOV, Y. A. (1969) *Nature (Lond.)* **224,** 695–699.
LI, C. H. and YAMASHIRO, D. (1970) *J. Am. Chem. Soc.* **92,** 7608–7609.
NIALL, H. D. (1971) *Nature New Biol.* **230,** 90–91.
NIALL, H. D., HOGAN, M. L., SAUER, R., ROSENBLUM, I. Y. and GREENWOOD, F. C. (1971) *Proc. Natl. Acad. Sci. (U.S.A.)* **68,** 866–869.
RAMACHANDRAN, L. K. and WITKOP, B. (1967) In *Methods in Enzymology*, vol. 11, pp. 283–298, ed. Hirs, C. H. W. New York: Academic Press.
SHERWOOD, L. M. (1967) *Proc. Natl. Acad. Sci. (U.S.A.)* **58,** 2307–2314.
SHERWOOD, L. M. (1969) In *Progress in Endocrinology*, pp. 394–401, ed. Gual, C. Amsterdam: Excerpta Medica Foundation.
SHERWOOD, L. M. and HANDWERGER, S. (1969) In *Proceedings of the Fifth Rochester Trophoblast Conference*, pp. 230–255, ed. Lund, C. J. and Choate, J. W. Rochester: The Rochester Press.

SHERWOOD, L. M. (1971) *New Engl. J. Med.* **284**, 774–776.
SHERWOOD, L. M., HANDWERGER, S., MCLAURIN, W. D. and LANNER, M. (1971a) *Nature New Biol.* **233**, 59–61.
SHERWOOD, L. M., HANDWERGER, S., MCLAURIN, W. D. and PANG, E. C. (1971b) In *Growth Hormone (Proceedings of the Second International Symposium)*, ed. Pecile, A. and Müller, E. E. Amsterdam: Excerpta Medica Foundation. In press.
TURKINGTON, R. W. and TOPPER, Y. J. (1966) *Endocrinology* **79**, 175–181.
WOODS, K. and WANG, K. T. (1967) *Biochim. & Biophys. Acta* **133**, 369–370.

DISCUSSION

Greenwood: In the sequence around 67 to 70, in HPL, Dr Niall has an additional glutamine. How many amino acids are there in HPL?

Sherwood: Both Dr Li and I find 190 amino acids. Dr Niall has found an additional glutamine residue at position 68. Minor differences in the sequence studies of the three groups have not yet been sorted out.

Greenwood: I find the existence of cyanogen bromide fragments of HPL and HGH with full activity in an *in vitro* assay incredible. You then have to postulate that the tissue receptor site is going to reassemble the molecule, if there is a requirement for a significant tertiary structure. It may simply be that there is a large HPL sequence after cyanogen bromide treatment which is active. Do you oxidize with performic acid first?

Sherwood: We are testing the cyanogen-bromide cleaved and performic-oxidized material now, to find out which fragment or fragments retains lactogenic activity. Whenever one modifies a protein chemically, there may be changes in the primary sequence as well as the secondary and tertiary structure. Clearly there are many precedents for fragments of protein hormones which retain biological activity, such as ACTH, parathyroid hormone and so forth. The pentapeptide of gastrin, which can't have very much tertiary structure, is fully active biologically. When HPL is cleaved with cyanogen bromide there is still one fragment as large as 50 amino acids in length. We are now trying to determine the biologically active site for both lactogenic and somatotropic effects.

Turkington: In view of the fact that you find very little difference even over a 10 or 100-fold range on your biological assay, can you be confident that your alterations of the structure did not affect biological activity? You might have affected only 80 per cent of the population of molecules and still have obtained the same amount of activity in the assay.

Sherwood: While amino acid analysis indicated complete conversion to the derivative form, the radioimmunoassay data showed complete loss of immunoreactivity, ruling out the presence of any native material. There was no immunoactivity in the cyanogen bromide fragments of HPL and less than 1 per cent in the HGH fragments. Performic-oxidized HPL even at a thousand times the concentration of native hormone caused no displacement in the radioimmunoassay.

Li: When we reduce and alkylate HCS (HPL) very carefully in the absence of denaturant, only one S-S bridge is broken, and yet over 50 per cent of the lactogenic activity (as estimated by the pigeon crop sac assay) is lost. However, when we reduced completely and alkylated both S-S bridges, we lost almost all lactogenic activity. Yet when you reduced and alkylated with the same reagent (iodoacetamide), you found no loss of lactogenic activity. We found that it lost both growth-promoting and lactogenic activity. We find that after performic acid oxidation almost every protein hormone loses all its activity.

Secondly, in the radioimmunoassay of the reduced and alkylated HGH, you found that less than 2 per cent activity was retained. However, Dr R. Luft in Stockholm and Dr M. M. Grumbach tested our HGH derivative and found 50 to 80 per cent of the activity was retained. Finally, have you really fully established the sequences?

Sherwood: The structural work we have done is based on over 1400 amino acid analyses using the subtractive Edman method. We have determined the sequence of all of the tryptic peptides and many of the chymotryptic and peptic peptides to complete the total sequence. I did not present the final determination of all of the amides. Firstly, we have not established all of them unequivocally, although in most instances they are similar to the ones in growth hormone. Secondly, it is difficult to be absolutely confident of the presence of glutamine versus glutamic acid, or asparagine versus aspartic acid, in all positions. The placental hormone, like growth hormone, is very easily deamidated, particularly during some of the *in vitro* procedures used to purify the peptides. One has to be very careful therefore about the assignment of the amides. In several places in the original growth hormone structure (Li, Liu and Dixon 1966), there were incorrect assignments.

We have repeatedly tested the performic-oxidized preparations. In our hands, oxidized HPL has absolutely no immunological activity, yet it is active biologically both *in vitro* and *in vivo* (Handwerger, Chan and Sherwood 1970). Although our results differ, we are using different assay systems. Drs C. Beck and K. J. Catt (1971) have also shown that the performic-oxidized placental lactogen and growth hormone have lost most of their immunological activity.

Forsyth: On the question of biological activity, most people have been talking about growth hormone activity and opposing it to prolactin-like activity, measured in various ways, and suggesting that different portions of the molecule are responsible for these two activities. I think there is no good reason to treat prolactin-like activity as a single entity; it's quite possible that different prolactin-like activities may themselves be related to different, though perhaps overlapping sequences. We have done bioassays of different preparations of human growth hormone using both the rabbit mammary intraductal test and a local pigeon crop assay (Forsyth and Folley 1970). One preparation, HS551F, from Dr Wilhelmi, showed quite a different ratio of these two activities from all the others that we tested. It seems quite possible, therefore, that prolactin-like activity isn't a single entity but that there may be different portions of the molecule responsible for the different activities which we associate with prolactin.

Greenwood: We have known for several years that you can get divergence of results from different immunoassays, depending upon different antibody specificities.

Cotes: I agree with Dr Greenwood. I find little difficulty in reconciling discrepant findings obtained with different radioimmunoassays if the discrepancies are attributable to differences in the specificities of the antisera used. Reports of such work would be much clearer if authors did not just define systems as radioimmunoassays, but also gave an identification of the antiserum.

Greenwood: I would like to propose a bridge between the mouse mammary gland and the tibia by postulating that the tissue receptors are different. If a hormone is a coded message, it may have a coded message for the tibia which is different from the coded message for the mammary gland.

I would also like to clarify our interest in HPL. Dr Niall came to our laboratory and helped Dr I. Y. Rosenblum to set up the Beckman sequenator. We decided that HPL was a good apprenticeship for structural work since Dr Niall with Drs Catt and Moffat (Catt, Moffat and Niall 1967) had been the first to attempt its sequence and because we had first picked up HPL by radioimmunoassay (Greenwood, Hunter and Klopper 1964). Homologies between our sequences of HPL and the published data on HGH and ovine prolactin were noted but Dr Niall recognized that the HGH sequence did not fit as well as might be expected. This led him to re-sequence the first 25 amino acids of HGH (Niall 1971). The revised sequence improved the homologies between these three peptides and enabled us to postulate a common primordial peptide for HPL, HGH and ovine prolactin (Niall *et al.* 1971). From these data one can almost write the structure of human prolactin.

I would like to pay tribute to the work of Dr Li and his group. We appreciate more than most the hard work involved in sequencing HGH, ovine prolactin and then HPL before the automated sequencer became available. Dr Niall and I got together with Dr Li just before the Prolactin Workshop at the National Institutes of Health in January, 1971 and discussed our results, and Dr Li was most generous in his formal presentation on the full sequence of HPL at that meeting.

Friesen: Placental lactogen is also found in monkeys. Dr B. Shome in our laboratory (Shome and Friesen 1971) has purified monkey placental lactogen and it's interesting that there are two molecular forms of MPL with slightly different amino acid compositions. When we were studying the biosynthesis of monkey placental lactogen (Friesen 1968), we identified MPL as the principal radioactive component. A second component, which contained a very substantial proportion of the radioactive proteins as well, remained as an unknown. It appears that this component was in fact the other component of monkey placental lactogen. Has anyone any information on placental lactogens in other species? I believe Dr Forsyth has preliminary evidence for one in the goat.

Forsyth: We have bioassay evidence of a placental lactogen in the goat (see pp. 151–167).

Bryant: Several years ago we picked up what we thought was ovine placental lactogen in a radioimmunoassay for ovine prolactin; we found a slope change as we followed plasmas taken throughout pregnancy. We haven't done the isolation work yet.

Sherwood: One of the interesting aspects of comparative endocrinology is the relative ease with which one finds prolactin in the pituitaries of lower forms and the great difficulty, until recently, of finding it in human pituitary extracts (Sherwood 1971). On the other hand, the placental hormone is so prominent in monkeys and man. One wonders about diverging pathways of evolution, as Dr Forsyth mentioned. Studies of placental extracts of lower mammals would be of great interest in trying to complete this picture.

Meites: Some years ago we (Meites and Turner 1948) and Lyons (1958) reported the presence of pigeon crop sac and mammary growth activity in placentas from rats, guinea pigs, rabbits and mice. Mammary-stimulating activity in human placenta was also reported many years before the present interest in HPL (see Lyons 1958). Thus there is biological evidence for placental activity in several species.

Something which puzzles me is that the sequences of human placental lactogen and human growth hormone are very similar, yet from the point of view of biological activity, human placental lactogen has about 85 per cent lactogenic activity, whereas human growth hormone has mainly growth-promoting activity.

Forsyth: We found that HPL and HGH are very similar in their lactogenic activity in the rabbit intraductal assay. Both have quite marked activity in this system, and also on rabbit and mouse mammary glands *in vitro*. The HPL and HGH preparations that we studied both have low activity on the pigeon crop (Forsyth and Folley 1970).

Greenwood: Can you express the activity in terms of a percentage of ovine prolactin activity in the rabbit assay?

Forsyth: I don't think that would be very meaningful, because potency varies from preparation to preparation, and I would have thought that activity varying from say about 80 to 120 per cent probably is not significant.

Frantz: We have done some work on that too and we find that most growth hormone preparations assay out at about 50–70 per cent of the prolactin activity of a highly purified ovine prolactin standard, NIH-S-8. The HPL that we have studied less thoroughly assays a bit lower, probably at about 30–50 per cent of the activity of ovine prolactin.

Forsyth: That certainly isn't our experience; essentially the three purified preparations are about the same.

Beck: But these are two entirely different bioassay systems.

Sherwood: Our findings are similar to Dr Frantz's, in that placental lactogen and growth hormone give comparable lactogenic responses. Ovine prolactin is perhaps 20 or 30 per cent more active in most of our studies.

In relation to Dr Meites' question, the difference in growth-promoting activity between HPL and HGH may lie either in primary structure or in three-dimensional structure. I would predict that it is due to differences in the primary sequence. Recent studies by Aloj and Edelhoch (1971) show minimal differences in the conformation of HPL and HGH by physicochemical techniques. Studies by Yamasaki, Kikutani and Sonenberg (1970) have shown that fragments of bovine growth hormone may have some growth-promoting activity.

Cotes: I wonder if either Dr Li or Dr Sherwood could tell us whether the dimer of HPL is simply a purification artifact or occurs under physiological conditions. I understood from Dr Li (p. 24) that the dimer and monomer show different biological activity (at least in respect of growth-promoting activity). Thus, if preparations of HPL do not all contain the same proportion of dimer, the ratio of growth-promoting to lactogenic activity of any preparation may be related to its content of monomer and dimer.

Li: We are working on the dimer of HGH as well as on the dimer of HCS (HPL), and also on the monomers of both. The biological profile of the HGH dimer is different from that of the HGH monomer; the growth-promoting activity is lower in the dimer. The lactogenic activity might also vary. On the other hand, the HCS (HPL) monomer has more growth-promoting activity than the dimer, but the lactogenic activity is the same. What it means biologically I can't say because we don't know whether HCS or HGH exist as the monomer or the dimer in the body.

Wilhelmi: Do you think the dimer of HCS (HPL) is a stable species when you inject it, or is there some tendency to revert to the monomer as it is diluted?

Li: We do not have much information on HCS (HPL) but we have some on HGH. You can convert the dimer of HGH to a monomer by 6M-guanidine hydrochloride or even 4 M. We can partially convert the monomer back to the dimer very easily, by leaving it at room temperature or lyophilizing it.

Friesen: On gel filtration of pregnant human serum and crude human placental extracts, as well as our purified preparation of HPL, the elution volumes are all identical, which suggests to me that the circulating form of HPL is probably the dimer. Certainly 90 per cent of the activity in the serum has the elution pattern of the dimer of HPL.

Wilhelmi: This could be a concentration phenomenon?

Friesen: Yes, I believe it is, but when human serum is fractionated on Sephadex the HPL behaves as a dimer and it is this form which I believe is circulating.

Greenwood: Dr Friesen, our 'ovine placental lactogen' cross-reacts with ovine prolactin rather than with ovine growth hormone. Does your monkey placental lactogen cross-react with monkey growth hormone?

Friesen: Yes, and with HPL to a lesser degree.

Greenwood: Is your monkey prolactin activity in pregnancy in fact monkey placental lactogen?

Friesen: No. Monkey placental lactogen fails to cross-react in our radioimmunoassay for monkey prolactin until we reach concentrations 10 000-fold greater than the minimal amount of monkey prolactin which cross-reacts in the assay.

Cowie: I think confusion is creeping in with the use of the term 'prolactin-like' activity. The activity as tested should be defined, for example pigeon crop activity or 'lactogenic' activity. 'Prolactin-like' activity is confusing when applied to preparations exhibiting low pigeon crop activity but high lactogenic activity.

REFERENCES

ALOJ, S. M. and EDELHOCH, H. (1971) *J. Biol. Chem.* **246,** 5047–5052.
BECK, C. and CATT, K. J. (1971) *Endocrinology* **88,** 777–782.

CATT, K. J., MOFFAT, B. and NIALL, H. D. (1967) *Science* **157**, 3786.
FORSYTH, I. A. and FOLLEY, S. J. (1970) In *Ovo-implantation: Human Gonadotropins and Prolactin*, pp. 266–278, ed. Hubinont, P. O., Leroy, F., Robyn, C. and Leleux, P. Basel: Karger.
FRIESEN, H. (1968) *Endocrinology* **83**, 744–753.
GREENWOOD, F. C., HUNTER, W. M. and KLOPPER, A. (1964) *Br. Med. J.* **1**, 22.
HANDWERGER, S., CHAN, E. and SHERWOOD, L. M. (1970) *Clin. Res.* **18**, 455.
LI, C. H., LIU, W.-K. and DIXON, J. S. (1966) *J. Am. Chem. Soc.* **88**, 2050–2051.
LYONS, W. (1958) *Proc. R. Soc. B* **149**, 303–325.
MEITES, J. and TURNER, C. W. (1948) *Res. Bull. Mo. Agric. Exp. Stn.* nos. 415, 416.
NIALL, H. D. (1971) *Nature New Biol.* **230**, 90–91.
NIALL, H. D., HOGAN, M. L., SAUER, R., ROSENBLUM, I. Y. and GREENWOOD, F. C. (1971) *Proc. Natl. Acad. Sci. (U.S.A.)* **68**, 866–869.
SHERWOOD, L. M. (1971) *New Engl. J. Med.* **284**, 774–777.
SHOME, B. and FRIESEN, H. (1971) *Endocrinology* in press (Oct.).
YAMASAKI, N., KIKUTANI, M. and SONENBERG, M. (1970) *Biochemistry* **9**, 1107–1114.

REGULATION OF THE SYNTHESIS AND RELEASE OF PROLACTIN

ROBERT M. MACLEOD AND JOYCE E. LEHMEYER

Department of Internal Medicine, University of Virginia School of Medicine, Charlottesville, Virginia

THE mechanisms which regulate the biosynthesis of pituitary hormones are varied and complex. Some factors which are active at the hypothalamic or pituitary level stimulate the production of one pituitary hormone yet inhibit the production of others. This paper will summarize some of our investigations of several factors which modify the *in vivo* and the *in vitro* synthesis of prolactin.

It is well established that the incubation *in vitro* of the anterior pituitary glands of rats with labelled amino acids results in incorporation of large quantities of radioactivity into both prolactin and growth hormone (Catt and Moffat 1967; MacLeod and Abad 1968). While the glands of female rats synthesize comparable amounts of prolactin and growth hormone, the pituitary cells regulate the release of these hormones quite differently. A typical experiment is illustrated in Fig. 1. During the first hour of

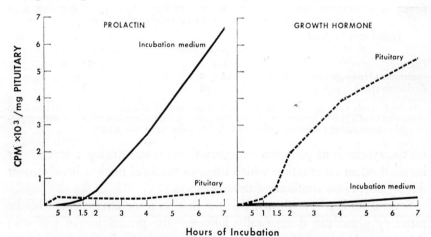

FIG. 1. The *in vitro* synthesis and release of prolactin and growth hormone. Anterior pituitary glands from female rats were incubated with 10 μCi [4,5-^{3}H]leucine in Medium 199 containing bicarbonate at 37°C under 95 per cent O_2–5 per cent CO_2. At the end of the incubation period, the glands were homogenized in phosphate buffer and aliquots of the homogenates and of the incubation media were subjected to polyacrylamide gel electrophoresis.

incubation with radioactive leucine most of the newly synthesized prolactin was retained by the pituitary gland. With continued incubation, however, no further accumulation of hormone within the gland occurred but, instead, the radioactive prolactin was released into the incubation medium. At the termination of a seven-hour incubation, approximately ten times more labelled prolactin was found in the incubation medium than in the pituitary gland. Addition of NIH anti-rat prolactin antibody to aliquots of the incubation medium precipitated 71 per cent of the radioactive prolactin which was detected by the polyacrylamide gel electrophoresis method. Conversely, growth hormone was retained by the gland throughout the incubation period, little being released into the medium. When glands of male rats were incubated *in vitro*, very little radioactive prolactin was produced but the glands synthesized two to three times more growth hormone than did glands of female animals (MacLeod, Abad and Eidson 1969).

When pituitary glands from female rats were incubated with puromycin, the incorporation of radioactive leucine into prolactin and growth hormone was completely inhibited. Thus it is readily apparent that the presence of radioactivity in the pituitary hormones after incubation with labelled amino acids in the absence of the antibiotic was due to *de novo* protein synthesis (Table I). Actinomycin D, however, was without effect

TABLE I

EFFECT OF ANTIBIOTICS ON THE *IN VITRO* INCORPORATION OF [4,5-^{3}H]LEUCINE INTO PROLACTIN AND GROWTH HORMONE BY THE PITUITARY GLAND

Treatment of incubated gland	Prolactin c.p.m. per mg gland	Growth hormone c.p.m. per mg gland
Control	2770±20	1283±78
Puromycin (100 μg)	18± 8	27± 3
Actinomycin D (1 μg)	2222±38	1518±67

In the controls, two pituitary glands from female rats were incubated in 1·4 ml of Medium 199 containing 10 μCi [4,5-^{3}H]leucine for 7 hours. Puromycin and actinomycin were present in the experimental cultures during the 7-hour culture period. The values are means ± S.E.M.

on the synthesis of prolactin and growth hormone during a seven-hour incubation, an observation which suggests minimal nuclear involvement during the *in vitro* synthesis of these hormones.

Several studies have shown that the *in vitro* ionic environment of the pituitary gland can dramatically influence the production of prolactin (MacLeod, Abad and Eidson 1969; Parsons 1969). In the first experiment presented in Fig. 2 (upper graph, left) pituitary glands, incubated in Krebs-Ringer bicarbonate buffer (KRB) and fortified with a complete amino acid mixture, synthesized and released large quantities of radioactive prolactin. When Krebs-Ringer phosphate buffer (KRP) was substituted for KRB a dramatic decrease in prolactin synthesis occurred and most of the newly synthesized hormone was retained within the pituitary gland. The second experiment (upper graph, right) once again demonstrates the inhibitory action of KRP buffer on prolactin synthesis, but also shows that the addition of bicarbonate ion to the KRP incubation medium restored synthesis and release of the hormone. It should be noted that in Krebs-Ringer phosphate buffer more newly synthesized prolactin was present

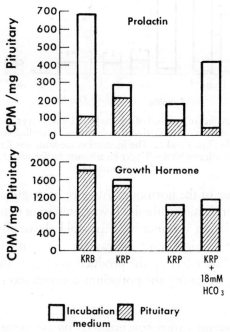

FIG. 2. The effect of bicarbonate on the synthesis and release of pituitary hormones. Pituitary glands from female rats were incubated in Krebs-Ringer bicarbonate or Krebs-Ringer phosphate buffer for 7 hours, as described in Fig. 1.

within the gland than in the incubation medium. In KRB or KRP containing bicarbonate, however, seven or eight times more radioactive prolactin was found in the medium than in the gland. The results of these experiments indicate that the primary effect of the bicarbonate ion is on the release of prolactin from the gland. No effect of bicarbonate ion on growth hormone synthesis or release was found (Fig. 2, lower graph).

The importance of the cationic environment for prolactin synthesis and release is demonstrated by the experiments illustrated in Fig. 3. The omission of potassium or, more importantly, of calcium, from either KRB or KRP buffer significantly decreased the synthesis of prolactin. The simultaneous omission of both Ca^{++} and K^+ did not further reduce the

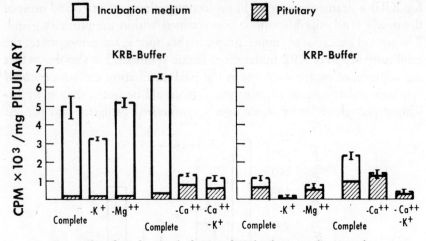

FIG. 3. The effect of removal of cations from buffers on prolactin synthesis. The methods of incubation and detection of labelled hormones are described in Figs. 1 and 2. The incubation medium was Krebs-Ringer phosphate buffer or Krebs-Ringer bicarbonate buffer, with K^+, Mg^{++} or Ca^{++} omitted.

synthesis or release of the hormone. Absence of magnesium ion had little, if any, effect on prolactin synthesis. It was interesting to note that when the calcium concentration was increased five-fold above physiological levels, the ion produced a significant increase in the synthesis and release of prolactin but had no effect on the production of growth hormone (Table II). Conversely, increasing the potassium concentration five-fold did not

TABLE II

EFFECT OF INCREASING THE CATION CONCENTRATION ON THE SYNTHESIS AND RELEASE OF PITUITARY GLAND HORMONES

Culture medium	Prolactin c.p.m. per mg gland		Growth hormone c.p.m. per mg gland	
	Pituitary	Medium	Pituitary	Medium
Control: Medium 199	456± 36	1978±309	1293±137	227±39
Medium 199+21 mM K^+	766±163	1914±186	1274±119	711±70*
Medium 199+5.1 mM Ca^{++}	495±110	3338± 77*	1330± 54	305±58
Medium 199+21 mM K^+ +5.1 mM Ca^{++}	836± 58*	3009±153*	1458±200	783±42*

Pituitary glands from female rats were incubated in Medium 199 containing 10 μCi [4,5-^{3}H]leucine. The indicated cation concentrations were obtained by using the chloride salts. Means± S.E.M.
* $P<0.01$.

influence prolactin production but caused a three-fold stimulation of growth hormone release. Simultaneously increasing the potassium and calcium concentrations significantly increased the synthesis and release of both hormones. Since bicarbonate and calcium affect only prolactin synthesis and release, while potassium stimulates only growth hormone release, it is evident that the synthesis and release of each hormone is regulated by a different mechanism.

Perturbations of the endocrine balance *in situ* are manifested by an increase or decrease in prolactin production. Several years ago we observed that implantation of hormone-secreting pituitary tumours into rats, or of pituitary isografts into mice, caused atrophy of the hosts' pituitary glands and decreased the concentration of pituitary hormones in the gland (MacLeod et al. 1966; MacLeod, Smith and DeWitt 1966; MacLeod 1970). When the pituitary glands of these animals were incubated *in vitro*, it was

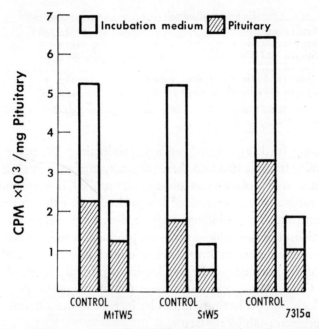

FIG. 4. The incorporation of [4,5-³H]leucine into prolactin by pituitaries from rats bearing pituitary tumours. Pituitary glands from normal female rats and females bearing transplanted hormone-secreting tumours were incubated for 7 hours and the radioactive prolactin detected as described in Fig. 1.

found that the synthesis and release of prolactin were greatly diminished when compared to production of the hormone in glands of normal animals (MacLeod and Abad 1968). Fig. 4 illustrates the effects of these tumours on prolactin production in the hosts' pituitary glands. The prolactin-secreting tumours MtTW5 and 7315a decreased the *in vitro* biosynthesis of prolactin by 55–70 per cent. Similarly, the pituitary tumour StW5, which secretes primarily growth hormone but detectable amounts of prolactin as well, also decreased the *in vitro* synthesis of prolactin.

Because of the well-known luteotropic effects of prolactin, a study was initiated to determine whether the tumour hormones exerted their feedback inhibition of prolactin synthesis via the ovary or through more central mechanisms. It was found that pituitary glands of spayed female rats when incubated *in vitro* synthesized only 17 per cent as much prolactin as did glands of intact animals (Table III). A combination of ovariectomy and

TABLE III

EFFECT OF PITUITARY TUMOURS ON THE *IN VITRO* SYNTHESIS OF PROLACTIN BY PITUITARY GLANDS FROM SPAYED AND ADRENALECTOMIZED RATS

Experimental animals	Prolactin c.p.m. per mg gland
Intact control	8770 ± 330
Spayed control	1480 ± 356
Spayed MtTW5	93 ± 25
Spayed StW5	516 ± 124
Spayed and adrenalectomized control	728 ± 39
Spayed and adrenalectomized 7315a	145 ± 37

Female rats were ablated 8 weeks before sacrifice, and their pituitary glands were incubated *in vitro* as previously described. Means $\pm$ S.E.M.

adrenalectomy further decreased prolactin production. When the pituitary tumours were transplanted into these surgically altered animals and the *in vitro* production of prolactin was subsequently studied, it was found that the glands of these animals synthesized less prolactin than the corresponding ablated controls. These findings suggested that the ovary was not the primary initiator of the reduced prolactin production and it was concluded that the pituitary tumour hormones inhibited the *in vivo* and *in vitro* synthesis of the hormone at the hypothalamo-pituitary level. Additional work by Chen, Minaguchi and Meites (1967) suggested that pituitary tumours decreased prolactin synthesis by increasing the amount of hypothalamic prolactin-inhibiting factor.

Although in this instance the ovary was not the primary initiator of decreased prolactin synthesis, oestrogens have an important function in governing the production of the hormone. Administration of a long-acting oestrogen, polyoestradiol phosphate, caused a significant increase in prolactin production in both male and female rats (Table IV). When animals bearing pituitary tumours were injected with the oestrogen, the steroid overcame the suppressive action of the tumour hormones and potentiated prolactin synthesis in the hosts' glands, a result similar to that found with glands of normal animals.

Among their many other effects, oestrogens increase the turnover of catecholamines in the hypothalamus (Coppola *et al*. 1965). Hence a study was begun to determine whether catecholamines and agents which influence their production have an effect on the synthesis and release of prolactin. Additionally, the biochemical mechanisms through which the

TABLE IV

THE EFFECT OF OESTROGEN ON THE *IN VITRO* SYNTHESIS OF PROLACTIN BY THE PITUITARY GLAND

Treatment of rats	Prolactin	
	Total c.p.m. per mg gland	Percentage change
Female control	4627 ± 288	—
Female control + 100 μg polyoestradiol phosphate	7446 ± 256	+61
Male control	660	
Male control + 200 μg polyoestradiol phosphate	1405	+111
Female MtTW5	778 ± 115	
Female MtTW5 + 100 μg polyoestradiol phosphate	4071 ± 388	+425
Female StW5	1731 ± 128	
Female StW5 + 100 μg polyoestradiol phosphate	2922 ± 167	+69
Female 7315a	1447	
Female 7315a + 200 μg polyoestradiol phosphate	7147	+375

Rats bearing pituitary tumours for 8 weeks were injected subcutaneously with oestrogen one week before sacrifice. Pituitary glands were incubated with 10 μCi [4,5-^{3}H]leucine for 7 hours, as previously described.

pituitary tumour hormones exert their inhibitory action at the hypothalamo-pituitary level were investigated. Recent studies have shown that the *in vitro* addition of noradrenaline or of dopamine profoundly inhibits

TABLE V

EFFECT OF CATECHOLAMINES AND METABOLITES ON THE INCORPORATION OF [4,5-^{3}H]LEUCINE INTO PROLACTIN BY THE PITUITARY GLAND

Compound added to culture medium	Prolactin c.p.m. per mg gland	
	Pituitary	Medium
1. Control	396 ± 60	3283 ± 247
2. L-Adrenaline	335 ± 51	386 ± 44
3. D-Adrenaline	389 ± 84	538 ± 35
4. L-Noradrenaline	489 ± 185	325 ± 21
5. DL-Metanephrine	425 ± 15	4175 ± 369
6. DL-3,4-Dihydroxymandelic acid	450 ± 17	3454 ± 446
7. DL-3-Methyl, 4-hydroxymandelic acid	381 ± 56	3710 ± 217

Compounds 2–4: 2×10^{-6}M.
Compounds 5–7: 2×10^{-5}M.
Pituitary glands of female rats were cultured for 7 hours in the presence of 10 μCi [4,5-^{3}H]leucine as previously described. Means ± S.E.M.

prolactin production by pituitary glands (MacLeod 1969; MacLeod, Fontham and Lehmeyer 1970; Birge *et al.* 1970). Table V shows that 2×10^{-6}M-adrenaline or noradrenaline caused a 90 to 95 per cent decrease in the amount of newly synthesized prolactin released into the incubation medium during a seven-hour incubation. The oxidized and methylated metabolites of the catecholamines were completely inactive. Incubation of glands with 10^{-6}M-dopamine hydrochloride almost completely inhibited the release of newly synthesized prolactin into the incubation medium and caused an accumulation of labelled hormone within the glands (Fig. 5). Incubation with 10^{-6}M-noradrenaline also decreased the release of

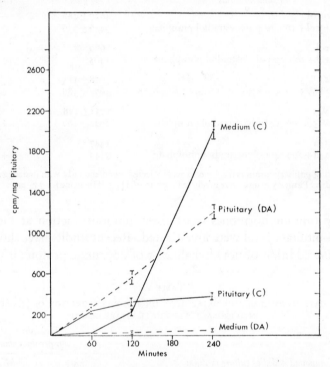

FIG. 5. The effect of dopamine on the incorporation of [4,5-³H]leucine into prolactin *in vitro*. The incubation procedure is described in Fig. 1. Dopamine hydrochloride (DA) was added at a concentration of 10^{-6}M and aliquots of the incubation medium were removed for analysis after 1, 2, 3 and 4 hours of incubation. C, control culture.

labelled prolactin and produced an accumulation of labelled hormone within the tissue. Subsequently, total prolactin synthesis was decreased approximately 50 per cent by noradrenaline. It is evident from these findings that the primary effect of dopamine and noradrenaline is the inhibition of release of newly synthesized prolactin, rather than inhibition of synthesis. Presumably, as prolactin accumulates within the gland other biochemical events are initiated which inhibit synthesis of the hormone. These results confirm the puromycin data and show that the mechanism governing the release of prolactin is independent of synthesis of the protein.

Distinct morphological changes are produced in the pituitary when it is incubated with noradrenaline. Electron microscopic examination of the pituitary glands of female rats by Dr Carlo Bruni (Department of Pathology) revealed that the prominent effect of adrenaline is on the two membranous systems of the cytoplasm, the rough endoplasmic reticulum (RER) and the Golgi system (Fig. 6).

The overall fine structure of the mammatrophs is similar in many basic respects to that observed in other cell types which secrete protein material for export. They have a well-developed RER which consists of long, flattened vesicles (cisternae) converging with one of their extremities toward the Golgi system. Secretory material within the cisternae of the RER is of low density, but it is of high density within the vacuoles in the Golgi area. This indicates that secretory material is first segregated within the cisternae of the RER and becomes condensed into secretory granules of higher density in the Golgi system and in the Golgi area. Granules of intermediate density are observed in the areas between the RER and the Golgi. The mature secretory granules, those of higher density and large size, are finally extruded into the extracellular environment. In the act of extrusion, the secretory granule loses the surrounding membrane as a result of its fusion with the plasma membrane.

The prominent effect of noradrenaline treatment was the dilatation of the RER and of the Golgi system—that is, the two systems involved in the segregation and in the transport of the secretory material from the site of synthesis, the ribosomes associated with the RER, to the extracellular environment. This dilatation of the two membranous components of the cytoplasm of the mammatrophs was not observed in all cells and varied in degree among different groups of cells. Mammatrophs free of deviations were found in the treated glands, but in remarkably lower number than in the control. Associated with the dilated condition of the ER and the Golgi was a decrease in the density of the ground substance of the cytoplasm. It should be pointed out that this less dense condition of the cytoplasm is not a result of a decrease in the number of ribosomes free in the ground substance. The number of these ribosomes in the treated glands was the same as in the controls. No differences were noticed in the quantity of ribosomes attached to the membranes of the ER, presumably the site of production of prolactin. Significant differences in the number of secretory granules between the two groups of animals were not detected, indicating that their activity in synthesizing prolactin was similar.

We are led to interpret these observations as suggesting that noradrenaline induces damage of the membranous components of the cytoplasm, which in turn interferes with their normal activity in the transport of secretory material into the extracellular environment.

It has been reported that adrenaline has a biphasic action on prolactin production and that lower concentrations of noradrenaline cause a significant increase in prolactin release (Koch, Lu and Meites 1970). Although experiments in our laboratory showed that only 2×10^{-7}M-noradrenaline increased prolactin release by 20 per cent, the result was not statistically significant (Table VI). Several other biological amines tested, such as

TABLE VI

EFFECT OF THE CONCENTRATION OF NORADRENALINE *IN VITRO* ON THE RELEASE OF PROLACTIN BY THE PITUITARY GLAND

	Release of prolactin as percentage of control	
Control	100	(N.S.)
With noradrenaline:		
2×10^{-8}M	110	(N.S.)
2×10^{-7}M	120	(N.S.)
1×10^{-6}M	30	$P < 0.001$
2×10^{-6}M	8	$P < 0.001$

Pituitary glands from female rats were incubated in 1·4 ml Medium 199 containing 10 µCi [4,5-^{3}H]-leucine and noradrenaline for 7 hours, as previously described.

cadaverine, putrescine, spermidine, 5-hydroxytryptamine and vasopressin, were without effect on prolactin release.

The previous experiments have demonstrated that pituitary glands from female rats are capable of initiating prolactin synthesis and release immediately upon incubation *in vitro* and that the addition of a catecholamine to the incubation medium inhibits the release of labelled prolactin. The following experiment was conducted to see whether catecholamines are effective in inhibiting the release of prolactin after the pituitary gland has initiated the *in vitro* synthesis and release of the hormone. It was found that after a four-hour preincubation with radioactive leucine, the addition of 10^{-5}M-noradrenaline caused a detectable inhibition of prolactin release as early as 30 minutes after the addition of the catecholamine and produced an almost complete inhibition after 60 minutes (Table VII).

Apparently, the tissue must be continuously exposed to the catecholamine for a period of time before prolactin release is inhibited. When pituitary glands were incubated with 10^{-5}M-noradrenaline for 15 minutes, then transferred to fresh medium without noradrenaline and re-incubated for various periods of time from 30 minutes to seven hours, prolactin synthesis and release were not affected by this short exposure to noradrenaline. If glands were pre-incubated with the catecholamine for 180

A

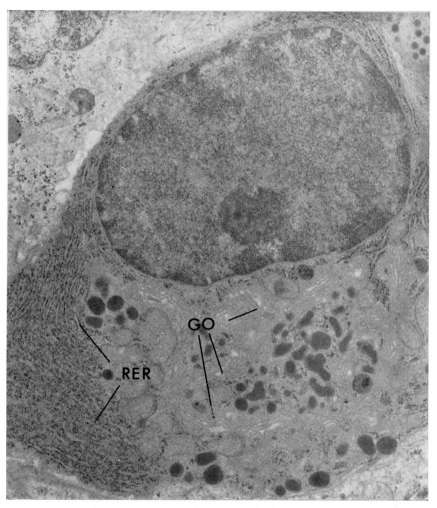

FIG. 6. Electron micrographs of mammatrophs from a control pituitary, A and a pituitary incubated for 2 hours with noradrenaline, B. In the latter the rough endoplasmic reticulum (RER) and the Golgi system (GO) are moderately but distinctly dilated. The ground substance of the cytoplasm is denser in the control mammatroph. × 13 300.

B

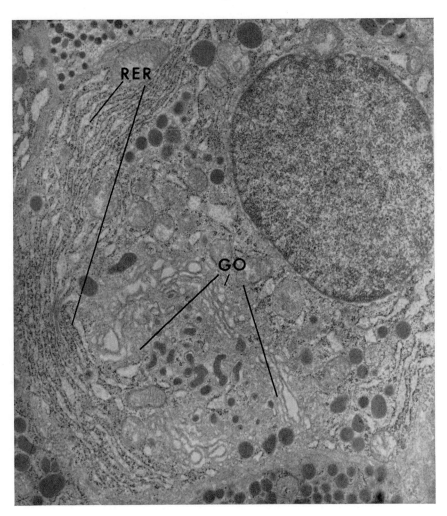

PROLACTIN REGULATION

TABLE VII
EARLY EFFECT OF NORADRENALINE ON THE RELEASE OF PROLACTIN BY THE PITUITARY GLAND
IN VITRO

Incubation period hours	Prolactin Total c.p.m. per mg gland		Percentage change
2	258		
4	1898		
Noradrenaline added:	Theoretical	Observed	
4·5	2380	2179	−8
5·0	2680	2299	−14
5·5	3000	2329	−22
6·0	3400	2287	−33

Pituitary glands from female rats were incubated in 3·0 ml Medium 199 containing 25 μCi [4,5-^{3}H]-leucine for 4 hours. Noradrenaline (10^{-5}M) was then added and the incubation allowed to proceed, with aliquots withdrawn as indicated.

minutes, prolactin release was completely inhibited, but after transfer to fresh medium without noradrenaline, prolactin release was resumed

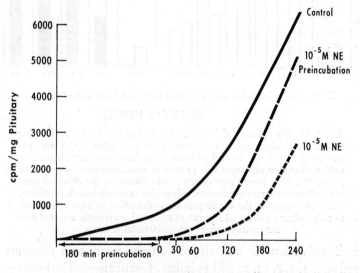

FIG. 7. Pituitary glands from female rats were pre-incubated in Medium 199 containing [4,5-^{3}H]leucine and noradrenaline (NE) for 180 minutes. The tissue was rinsed and re-incubated in fresh medium without catecholamine and aliquots removed at intervals up to 240 minutes. Other glands were incubated with or without noradrenaline for the entire 240-minute period.

immediately and at a rate comparable to that in control glands (Fig. 7). These results suggest that the catecholamine is either very loosely bound to the pituitary cells or is rapidly metabolized by the monoamine oxidase in the gland. We have previously demonstrated that the amount of this enzyme in the pituitary gland is capable of metabolizing two nmoles of catecholamine per minute (MacLeod, Fontham and Lehmeyer 1970).

The long-term effect of noradrenaline on the release of labelled prolactin *in vitro* was next studied. Pituitary glands from female rats were incubated

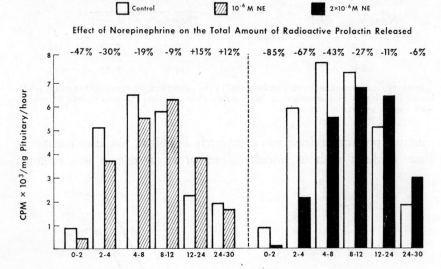

FIG. 8. Pituitary glands of female rats were incubated with radioactive leucine in Medium 199 and aliquots of the incubation media were removed at 2, 4, 8, 12, 24 and 30 hours after the start of the incubation. L-Noradrenaline bitartrate was present in concentrations of 10^{-6}M or 2×10^{-6}M. The lower part of the figure illustrates the effect of noradrenaline on the rate of release of radioactive prolactin into the medium. The percentages in the upper part show the effect of incubation with noradrenaline on the total amount of radioactive prolactin released into the medium after various incubation times.

with radioactive leucine and aliquots of the incubation medium were removed after 2, 4, 8, 12, 24 and 30 hours of incubation. The lower portion of Fig. 8 illustrates the rate at which radioactive prolactin was released into the incubation medium and indicates that the maximum rate of release of labelled prolactin by control tissue was attained between the fourth and eighth hours of incubation. Noradrenaline at 10^{-6}M exerted its maximum inhibitory effect on the rate of prolactin release during the early time intervals. After eight hours of incubation, the rate of release of labelled prolactin by control pituitary glands gradually decreased. Glands incubated with 10^{-6}M-noradrenaline, however, reached their maximum rate of prolactin release between eight and 12 hours of incubation. After 12 or more hours the rate of release of prolactin by treated glands exceeded the rate in control glands. The data in the upper portion of Fig. 8 show that 10^{-6}M-noradrenaline produced a 47 per cent decrease in the total amount of labelled prolactin released after two hours of incubation and that the effectiveness of the catecholamine decreased as incubation proceeded. At some point between 12 and 24 hours, control and treated glands had released equal amounts of labelled hormone and after 24 hours the production of prolactin by treated glands exceeded that of controls. The higher concentration of noradrenaline (2×10^{-6}M) produced a greater degree of inhibition initially, but its effectiveness also decreased with time, so that after 30 hours of incubation, only a 6 per cent decrease in the total amount of prolactin produced was observed. These findings indicate that the very pronounced inhibitory effect on prolactin release produced by the *in vitro* addition of noradrenaline is transient in nature and that no permanent effects on hormone release occur. The fact that the pituitary gland contains sufficient monoamine oxidase activity for metabolizing the catecholamine could explain these results.

Although there is no direct evidence that catecholamines are the agents through which the physiological control of prolactin synthesis and release is mediated, the possibility must be considered because of the dramatic *in vitro* effects of catecholamines and because drugs which decrease the dopamine and noradrenaline concentrations in the hypothalamus potentiate the production of prolactin. We have previously reported (MacLeod, Fontham and Lehmeyer 1970) that 16 hours after injection of 2 mg/kg body weight of reserpine, the tranquillizer produced a 100 per cent increase in the *in vitro* release of newly synthesized prolactin from the pituitary gland. Although reserpine depletes the catecholamine and indoleamine stores in most tissues, the particular importance of the brain stores is demonstrated by the fact that the administration of guanethidine, an agent which reduces the peripheral but not the central stores of catecholamines, did not increase prolactin production and, in fact, slightly decreased synthesis and release of the hormone. The administration of perphenazine, another tranquillizing drug, or α-methyl-*p*-tyrosine, an inhibitor of dopamine synthesis, caused a large increase in prolactin production.

Thus, in spite of very different chemical structures, oestradiol, reserpine and perphenazine are all capable of stimulating prolactin synthesis and release when administered *in vivo*. As a means of further elucidating the function of catecholamines in prolactin production it was decided to see whether pituitary glands from rats treated with these three drugs would retain their sensitivity to noradrenaline *in vitro*.

The data presented in Table VIII show, once again, that the addition *in vitro* of 2×10^{-6}M-noradrenaline to pituitary glands of untreated female rats completely inhibited prolactin release and subsequently inhibited synthesis. The catecholamine was without effect, however, when incubated with pituitary glands from female rats injected with perphenazine.

TABLE VIII

EFFECT OF PERPHENAZINE, OESTRADIOL OR RESERPINE ON THE *IN VITRO* INCORPORATION OF [4,5-^{3}H]LEUCINE INTO PROLACTIN IN THE PRESENCE OF NORADRENALINE

Group	Prolactin c.p.m. per mg gland	
	Pituitary	Medium
Female rats:		
Control	781 ± 123	2719 ± 167
2×10^{-6}M-noradrenaline	1284 ± 166	40 ± 16
Perphenazine	1609 ± 150	2243 ± 141
Perphenazine + noradrenaline	2032 ± 99	2289 ± 51
Oestradiol	1113 ± 72	2482 ± 182
Oestradiol + noradrenaline	1694 ± 146	892 ± 115
Reserpine	706 ± 81	3464 ± 248
Reserpine + noradrenaline	1783 ± 372	95 ± 33
Male rats:		
Control	225 ± 3	454 ± 36
2×10^{-6}M-noradrenaline	345 ± 95	13 ± 13
Perphenazine	1300 ± 199	2437 ± 36
Perphenazine + noradrenaline	1150 ± 57	1998 ± 108

Treatments: 5 mg/kg perphenazine injected daily for 5 days; 200 μg polyoestradiol phosphate injected 5 days before sacrifice; 2 mg/kg reserpine injected daily for 5 days.

When noradrenaline was added to glands from oestrogen-treated female rats, a 65 per cent inhibition of release of newly synthesized prolactin occurred. Glands of reserpine-treated females, however, were as responsive as control glands to the inhibitory effect of noradrenaline on prolactin production. Treatment of male rats with perphenazine stimulated prolactin synthesis *in vitro* by 450 per cent and the addition of noradrenaline to the glands decreased synthesis of the hormone only 16 per cent. These results suggest that each of these drugs stimulates prolactin synthesis by a different mechanism. Additionally, it is the catecholamine depletor which is most effective in producing pituitary glands that are sensitive to the inhibitory effect of noradrenaline.

It is known that chlorpromazine increases the turnover of brain catecholamines, causes the accumulation of pyridoxal phosphate, and stimulates prolactin production. An experiment was designed to determine whether pyridoxal phosphate could influence the *in vitro* synthesis of prolactin (Fig. 9). Surprisingly, the vitamin caused a slight inhibition of the release of labelled prolactin and, even more surprisingly, prevented the decrease in prolactin release induced by noradrenaline. Although this action of pyridoxal phosphate is very similar to the *in vivo* effect of perphenazine on prolactin production, it is not suggested that the mechanisms of action of the two compounds are identical. These results may merely reflect the known ability of pyridoxal phosphate to complex with noradrenaline (Gey and Burkard 1969), thus rendering it inactive.

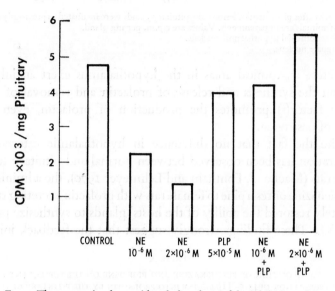

FIG. 9. The prevention by pyridoxal phosphate of the inhibition of prolactin release produced by noradrenaline. Anterior pituitary glands from female rats were incubated for 7 hours in the usual manner (Fig. 1). Noradrenaline (NE) was present at the concentrations indicated. The concentration of pyridoxal phosphate (PLP) was 5×10^{-5}M. The bars represent the amount of radioactive prolactin released into the incubation medium by the glands.

In collaboration with Dr Richard Weiner (Department of Anatomy, UCLA), electrolytic lesions were placed bilaterally in the hypothalamus of male rats. Twenty-one days later the animals were killed by decapitation and the pituitary glands removed and incubated in Medium 199 containing radioactive leucine. Although male rats synthesize relatively small amounts of prolactin *in vitro*, discrete bilateral lesions in the ventromedial hypothalamic nucleus (VMH) significantly increased the *in vitro* synthesis and release of prolactin (Table IX). When the VMH was incompletely destroyed, or when a portion of the arcuate nucleus (AHA) was destroyed, no effect on prolactin production was observed. It is evident

TABLE IX

INFLUENCE OF BRAIN LESIONS ON THE *IN VITRO* INCORPORATION OF [4,5-^{3}H]LEUCINE INTO PROLACTIN AND GROWTH HORMONE IN PITUITARY GLANDS OF MALE RATS

Operation	Prolactin in medium	Growth hormone		
		Gland	Medium	Total
Control (10)	585 ± 40	3305 ± 218	1569 ± 105	4874
Bilateral lesion in VMH (7)	956 ± 76	1995 ± 127	1266 ± 106	3261
Partial lesion in VMH (4)	630 ± 75	2267 ± 221	1281 ± 73	3548
Lesions in VMH + AHA (4)	587 ± 127	1085 ± 212	819 ± 65	1904
Lesion in AHA (1)	647	2278	1698	3976

Three weeks after placement of lesions, the pituitary glands were incubated as previously described. Number of animals are in parentheses. Values are c.p.m. per mg gland.
VMH, ventromedial hypothalamic nucleus.
AHA, arcuate nucleus.

that discrete anatomical areas in the hypothalamus exert an inhibitory action on the synthesis and release of prolactin and removal of this inhibitory factor(s) promotes the production of prolactin, even in the absence of oestrogen.

Despite the fact that no difference in hypothalamic catecholamine concentration has been observed between normal and pituitary tumour-bearing rats (MacLeod, Fontham and Lehmeyer 1970), the administration of perphenazine or reserpine to female rats with prolactin-secreting tumours completely restored the ability of the hosts' glands to synthesize prolactin (Table X). These findings strongly suggest that the feedback inhibition

TABLE X

EFFECT OF ADMINISTRATION OF PERPHENAZINE AND RESERPINE ON THE SUBSEQUENT *IN VITRO* INCORPORATION OF [4,5-^{3}H]LEUCINE INTO PROLACTIN BY THE PITUITARY GLAND

Treatment of rats	Number of flasks	Prolactin c.p.m. per mg gland	
		Pituitary	Medium
Control (no tumour, no drug)	3	340 ± 58	3123 ± 94
No tumour + perphenazine	3	1197 ± 194*	5466 ± 47*
MtTW5	8	110 ± 91*	1552 ± 155**
MtTW5 + perphenazine	3	897 ± 216*	5371 ± 676*
MtTW5 + reserpine	4	317 ± 55	5865 ± 150*

Control female rats and rats bearing pituitary tumours for 8 weeks were injected daily for 5 days with perphenazine (5 mg/kg) or reserpine (2 mg/kg). Means ± s.e.m.
* $P < 0.01$ vs. non-tumour-bearing control.
** $P < 0.05$ vs. non-tumour-bearing control.

exerted by prolactin from the tumour on prolactin production by the hosts' pituitary gland is mediated by a catecholamine-dependent mechanism.

Of the many drugs reported to alter the secretion of prolactin, the derivatives of ergot are among the most interesting. Ergocornine (Nagasawa and Meites 1970) and ergocristine (Wuttke, Cassell and Meites 1971) have recently been shown to decrease prolactin in the pituitary gland and in serum and to inhibit the growth of hormone-dependent mammary tumours. We have studied the effect of these and other ergot derivatives on rat pituitary glands and on hormone-secreting pituitary tumours. The daily injection of 0·2 mg/kg body weight of ergotamine had no effect on the *in vitro* synthesis and release of prolactin (Fig. 10) but administration

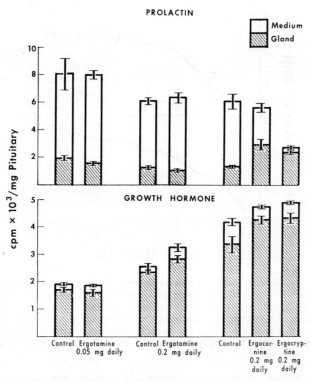

FIG. 10. The effect of ergot derivatives on the synthesis and release of prolactin and growth hormone by pituitary glands *in vitro*. Female rats were injected subcutaneously with an ergot once daily for 7 days before sacrifice. The pituitary glands of these animals and those of control animals were incubated with radioactive leucine for 7 hours in the usual manner.

of ergocornine or ergocryptine significantly decreased prolactin release and caused the newly synthesized prolactin to accumulate within the gland. It is interesting to note, however, that all three ergot derivatives stimulated the synthesis of growth hormone *in vitro*.

Despite the fact that injection of ergotamine produced only minimal changes in the pituitary gland, administration of the ergot to female rats bearing pituitary tumours inhibited further tumour growth. Fig. 11 illustrates the effect of ergotamine injection on the growth of the StW5 and 7315a tumours. In other experiments, ergocornine injection also stopped StW5 growth. Growth hormone-secreting pituitary tumours are known to cause marked spleno-hepatomegalia (Yokoro, Furth and Haran-Ghera 1961; MacLeod, Allen and Hollander 1964). Injecting

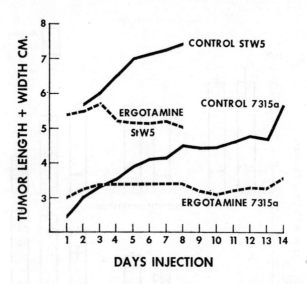

FIG. 11. The effect of ergotamine on the growth of transplanted pituitary tumours. Female Buffalo strain rats bearing the prolactin and ACTH-secreting tumour 7315a were injected subcutaneously daily with 0·05 mg ergotamine tartrate. Wistar-Furth females bearing the prolactin and growth hormone-secreting tumour StW5 received 0·2 mg ergotamine tartrate daily. Corresponding control animals were not injected.

ergotamine or ergocornine into rats with these tumours reversed the *in situ* effects of the hormone on liver and spleen weights and also reversed the effect of the ACTH-secreting tumour 7315a on the adrenal (Table XI). Thus it would appear that the drugs have a greater effect on hormone production by the pituitary tumour than by the pituitary gland.

The ergot derivatives also reversed the pituitary tumour-induced inhibition of prolactin and growth hormone production by the hosts' pituitary gland (Table XII). The drugs did not completely inhibit the secretion of hormones by the tumours, however, since the mammary glands of the treated tumour-bearing rats were still well developed and the tumour hormones presumably still exerted some inhibitory influence on the synthesis of prolactin by the hosts' pituitary glands.

TABLE XI

EFFECT OF ERGOTAMINE AND ERGOCORNINE INJECTIONS ON TISSUE WEIGHTS OF FEMALE RATS BEARING PITUITARY TUMOURS

Treatment of animals	Pituitary gland mg	Pituitary tumour g	Liver g	Spleen g	Adrenals mg
Control	8·81	—	5·51±0·32	0·62±0·10	51± 4
StW5	6·60	23·8±6·7	12·75±1·17	1·44±0·77	53± 3
StW5+0·2 mg ergotamine tartrate*	8·21	10·4±2·6	6·84±0·29	0·62±0·02	57± 1
7315a	7·72	12·7±3·4	9·95±0·74	—	176±24
7315a+0·05 mg ergotamine tartrate**	8·55	1·9±0·3	7·90±0·22	—	69± 5
StW5	6·85	7·1±1·2	11·98±0·44	1·33±0·36	58± 2
StW5+0·2 mg ergocornine maleate*	7·30	2·3±0·2	8·40±0·24	0·57±0·02	58± 3

* Injected subcutaneously daily for 7 days.
** Injected subcutaneously daily for 13 days.

It is not known whether the injected ergocornine and ergocryptine act directly on the pituitary gland or through the hypothalamus to decrease prolactin synthesis but incubation of the alkaloids with pituitary glands significantly decreased prolactin production (Fig. 12). Ergocryptine and ergocornine completely inhibited the release of labelled prolactin and produced a slight retention of the hormone within the tissue. Ergotamine was less effective than the other derivatives in decreasing prolactin synthesis but it caused an even greater accumulation of newly labelled hormone within the gland. Although ergotamine had no effect on the incorporation of isotope into growth hormone, the other ergot derivatives decreased growth hormone synthesis 30–35 per cent.

TABLE XII

EFFECT OF ADMINISTRATION OF ERGOT DERIVATIVES ON THE *IN VITRO* SYNTHESIS OF PROLACTIN BY PITUITARY GLANDS OF FEMALE RATS BEARING PITUITARY TUMOURS

Treatment of animals	Pituitary gland	Medium	Total synthesis
Control	425±114	5489±473	5914
7315a	19± 19	2374±177	2393
7315a+ergotamine tartrate*	185± 61	3586±943	3771
Control	1230±124	4838±241	6068
MtTW5-S	131± 47	1720± 23	1851
MtTW5-S+ergotamine tartrate**	551±104	3223±210	3774
MtTW5-S	182± 18	2566±393	2748
MtTW5-S+ergocornine maleate**	326±182	1648±213	1974

* 0·05 mg injected subcutaneously daily for 13 days.
** 0·20 mg injected subcutaneously daily for 7 days.

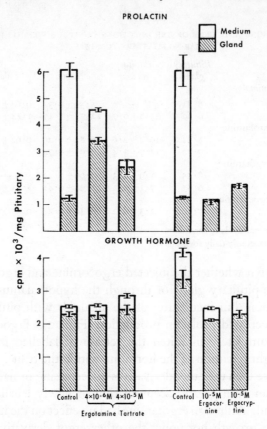

FIG. 12. The effect of ergot derivatives *in vitro* on the synthesis and release of prolactin and growth hormone. Ergot derivatives were added as indicated to incubation flasks containing pituitary glands of female rats and [4,5-^{3}H]leucine in Medium 199. The incubation time was 7 hours.

DISCUSSION

The pioneering work of Meites, Nicoll and Talwalker (1963) and Pasteels (1961) demonstrated that pituitary glands synthesized large quantities of prolactin when removed from the inhibitory influence of the hypothalamus. Investigations in our laboratory have shown that the pituitary glands of female rats are capable of *de novo* prolactin synthesis and release immediately upon incubation *in vitro*. The rapidity with which this process begins and the fact that actinomycin D is without effect on it indicate that the genetic transcription process is functional *in situ* but its translation is repressed by a hypothalamic hypophysiotropic factor. The validity of this hypothesis is further demonstrated by the fact that placing discrete lesions in the ventromedial hypothalamic nucleus significantly increased the *in vitro* production of prolactin.

The function of the brain catecholamines in the regulation of pituitary hormone secretion has intrigued investigators for some time. The observation of Kanematsu and Sawyer (1963) that milk secretion in rabbits was stimulated by a reserpine-induced decrease in hypothalamic catecholamines was confirmed by Van Maanen and Smelik (1968), and the findings of Fuxe and Hökfelt (1969) are in agreement with the concept of a dopaminergic mechanism in the hypothalamus inhibiting prolactin production. There is a consensus of opinion among investigators that a prolactin-inhibiting factor (PIF) exists in the hypothalamus (Meites, Nicoll and Talwalker 1963). It is uncertain, however, if the hypothalamic catecholamine is the long-sought PIF or is a transmitter which stimulates the release of PIF, thereby blocking prolactin release. The latter concept is favoured by Kamberi, Mical and Porter (1971) who were unable to decrease plasma prolactin levels in male rats by infusing catecholamines into the pituitary gland. Since recent work (Neill 1970) has demonstrated that stress greatly increases plasma prolactin, these infusion experiments should be interpreted with caution because of the severe stress imposed by the anaesthetic and the elegant surgical techniques and because of the sex of the animals.

The results presented here are consistent with the hypothesis that dopamine functions as the physiological PIF. The dopaminergic secretory granules are located adjacent to the portal vessels leading to the pituitary gland and the number of granules and the hypothalamic catecholamine concentration are inversely related to prolactin production. In addition, pharmacological agents which decrease brain catecholamine content increase prolactin synthesis (MacLeod, Abad and Eidson 1969; MacLeod, Fontham and Lehmeyer 1970), and increase plasma prolactin concentrations (Lu *et al.* 1970). The likelihood of a direct effect of catecholamines on prolactin release *in vivo* should be emphasized, because the amines act immediately to inhibit prolactin release and the duration of the influence of catecholamines *in vitro* is proportional to their concentration. Presumably, the transient nature of this effect is due to the action of monoamine oxidase in the pituitary gland (MacLeod, Fontham and Lehmeyer 1970).

It is apparent that the mechanisms which govern the synthesis of prolactin are distinct from those governing its release. Normally, newly synthesized prolactin was quickly released into the incubation medium. No accumulation of prolactin within the gland accompanied the increased synthesis of prolactin following treatment with reserpine, perphenazine or oestradiol, indicating the ease with which prolactin is released. Release, however, was inhibited by catecholamines without initially affecting hormone synthesis. The results of electron microscopic examination of the pituitary glands incubated with catecholamine were completely consistent with the biochemical data.

The specific biochemical mechanism through which pituitary tumour prolactin inhibits the synthesis of the host's pituitary hormone is unknown. The shortloop feedback, however, apparently inhibits prolactin synthesis through some mechanism which is responsive to catecholamines, because reserpine, perphenazine and oestradiol reverse the inhibitory effects of the tumour prolactin.

The important finding of Meites and co-workers (Nagasawa and Meites 1970; Wuttke, Cassell and Meites 1971) that the growth of mammary tumours was inhibited and the plasma prolactin concentration was decreased by injecting ergocornine prompted a study of the effects of ergot drugs on prolactin synthesis and pituitary tumour function. It was interesting to note that hormone production by transplanted pituitary tumours was greatly decreased by ergotamine whereas the drug had no effect on prolactin synthesis by the pituitary gland. Whether this represents vascular differences or a more fundamental cellular response of the hypophysis is unknown. The effects of this drug are in contrast to the inhibitory effects of ergocornine and ergocryptine on prolactin synthesis in both the pituitary gland and tumour. We have extended the work of Wuttke, Cassell and Meites (1971) and shown that these drugs completely inhibit the *in vitro* release of newly synthesized prolactin while ergotamine is less effective. The fact that prolactin release was greatly decreased while growth hormone release was stimulated by the ergots demonstrates, once again, the selectivity and independence of the mechanisms governing the production of pituitary hormones.

SUMMARY

Some of the factors which regulate the incorporation of $[4,5-^3H]$leucine into rat pituitary gland prolactin *in vitro* and the subsequent release of the hormone were studied. Labelled prolactin appeared in the incubation medium promptly and little remained in the gland, whereas growth hormone was largely retained in the tissue. Puromycin, but not actinomycin D, completely inhibited incorporation of isotope into the proteins, indicating that true protein synthesis occurred. In female rats, prolactin and growth hormone were synthesized at comparable rates, but in males prolactin synthesis was decreased over 80 per cent. Implantation of female rats with a prolactin-secreting pituitary tumour decreased the host's *in vitro* synthesis 50–75 per cent. Although ovariectomized–adrenalectomized rats synthesized greatly decreased amounts of prolactin, pituitary tumour implantation further decreased synthesis, indicating that the tumour hormones inhibited the host's synthesis of prolactin at the hypothalamo-pituitary level. The administration of oestrogen significantly increased prolactin synthesis in glands of normal animals and restored synthesis to control values in rats bearing prolactin-secreting tumours. Like oestradiol, reserpine and perphenazine also decreased the hypothalamic catecholamine concentration and had similar potentiating effects on prolactin synthesis in normal and tumour-bearing rats. Catecholamines may be important in regulating prolactin synthesis and release since the *in vitro* addition of dopamine, noradrenaline or adrenaline promptly decreased the release of labelled prolactin and subsequently inhibited its synthesis. Since oestradiol, reserpine and perphenazine had no significant effect on the growth of the pituitary tumours, their ability to increase prolactin synthesis by the gland was a result of direct action. In contrast, ergotamine injection completely inhibited growth of prolactin-secreting tumours but had little effect on the synthesis and release of pituitary gland prolactin. These results suggest that bioamines and agents which affect their distribution have a physiological function to regulate the synthesis and release of prolactin.

Acknowledgements

This investigation was supported in part by United States Public Health Service grant CA-07535 from the National Cancer Institute. R.M.M. is a USPHS Research Career Development Awardee.

The authors thank several pharmaceutical companies for donating the drugs used: Schering Corporation (perphenazine), CIBA Ltd. (reserpine), and Sandoz Pharmaceuticals (ergot derivatives).

REFERENCES

BIRGE, C. A., JACOBS, L. S., HAMMER, C. T. and DAUGHADAY, W. H. (1970) *Endocrinology* **86**, 120–130.
CATT, K. and MOFFAT, B. (1967) *Endocrinology* **80**, 324–328.
CHEN, C. L., MINAGUCHI, H. and MEITES, J. (1967) *Proc. Soc. Exp. Biol. & Med.* **126**, 317–325.
COPPOLA, J. A., LEONARDI, R. G., LIPPMANN, W., PERRINE, J. W. and RINGLER, I. (1965) *Endocrinology* **77**, 485–490.
FUXE, K. and HÖKFELT, T. (1969) In *Frontiers in Neuroendocrinology*, pp. 47–96, ed. Ganong, W. F. and Martini, L. New York: Oxford University Press.
GEY, K. F. and BURKARD, W. P. (1969) *Ann. N.Y. Acad. Sci.* **166**, 213–224.
KAMBERI, I. A., MICAL, R. S. and PORTER, J. C. (1971) *Endocrinology* **88**, 1012–1020.
KANEMATSU, S. and SAWYER, C. H. (1963) *Proc. Soc. Exp. Biol. & Med.* **113**, 967–969.
KOCH, Y., LU, K. H. and MEITES, J. (1970) *Endocrinology* **87**, 673–675.
LU, K. H., AMENOMORI, Y., CHEN, C. L. and MEITES, J. (1970) *Endocrinology* **87**, 667–672.
MACLEOD, R. M. (1969) *Endocrinology* **85**, 916–923.
MACLEOD, R. M. (1970) *Proc. Soc. Exp. Biol. & Med.* **133**, 339–341.
MACLEOD, R. M. and ABAD, A. (1968) *Endocrinology* **83**, 799–806.
MACLEOD, R. M., ABAD, A. and EIDSON, L. L. (1969) *Endocrinology* **84**, 1475–1483.
MACLEOD, R. M., ALLEN, M. S. and HOLLANDER, V. P. (1964) *Endocrinology* **75**, 249–258.

MacLeod, R. M., Bass, M. B., Buxton, E. P., Dent, J. N. and Benson, Jr, D. G. (1966) *Endocrinology* **78**, 267–276.

MacLeod, R. M., Fontham, E. H. and Lehmeyer, J. E. (1970) *Neuroendocrinology* **6**, 283–294.

MacLeod, R. M., Smith, M. C. and DeWitt, G. W. (1966) *Endocrinology* **79**, 1149–1156.

Meites, J., Nicoll, C. S. and Talwalker, P. K. (1963) In *Advances in Neuroendocrinology*, pp. 238–277, ed. Nalbandov, A. V. Urbana, Ill.: University of Illinois Press.

Nagasawa, H. and Meites, J. (1970) *Proc. Soc. Exp. Biol. & Med.* **135**, 469–472.

Neill, J. D. (1970) *Endocrinology* **87**, 1192–1197.

Parsons, J. A. (1969) *Am. J. Physiol.* **217**, 1599–1603.

Pasteels, J. L. (1961) *C.R. Hebd. Séance Acad. Sci., Sér. D* **253**, 2140–2142.

Van Maanen, J. H. and Smelik, P. G. (1968) *Neuroendocrinology* **3**, 177–186.

Wuttke, W., Cassell, E. and Meites, J. (1971) *Endocrinology* **88**, 737–741.

Yokoro, K., Furth, J. and Haran-Ghera, N. (1961) *Cancer Res.* **21**, 178–186.

DISCUSSION

Sherwood: In view of evidence for the role of adenyl cyclase in the regulation of growth hormone secretion, how do you think adenyl cyclase fits into the regulation of prolactin release by catecholamines?

MacLeod: Adenyl cyclase probably has minimal or no effect on the regulation of prolactin synthesis or release since catecholamines do not stimulate adenyl cyclase activity in the pituitary gland, in contrast to other tissues. The incubation of pituitaries with dibutyryl cyclic AMP has no effect on prolactin synthesis or release, so we don't think the nucleotide has an important function. The regulation of growth hormone secretion, however, is controlled by this enzyme.

Turkington: I was interested that just by electrophoresis you can get out a labelled protein with such a high specific activity. Do you feel that all those gel fractions represent highly purified hormone, or do you take another step to purify them?

MacLeod: At first we put it through a Sephadex column to eliminate most of the bulk protein. Later studies have shown this step to be unnecessary, because we get the same results with or without the column step. We make no claim that the bands which we show on the polyacrylamide gel are exclusively prolactin or exclusively growth hormone. Since 75 per cent of the radioactivity was precipitated by the specific prolactin antibody, however, it suggests that the prolactin band on the acrylamide gel is relatively pure hormone.

Nicoll: We are doing similar studies to those reported by Dr MacLeod and have been concerned with the possible heterogeneity of prolactin. In studying the kinetics of prolactin turnover *in vivo* and *in vitro* Dr Karen C. Swearingen (1971) has found that synthesis is heterogeneous (that is, there is at least one rapidly turning over component). We felt that this may have been a result of another protein running with prolactin. Accordingly, Dr Swearingen isolated the prolactin from the 7.5 per cent acrylamide gel and then re-ran the same fraction at two different pH's and at several different gel concentrations. No indication of any significant contamination of the prolactin band was obtained. Of course, this doesn't rule out the possibility of very slight contamination, but it looks as though the prolactin separated by this means is very 'clean', both in the medium and in tissue homogenates.

Li: Are you sure that the two protein bands are growth hormone and prolactin? Have you assayed them? Can you estimate how active these bands are, per mg of nitrogen or mg of protein?

Nicoll: We have assayed the prolactin band and shown that it has activity in the crop sac assay. Others have shown this also and have identified the growth hormone using the tibia test. No one to my knowledge has been able adequately to quantify the amount of protein extracted from those bands in the gels. Accordingly, a good test of the correlation between the amount of protein extracted and its specific activity has not been possible.

Li: But doesn't this make the discussion rather meaningless, because it is possible that you simply have radioactive proteins in the band and the active protein is just a tiny bit?

Nicoll: When we subjected the protein eluted from the band to electrophoresis under different conditions we expected to see multiple forms. We didn't see that. Also, as Dr MacLeod stated, 75 per cent of the protein was precipitated by specific antiserum.

MacLeod: I would doubt whether you could put one unit of prolactin or growth hormone on to a polyacrylamide gel and hope to get anything like one unit of biological activity off. Since this method subjects the protein to at least three different pH's, and the electrophoretic current, the biological activity may be significantly altered. I don't think we should expect to be able to correlate the amount put on and the amount taken off. These bands have been assayed for biological activity, however, by Jones and co-workers (1965).

Turkington: Are any of the effects influenced by changes in the intracellular concentration of the tritiated leucine precursor? I wonder whether bicarbonate or calcium or the catechols are changing the ability of the isotopic precursor to enter the cell.

MacLeod: We haven't investigated this *per se*, but the fact that growth hormone synthesis was not affected by these agents suggests that amino acid transport was not modified by the conditions of incubation.

Beck: Is there any evidence in any other tissue that these ions alter amino acid transport?

Sherwood: There is only one tissue, as far as I know, in which calcium decreases amino acid uptake. In the parathyroid glands, calcium inhibits secretion, whereas in most other secretory systems it stimulates hormone secretion.

Pasteels: I am quite convinced by your beautiful results on the action of catecholamines on the hypophysis, Dr MacLeod, but I am puzzled by the electron micrographs. You described an enlargement of the ergastoplasmic reticulum as a result of catecholamine administration. However, the best evidence that the endoplasmic reticulum is a system of vacuoles surrounded by membranes is that it is highly sensitive to osmotic change. So it is possible that the enlargement of the cisternae was the consequence of an osmotic change, perhaps due to a specific action of the catecholamines on the membranes.

MacLeod: I agree completely with this possibility. We are uncertain whether the catecholamines act directly to inhibit the release of prolactin or whether they modify intracellular biochemical mechanisms which control the secretion of the hormone.

Beck: Have you any information on the intracellular water transport in catecholamine-treated tissues?

MacLeod: No, but we have some preliminary data suggesting that calcium transport is decreased by the presence of catecholamine.

Meites: We have been working on similar problems, and differ somewhat in our interpretation of these results. One point pertains to the catecholamines present in high concentrations in the hypothalamus (Vogt 1954). The evidence is growing that these hypothalamic catecholamines have an important role in regulating the secretion of anterior pituitary hormones, not by acting directly on the pituitary but rather by acting on the release of hypothalamic factors. The evidence is quite convincing that hypothalamic catecholamines stimulate the release of prolactin-inhibitory factor (PIF), thereby inhibiting release of prolactin. Hypothalamic catecholamines also stimulate release of FSH-RF and LRF and their corresponding pituitary hormones.

Dr MacLeod's demonstration that catecholamines can directly inhibit prolactin release *in vitro* has been confirmed by us and by others at the doses Dr MacLeod used, but we also found that catecholamines at smaller doses *stimulate* prolactin release *in vitro* (Koch, Lu and Meites 1970). Drs Kamberi, Mical and Porter (1970) have shown with their technique for perfusing single portal vessels that infusion of catecholamines, particularly dopamine, has no effect on the release of prolactin. In addition, Porter, Goldman and Wilber (1970) reported that no catecholamines were detected in the portal circulation, so there is a question of how catecholamines reach the pituitary.

I believe Dr P. G. Smelik first suggested the possibility that dopamine might be PIF itself, but although we can't completely rule out the possibility that catecholamines may act directly on the pituitary, I am inclined to believe that the *in vitro* effects Dr MacLeod and others have reported are pharmacological effects and are not related to the normal physiological control of prolactin secretion.

MacLeod: The work of Kamberi, Mical and Porter is certainly a very good argument for the indirect effect of catecholamines. To my knowledge, however, they have studied the *in vivo* effects of catecholamines exclusively in male animals, whereas our work has used female rats. Perhaps the catecholamine in the hypothalamus of male rats suppresses maximally the plasma prolactin concentration and any further decrease is obtained only with difficulty. It would be interesting to see if the plasma prolactin concentration is decreased more significantly in female rats after perfusion of the pituitary with catecholamines. We know that stress increases plasma prolactin, and in Kamberi's experiments catecholamine may have been unsuccessful in decreasing plasma prolactin because of the severe stress experienced by the animals due to the anaesthesia and surgery. Certainly 10^{-6} or 10^{-7} M-catecholamine is a high concentration, but this cannot be used as an argument against a genuine physiological effect, because to my knowledge all the effects of cyclic AMP *in vitro* are demonstrated at 10^{-4} and 10^{-3} M, and this is presumably one of the major mechanisms through which many of the hormones act *in vivo*. I think we must keep an open mind on this topic and only time will tell whether catecholamines exert their action on prolactin release through a direct or indirect effect.

Nicoll: I am still rather fond of the notion that one of the catecholamines is PIF because it would simplify life for us. However, there is other evidence contrary to catecholamine being PIF. One can prepare an extract of hypothalamic tissue from rats and obtain inhibition of prolactin secretion *in vitro*. If one uses the same amount of catecholamine that would be expected to be in the extract, secretion is not affected. On the other hand, if one examines the data that Dr Meites (1970) reported showing a dose-related inhibitory effect of hypothalamic extract on prolactin release, as determined by immunoassay, and compares them with data published by Birge and colleagues (1970), who tested the inhibitory effects of various doses of catecholamines, one finds that the inhibitory curves are perfectly parallel. This raises the possibility that one of the catecholamines may be part of PIF, or that at least their action must be very similar on the lactotroph.

Spellacy: Dr MacLeod, I am not clear how you explain the inhibition of the effect of noradrenaline by phenothiazines when reserpine had no effect.

MacLeod: I don't have a good explanation for it. Our working hypothesis is that perphenazine is stimulating prolactin synthesis and/or release through a very different mechanism. I don't know how perphenazine acts *in vivo*. That it may work through a dopamine-dependent mechanism is a possibility but I don't think this has been demonstrated. The fact that it produces a pituitary gland which is refractory to noradrenaline doesn't support that. It seems to turn on protein synthesis and prolactin synthesis independently of the catecholamines.

Friesen: Have you any evidence for a precursor of prolactin or of growth hormone in your studies? Secondly, were you suggesting that the partial or the transient inhibition of the release of prolactin by noradrenaline was related to the metabolism of noradrenaline, and if so, could you create a sustained inhibition of prolactin release by adding fresh noradrenaline?

MacLeod: We have no evidence for or against a precursor of prolactin or growth hormone. With regard to the transient nature of the catecholamine effect, this can be explained by the rather high monoamine oxidase activity present in the pituitary gland. We can repeatedly prevent the gland from synthesizing prolactin if we add fresh catecholamine to it.

Pasteels: How do you think the ergot alkaloids act, in relation to the possible action of a specific PIF and of catecholamines? I can confirm your results with ergocornine and ergocryptine. We found that the primary action of ergot drugs was on the release of prolactin.

MacLeod: I certainly do not know how the ergot alkaloids function. I would agree that your morphological evidence and our biochemical evidence certainly suggests that they inhibit release of prolactin rather than synthesis, but I don't know through what biochemical mechanisms this is occurring.

Meites: I believe we understand something about the loci of action of the ergot drugs. We found that the ergot drugs increase PIF activity in the hypothalamus and we have recently confirmed this (Wuttke, Cassell and Meites 1971). An increase in PIF activity would of course decrease prolactin secretion. We have also shown that the ergot drugs act directly on the anterior pituitary (Lu, Koch and Meites 1971). I am inclined to believe that this is their major site of action. We have no information on the biochemical mechanisms involved.

Greenwood: I'd like to ask Dr Meites or Dr MacLeod what the relationship is between their different approaches to *circulating* catecholamines. Adrenaline and noradrenaline do not cross the blood-brain barrier, so their effects must be due to catecholamines synthesized within the hypothalamus. Dr Meites says that catecholamines act on the pituitary *via* releasing factors, whereas Dr MacLeod suggests that they act directly.

MacLeod: You stated the situation correctly. Although the pituitary receives some blood from the extraportal system it's more likely to be hypothalamic catecholamine that regulates prolactin release, but whether it is working as an intermediary to cause the production of PIF or is itself PIF is a question for discussion. That catecholamines have not been demonstrated in the portal blood is not too surprising. The concentration in the systemic circulation is certainly very low; a fraction of a microgramme per litre, I believe.

Meites: It is true that if you inject dopamine, adrenaline or noradrenaline systemically, either into the carotid artery or intraperitoneally, they have practically no effect (Lu *et al.* 1970). On the other hand, if you administer L-dopa to the rat (and it is interesting that Dr Friesen has also observed this in man), there is a marked decrease in serum prolactin levels. Also if we give monoamine oxidase inhibitors, which increase brain catecholamines, we again see a marked decrease in serum prolactin. These drugs increase hypothalamic PIF activity and they do enter the brain. On the other hand, if we give drugs systemically that decrease hypothalamic catecholamines, such as reserpine, chlorpromazine or haloperidol, there is a pronounced increase in blood prolactin. These drugs act by decreasing hypothalamic catecholamines and thereby decrease PIF activity.

Friesen: Dr MacLeod, how do you explain the elevated prolactin levels after pituitary stalk section, on the basis of inhibition by catecholamine?

MacLeod: The absence of portal plasma containing hypothalamic catecholamines would remove the chronic inhibitory influence that the hypothalamus exerts on pituitary prolactin synthesis. We believe that the pituitary has potential ability to synthesize and release prolactin but that it is normally repressed by the hypothalamic catecholamines, and that stalk-sectioning derepresses the pituitary prolactin cells and prolactin synthesis is initiated.

Friesen: Does your scheme allow for a releasing factor as well?

MacLeod: We have no evidence that there is a prolactin-releasing factor.

Nicoll: Dr Friesen, I believe you have some data on serum prolactin levels in women after pituitary stalk section?

Friesen: Our highest levels were 50–80 ng/ml of prolactin, measured some months after stalk section. I think Dr Turkington's values were even higher?

Turkington: Our range for this was about 40–500 ng/ml.

Nicoll: Dr Greenwood, could you clarify your data with Dr Bryant and Dr Denamur on prolactin levels in sheep after stalk section?

Greenwood: We finally resolved this problem. It's very simple. Originally in 1967 we were looking for high levels based on the assumption that there would be a flood of prolactin after stalk section, so we were setting up the assay at about 4 or 8 ng/ml, and didn't find it, except in one sheep which turned out to be an incomplete section. Our expectation was wrong because most of the sheep pituitary infarcts after stalk section. Dr Denamur has operated on five or six more sheep and applied a more sensitive radioimmunoassay and found those amounts of prolactin in plasma which he and Dr Short require to maintain the corpus luteum. The joint data are to be published (Bryant *et al.* 1971).

REFERENCES

BIRGE, C. A., JACOBS, L. S., HAMMER, C. T. and DAUGHADAY, W. H. (1970) *Endocrinology* **86**, 120.

BRYANT, G. D., GREENWOOD, F. C., KANN, G., MARTINET, J. and DENAMUR, R. (1971) *J. Endocrinol.* in press.

JONES, A. E., FISHER, J. N., LEWIS, U. J. and VANDERLAAN, W. P. (1965) *Endocrinology* **76**, 578–583.

KAMBERI, I. A., MICAL, R. S. and PORTER, J. C. (1970) *Fed. Proc. Fed. Am. Soc. Exp. Biol.* **29**, 378 (abst. 751).

KOCH, Y., LU, K. H. and MEITES, J. (1970) *Endocrinology* **87**, 673–675.

LU, K. H., AMENOMORI, Y., CHEN, C. L. and MEITES, J. (1970) *Endocrinology* **87**, 667–672.

LU, K. H., KOCH, Y. and MEITES, J. (1971) *Endocrinology* **89**, 229–233.

MEITES, J. (1970) In *Hypophysiotropic Hormones of the Hypothalamus: Assay and Chemistry*, p. 145, ed. Meites, J. Baltimore: Williams and Wilkins.

PORTER, J. C., GOLDMAN, B. D. and WILBER, J. F. (1970) In *Hypophysiotropic Hormones of the Hypothalamus: Assay and Chemistry*, pp. 282–297, ed. Meites, J. Baltimore: Williams and Wilkins.

SWEARINGEN, K. C. (1971) *Endocrinology* in press.

VOGT, M. (1954) *J. Physiol. (Lond.)* **123**, 451–481.

WUTTKE, W., CASSELL, E. and MEITES, J. (1971) *Endocrinology* **88**, 737–741.

THE SYNTHESIS AND SECRETION OF PLACENTAL LACTOGEN AND PITUITARY PROLACTIN

H. Friesen, C. Belanger, H. Guyda and P. Hwang

McGill University Clinic, Royal Victoria Hospital, Montreal

This paper presents experiments conducted in our laboratory on the biosynthesis and secretion of proteins *in vitro* by placental and pituitary tissue. These studies have been selected to demonstrate the usefulness of this experimental approach in searching for mechanisms which control the synthesis and secretion of placental and pituitary hormones. In the pituitary incubation experiments which will be outlined, we used many of the same techniques as in our earlier studies on placental hormone synthesis, but in this case the object of the experiment was to determine whether the pituitary gland in primates, as in other species, synthesizes and secretes growth hormone and prolactin independently.

In the studies on the placenta, human placental lactogen (HPL) was identified as one of the principal radioactive proteins in the incubation medium, where it accounted for as much as 60 per cent of the trichloroacetic acid precipitable proteins in some experiments (Friesen, Suwa and Pare 1969; Suwa and Friesen 1969a, b). When we incubated human and monkey pituitaries in the same manner, we consistently found a single large radioactive protein peak after gel filtration of the incubation media (Friesen, Guyda and Hardy 1970). To our surprise we found that a much larger percentage of the radioactive proteins in this fraction was precipitated by antiserum to sheep prolactin than by antiserum to human placental lactogen or growth hormone (Friesen and Guyda 1971; Friesen, Guyda and Hwang 1971). As a result of these and other observations we inferred that primate pituitaries, like pituitaries of other species, synthesize and secrete two hormones, growth hormone and prolactin; that the two hormones can be distinguished immunologically; and that primate prolactin is immunologically related to prolactin from other species. These hypotheses have been tested experimentally and the data we present here provide unequivocal evidence for the existence of human and monkey prolactin separate and distinct from growth hormone.

The studies on the synthesis of HPL *in vitro* were initiated after we had observed that when [³H]leucine was injected into the monkey placenta, [³H]monkey placental lactogen had the highest specific activity of any proteins which we identified (Friesen 1968). Therefore we were interested to see whether we could demonstrate HPL synthesis *in vitro* and finally whether we could elucidate some of the mechanisms which control the synthesis of HPL. The studies initiated by Dr S. Suwa and myself have been reported in some detail (Suwa and Friesen 1969a, b) and I will simply review some of their highlights.

The methods used are summarized in Table I. Fresh placental tissue

TABLE I

PROCEDURE FOR INCUBATION OF PLACENTAL TISSUE

A. Placental tissue incubated in Krebs-Ringer bicarbonate buffer with L-[4,5-³H]leucine.
B. Aliquots of medium or tissue extract analysed for:
    (1) Trichloroacetic acid precipitable proteins—³H-labelled proteins.
    (2) Anti-HPL precipitable proteins—[³H]HPL.
        (a) Small molecular weight component.
        (b) Large molecular weight component.
    (3) Protein—Lowry estimation.
    (4) HPL—radioimmunoassay.
C. Gel filtration on Sephadex G-100; same analysis as in B.

from different sections of the placenta was divided into many fragments and rinsed briefly in Krebs-Ringer bicarbonate buffer; 0·5 g tissue was incubated in 5 ml of buffer containing 200 mg per cent of glucose, penicillin, streptomycin and 5 μCi of L-[4,5-³H]leucine (5·5 Ci/mmole, New England Nuclear Corporation). The cultures were incubated in a Dubnoff incubator at 37°C in an atmosphere of 95 per cent O_2 and 5 per cent CO_2. The incubation medium was replaced with fresh medium at four-hour and subsequently at 24-hour intervals. At the end of each incubation period the medium was centrifuged immediately at 3000 rev/min at 4°C for ten minutes and the clear supernatant was kept frozen. Tissue fragments were homogenized in 0·1M-NH_4HCO_3 buffer and centrifuged at 60 000 × **g** for 30 minutes to obtain a supernatant and pellet which were frozen for subsequent analysis. Aliquots of the tissue supernatant and medium were analysed for (1) ³H-labelled proteins (trichloroacetic acid precipitable proteins) (2) [³H]HPL (³H-labelled proteins precipitated by antiserum to HPL), (3) total proteins (method of Lowry) and (4) HPL (measured by double antibody radioimmunoassay). The same analyses were made of fractions after gel filtration of the incubation medium and tissue supernatant on Sephadex G-100.

Using these methods we studied the synthesis and release of human placental proteins and HPL in eight placentas obtained at term. The amount of proteins and ^{3}H-labelled proteins released into the medium was fairly constant among different placentas, whereas the release of HPL varied to a considerably greater extent. The percentage ratio of [^{3}H]HPL to ^{3}H-labelled proteins in the medium of four placentas was between 6 and 12 per cent, which was considerably higher than in the tissue, where it was 2-4 per cent, indicating a preferential release of HPL. However, in an additional four placentas studied, the percentage ratio of [^{3}H]HPL to ^{3}H-labelled proteins was considerably greater in the medium (range 23–55 per cent), as shown in Table II. In two of the placentas studied, tissue

TABLE II

[^{3}H]HPL AS A PERCENTAGE OF TOTAL ^{3}H-LABELLED PROTEIN AFTER INCUBATION OF PLACENTAL TISSUE

Experiment no.	Incubation medium at 24 hours %	Tissue extract at 24 hours %
1	6	2
2	8	4
3	10	3
4	12	2
5	23	3
6	39·4	4
7	45	2·2
8	55	3·4

In each case the mean was calculated from 3 to 6 incubation flasks, containing 10 to 20 fragments of placental tissue.

fragments taken from different parts of the placenta were incubated separately to see whether HPL synthesis was uniform throughout the placenta or whether certain areas of the placenta synthesized proportionately more or less HPL relative to other proteins. Table III shows that protein and HPL

TABLE III

[^{3}H]HPL AS A PERCENTAGE OF TOTAL ^{3}H-LABELLED PROTEIN IN FIVE DIFFERENT AREAS OF THE PLACENTA

Site no.	Incubation medium				Tissue			
	4 hours		24 hours		4 hours		24 hours	
	1*	2†	1	2	1	2	1	2
1	41	26	59	33	10	15	6·0	9·0
2	42	35	50	47	13	19	9·0	13·0
3		43	35	40	23	21	5·0	8·0
4		18	48	34	15	14	8·0	8·0
5		18	31	43	11	13	5·0	13·0

* Experiment 1.
† Experiment 2.

synthesis in five different areas of the placenta was very similar. These studies suggested that no specialized areas for HPL synthesis exist and that the rate of HPL synthesis is fairly homogeneous throughout the placenta.

After filtration on Sephadex G-100 of the tissue supernatant and incubation media, aliquots of each fraction were precipitated with antiserum to

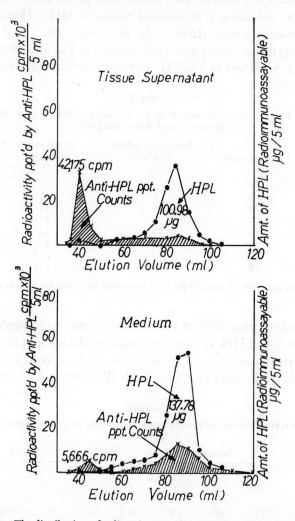

FIG. 1. The distribution of radioactive proteins precipitated by antiserum to HPL after gel filtration of a placental tissue extract (*upper*) and incubation medium (*lower*). The incubation time was 24 hours. Note that in the media the elution volumes of [³H]HPL and HPL (measured by radioimmunoassay) are identical whereas in the tissue [³H]HPL emerges mainly with the void volume.

HPL to determine the elution volume of the ³H-labelled proteins precipitated (Fig. 1). We expected to find that the elution volumes of the HPL standard and of radioactive proteins precipitated by anti-HPL were identical. While this was true in the case of proteins in the medium

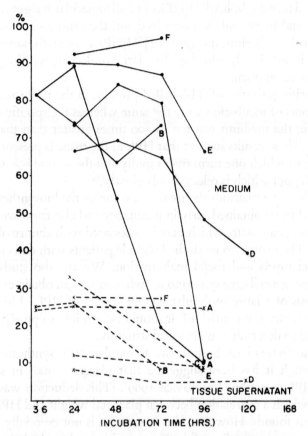

FIG. 2. The small molecular weight species of HPL([³H]HPL) is expressed as a percentage of all the radioactive proteins precipitated by anti-HPL in the incubation medium and tissue supernatant at different times of incubation in six experiments (A–F). The continuous and interrupted lines indicate the results in the medium and tissue respectively. The medium and tissue proteins were fractionated in the same way as shown in Fig. 1 and ³H-labelled proteins in aliquots of each fraction were precipitated with anti-HPL.

during the early incubation period, in the tissue most of the ³H-labelled proteins precipitated by anti-HPL serum appeared in the void volume. With more prolonged incubation times the proportion of large to small molecular weight proteins precipitated by anti-HPL in the medium also increased, as shown in Fig. 2.

The nature of the large molecular weight proteins precipitated by anti-HPL remains unclear. We have been unable to convert the large molecular weight fraction into a small molecular weight HPL by a variety of procedures including treatment with 6M-guanidine HCl, 8M-urea, ribonuclease and trypsin. When [^{3}H]HPL of small molecular weight is incubated with placental tissue and homogenized and the proteins are fractionated upon Sephadex, the elution volume of [^{3}H]HPL remains the same, showing that HPL is not simply adsorbed to a large molecular weight protein or other tissue component.

The specific activities of ^{3}H-labelled proteins in the tissue and medium after 24 hours of incubation were the same whereas the specific activity of [^{3}H]HPL in the medium was ten to 300 times greater than that of tissue [^{3}H]HPL. These results suggest that HPL in the tissue is present in at least two pools, of which one turns over rapidly and the second is a more stable tissue HPL pool which is released only slowly.

Using the same methods we are now exploring the biosynthesis of HPL in placental tissue obtained early in pregnancy and also from women with a variety of disease states which may be associated with abnormal placental function. The groups to be studied include patients with diabetes, hypertension, eclampsia and foetal malnutrition. We are also studying HPL synthesis using a cell-free system to see whether we can obtain evidence for the synthesis of a large molecular weight species of HPL. Using affinity chromatography with anti-HPL it should be possible to purify this component and to determine its chemical structure.

As yet no factors have been shown to regulate HPL synthesis and secretion, although it has been suggested that placental mass in some way controls HPL secretion (Josimovich 1969). This deduction was based on studies in which a correlation between placental weight and HPL concentrations was found. However, this suggestion is not especially helpful in providing an understanding of the exact mechanism by which the syncytiotrophoblast regulates the synthesis and secretion of HPL. Furthermore, in our studies on placental lactogen synthesis *in vitro*, we observed very wide variations in the relative rate of HPL synthesis after correcting for differences in the weight of the placental fragments incubated. If the studies of HPL synthesis *in vitro* have any relevance to HPL secretion *in vivo*, one may justifiably speculate about the existence of intracellular mechanisms which regulate HPL synthesis. In addition it is possible that extracellular factors may influence the secretion of HPL. Unfortunately at present we have almost no knowledge of any homeostatic mechanisms which regulate the endocrine function of the placenta.

The first exciting suggestion that a control mechanism exists may be inferred from the preliminary observations of Tyson, Austin and Farinholt (1971), who reported that with prolonged fasting, HPL concentrations increased by 30 per cent. The signal which triggers the presumed increased rate of secretion of HPL has not yet been identified. However, we have observed that with prolonged incubation of placental fragments *in vitro* there is a sudden increase in the HPL released into the medium. Fig. 3

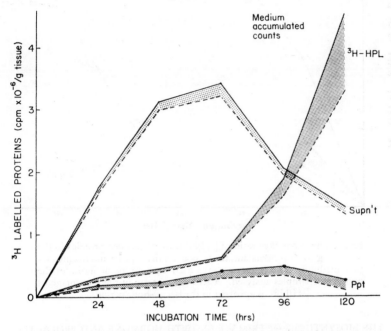

FIG. 3. The incorporation of [^{3}H]leucine into proteins by placental tissue during a 120-hour incubation period. The incubation medium was replaced at 24-hour intervals and the placental tissue was homogenized, extracted and aliquots were fractionated and precipitated with antiserum to HPL. The stippled areas indicate the ^{3}H-labelled proteins precipitated by anti-HPL. For the medium the cumulative counts in protein are indicated. It is apparent that between 72 and 120 hours there is a striking increase in the release of ^{3}H-labelled protein into the medium.

shows the results of one such experiment, where the ^{3}H-labelled proteins and [^{3}H]HPL in the tissue and medium are plotted against time, and Fig. 4 shows the percentage ratio of [^{3}H]HPL to ^{3}H-labelled proteins at each stage of incubation. Between 72 and 120 hours the [^{3}H]HPL as a percentage of ^{3}H-labelled proteins increases from 5 to 25 per cent. Therefore both *in vivo* and *in vitro* we have some evidence that a regulatory mechanism exists which in some, as yet ill-defined, manner controls HPL synthesis and secretion.

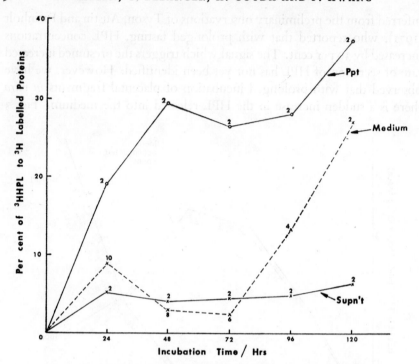

Fig. 4. The percentage ratio of [³H]HPL to ³H-labelled proteins in the 60 000 × **g** tissue supernatant, the 60 000 × **g** tissue pellet and the incubation medium. The number above each 24-hour point indicates the number of samples analysed. A striking increase in the percentage of [³H]HPL is apparent in the last 48-hour period.

THE BIOSYNTHESIS OF PRIMATE GROWTH HORMONE AND PROLACTIN

Pituitary fragments from monkey or humans were incubated in the same manner as outlined for the placenta with the exception that the medium contained 50 μCi/ml L-[4,5-³H]leucine. A flow-sheet outlining the analyses is given in Table IV and a more complete description of the

TABLE IV

PROCEDURE FOR ANALYSIS OF TISSUES AND MEDIA AFTER INCUBATION OF PITUITARY GLANDS IN KREBS-RINGER BICARBONATE BUFFER

A. Medium or tissue extract:
   (1) Trichloroacetic acid precipitable protein—³H-labelled protein.
   (2) Anti-HGH precipitable protein—[³H]HGH or [³H]MGH.
   (3) Anti-ovine prolactin precipitable protein—[³H]human prolactin or [³H]monkey prolactin.
   (4) Protein—fluorometric estimation.
   (5) HGH or MGH—radioimmunoassay.
   (6) Human or monkey prolactin—radioimmunoassay.
B. Gel filtration of tissue or medium.
C. Electrophoresis of fractions from B.

methods has been documented elsewhere (Friesen, Guyda and Hardy 1970). After the incubation, radioactive proteins were precipitated by trichloroacetic acid and immunoprecipitation studies were made on aliquots of the tissue and medium using antiserum to ovine prolactin and antiserum to HGH to precipitate radioactive prolactin and radioactive growth hormone respectively. Growth hormone and prolactin in the tissue extracts or media were measured by radioimmunoassay and total protein was measured by fluorometric analysis.

Table V summarizes two studies on monkey pituitary glands obtained

TABLE V

RADIOACTIVE PROTEINS IN FOETAL MONKEY PITUITARY INCUBATED *IN VITRO*

	Incubation media				Tissue	
	4 hours		24 hours		24 hours	
	Monkey no. 1*	Monkey no. 2**	Monkey no. 1	Monkey no. 2	Monkey no. 1	Monkey no. 2
^{3}H-labelled proteins $\times 10^3$ c.p.m.†	19·5	5·8	250·0	97·0	620·0	637·0
[^{3}H]MGH $\times 10^3$ c.p.m.	0·5	0·15	63·8	17·2	90·9	67·3
[^{3}H]human prolactin $\times 10^3$ c.p.m.	12·2	1·4	102·2	34·4	121·3	120·5
Percentage ratio [^{3}H]MGH/ ^{3}H-labelled protein	2·6	2·7	25·0	18·0	14·0	11·0
Percentage ratio [^{3}H]monkey prolactin/^{3}H-labelled protein	63·0	25·0	41·0	35·0	19·0	19·0

* Foetus alive in pregnant monkey no. 1.
** Foetus dead in pregnant monkey no. 2.
† c.p.m./mg wet weight pituitary.

from pregnant monkeys. One had a macerated dead foetus, but in the second the foetus was alive. It is apparent that the amount of radioactive protein released into the incubation media by the pituitary from the monkey with the dead foetus was considerably less than that released by the pituitary from the monkey with the live foetus. The difference is almost wholly accounted for by the increased amounts of radioactive protein precipitated by antiserum to prolactin. At 24 hours the ^{3}H-labelled proteins, [^{3}H]growth hormone and [^{3}H]prolactin contained in the extract of the two pituitary tissues are very similar, but the release into the medium of [^{3}H]growth hormone and especially [^{3}H]prolactin is considerably greater in the second pituitary. The total amount of protein released into the medium by this pituitary is also considerably greater at both 4 and 24 hours. After gel filtration of the combined media from both experiments the distribution of ^{3}H-labelled proteins, [^{3}H]prolactin and [^{3}H]growth hormone was determined (Fig. 5). There is a single radioactive protein peak which coincides with a protein peak emerging with a ratio of elution volume to void volume of 2·0. The [^{3}H]prolactin peak (tube 28) emerges slightly ahead of the [^{3}H]growth hormone peak (tube 29) and the latter is coincident with the growth hormone peak measured by radioimmunoassay. Sixty-three per cent of the ^{3}H-labelled proteins in tubes 22 to 32

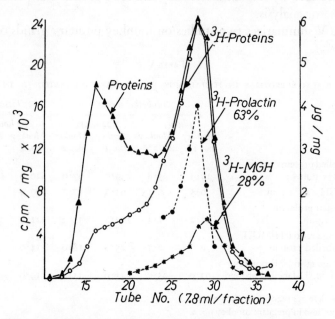

FIG. 5. Gel filtration of the incubation media from experiments on monkey pituitary glands outlined in Table V. The distribution of ^{3}H-labelled proteins and proteins is shown together with the distribution of ^{3}H-labelled proteins precipitated by anti-HGH (^{3}H-MGH) and anti-ovine prolactin (^{3}H-prolactin). Note the slightly smaller elution volume of ^{3}H-prolactin compared to ^{3}H-MGH.

were precipitated by antiserum to ovine prolactin and only 28 per cent by antiserum to growth hormone. We concluded from studies of this type in monkeys, that growth hormone and prolactin are separate hormones which are synthesized independently and can be distinguished immunologically, and that monkey and sheep prolactin share a common antigenic determinant which is absent from monkey and human growth hormone.

The same type of studies were performed with normal human pituitary glands (from both male and female subjects) and pituitary adenomas from a patient with acromegaly and two patients with galactorrhoea (Hwang et al. 1971b; Friesen et al. 1971). Fig. 6 shows the distribution of ^{3}H-labelled proteins, [^{3}H]HGH and [^{3}H]prolactin after gel filtration of the 24-hour media from incubated pituitaries in two experiments. Table VI sum-

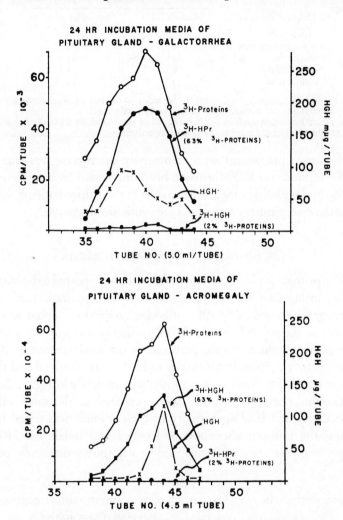

FIG. 6. Gel filtration of the 24-hour incubation media of the pituitaries from a patient with galactorrhoea and from one with acromegaly. Note the difference in the amount of ^{3}H-labelled proteins precipitated by anti-HGH and anti-ovine prolactin in the two media.

marizes the results of immunoprecipitation studies with antiserum to sheep prolactin and antiserum to HGH. In the two patients with galactorrhoea, antiserum to sheep prolactin precipitated 30-fold and 200-fold more ^{3}H-labelled proteins than antiserum to growth hormone

TABLE VI
IMMUNOPRECIPITATION OF ^{3}H-LABELLED PROTEINS WITH ANTISERA TO SHEEP PROLACTIN AND HGH in 24-HOUR INCUBATION MEDIUM

	Percentage of ^{3}H-labelled proteins precipitated by	
Origin of Pituitary	Anti-HGH	Anti-ovine prolactin
Acromegaly	63	2
Galactorrhoea, case 1	2	63
Galactorrhoea, case 2	13*	95
Normal female	9	28
Normal male	37	12

The incubation medium was fractionated on Sephadex G-100 and immunoprecipitation studies were confined to aliquots of the radioactive protein peak, Ve/Vo 1·8–2·2.

* In the 4-hour incubation medium 0·5 per cent and 99 per cent of the ^{3}H-labelled proteins were precipitated by anti-HGH and anti-ovine prolactin respectively.

did; whereas in the patient with acromegaly the reverse was true. The ratio of [^{3}H]prolactin to [^{3}H]growth hormone varied by a factor greater than 1000 in the pituitary incubation media in these three patients, suggesting that the two pituitary hormones are synthesized separately.

THE PURIFICATION OF PRIMATE PROLACTIN

Because primate prolactin and growth hormone appeared to be immunologically distinguishable, we reasoned that the application of affinity chromatography using anti-HPL antibodies coupled to Sepharose should be helpful in the separation of primate prolactin and growth hormone. HPL–Sepharose columns were prepared as outlined previously (Guyda and Friesen 1971). Specific antibodies to HPL were obtained and they in turn were coupled to Sepharose as shown schematically in Fig. 7. Sepharose anti-ovine prolactin columns were prepared in the same manner. The specificity and binding capacity of each column was tested by determining the immunoadsorption of labelled and unlabelled HGH or sheep prolactin on each column. Table VII shows that each column

TABLE VII
IMMUNOADSORPTION OF RADIOACTIVE TRACERS BY AFFINITY CHROMATOGRAPHY

	Percentage of tracer adsorbed by	
	Anti-HPL Sepharose	Anti-ovine prolactin Sepharose
[^{131}I]HGH	91	5
[^{131}I]ovine prolactin	15	93
[^{131}I]monkey prolactin	8	82
[^{131}I]human prolactin	—	85

adsorbed almost exclusively only one tracer. Incubation media from cultured pituitary glands from pregnant monkeys were passed through

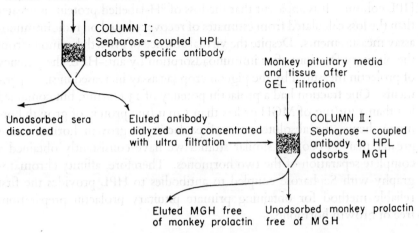

FIG. 7. Sheep antisera to HPL were pooled, diluted 1/3 with water and passed through a Sepharose column to which 0·5 g of HPL had been coupled. Anti-HPL antibodies were adsorbed and all other proteins passed through unretarded. The column was washed and specific anti-HPL antibodies were eluted with sodium thiocyanate. The antibodies were dialysed, concentrated and coupled to Sepharose. Human or monkey pituitary tissue was passed through the Sepharose columns to remove growth hormone; prolactin was not adsorbed. The same method was employed to couple anti-ovine prolactin antibodies to Sepharose which in turn could be used to bind primate prolactin.

Sepharose anti-HPL columns. The results of one experiment are shown in Table VIII. It is apparent that with a single passage of the media through the

TABLE VIII

IMMUNOADSORPTION OF MGH AND MONKEY PROLACTIN IN PITUITARY INCUBATION MEDIA BY SEPHAROSE ANTI-HPL COLUMNS

	Protein µg	MGH µg	Monkey prolactin µg	^{3}H-labelled proteins c.p.m. × 10^{-4}	[^{3}H]MGH c.p.m. × 10^{-4}	^{3}H-monkey prolactin c.p.m. × 10^{-4}
Medium applied	2700	680	990	1576	388	411
Unadsorbed	1724	<1·0	840	981	8	323
Eluted with NaSCN	945	892*	58	331	294	3
Percentage unadsorbed†	64	<0·05	85	62	2	79
Percentage eluted†	35	131	6	21	76	0·7
Percentage loss†	1	0	9	17	22	21

The media from several cultures were pooled after the incubation of pituitary glands from pregnant monkeys.
* The MGH content was analysed on two separate occasions.
† In these three rows the units under the column headings do not apply—the values are percentages.

Sepharose anti-HPL column more than 99 per cent of monkey growth hormone (MGH) and less than 15 per cent of monkey prolactin was removed. Only 2 per cent of the [^{3}H]MGH and 80 per cent of the [^{3}H]-

monkey prolactin in the media were not adsorbed by the Sepharose anti-HPL column. It is apparent that the loss of ^{3}H-labelled proteins is greater than the loss calculated from estimates of recovery based on radioimmunoassay measurements. Despite the virtual absence of growth hormone from the pituitary fractions after immunoadsorption by anti-HPL, the potency of prolactin measured by the pigeon crop sac assay increased in six experiments. One fraction had a prolactin potency of 13 i.u./mg, but contained less than 1 µg/mg of MGH or less than 1 m u./mg protein of growth hormone, hence the absolute ratio of prolactin to growth hormone was greater than 13 000. In similar studies we have consistently obtained a complete separation of the two hormones. Therefore, affinity chromatography with Sepharose coupled to antibodies to HPL provides the first reliable method for obtaining primate pituitary prolactin preparations free of growth hormone.

RADIOIMMUNOASSAY FOR HUMAN AND MONKEY PROLACTIN

Two rabbits were immunized with 300 µg of these growth hormone-depleted pituitary fractions containing prolactin. The antisera obtained bound ^{131}I-labelled sheep prolactin and in incubation media of pregnant monkey pituitaries the antisera precipitated the same percentage of ^{3}H-labelled proteins as antiserum to sheep prolactin did (Friesen, Guyda and Hwang 1971). Hence we concluded that the antisera contained antibodies against primate prolactin. Using one of the antisera we developed a radioimmunoassay for primate prolactin which distinguished between primate growth hormone and prolactin (Guyda, Hwang and Friesen 1971). Unfortunately the assay was not sensitive enough to measure circulating levels of serum prolactin. However, the sensitivity of the assay was improved greatly when monkey or human ^{131}I-labelled prolactin was used as tracer instead of ^{131}I-labelled sheep prolactin (Hwang, Guyda and Friesen 1971). The primate prolactin tracer was purified by affinity chromatography using columns of Sepharose coupled to anti-ovine prolactin. More than 80 per cent of the primate prolactin tracer was bound by anti-ovine or monkey prolactin sera and less than 10 per cent by anti-HGH serum. The same result was obtained when the primate prolactin tracer was passed through Sepharose columns to which antibodies to HPL or prolactin had been coupled (Table VII). In the homologous radioimmunoassay for primate prolactin, we have used antisera to human or monkey prolactin and human or monkey ^{131}I-labelled prolactin. The assay was standardized using a pool of postpartum serum in which prolactin had been measured by bioassay by Dr A. Frantz of Columbia

University, using a pregnant mouse mammary gland assay (Frantz and Kleinberg 1970). Fig. 8 shows that the sensitivity of our assay was 2 ng/ml and that HGH and HPL fail to cross-react in the assay at a concentration of 1000 ng/ml. HGH cross-reacts only when the concentration exceeds 1 μg ml and greater amounts of HGH inhibit the binding of the tracer in a manner identical to that of prolactin, suggesting that HGH is contaminated to a minor extent with prolactin.

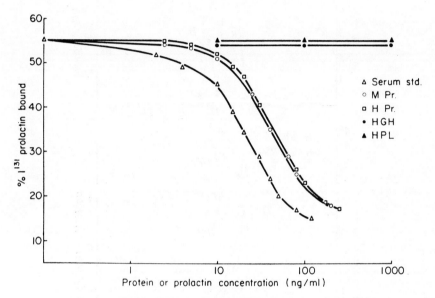

FIG. 8. A radioimmunoassay for prolactin. The sensitivity of the assay is 2·0 ng/ml. HGH and HPL do not inhibit the binding of the tracer [^{131}I]monkey prolactin (MPr) at concentrations of 1 μg and 10 μg/ml respectively.

The assay has been used to measure prolactin in pituitary incubation media, in eluates obtained after gel electrophoresis of pituitary fractions and finally in serum samples. A few interesting results obtained with the assay will be summarized briefly. In the incubation media of the monkey pituitaries shown in Fig. 5, the prolactin content in a pool of the tubes containing prolactin (tubes 22 to 31) is 163 μg. In seven human pituitary glands obtained at autopsy and homogenized with 0·1M-NH_4HCO_3, the prolactin content ranged between 0·20 and 0·75 μg/mg wet weight and in the 24-hour incubation media of human pituitaries obtained at hypophysectomy, it was approximately 0·30 μg/mg wet weight. Fig. 9 shows the electrophoretic distribution of human prolactin in eluates obtained after polyacrylamide gel electrophoresis of media from cultures from a case of pituitary adenoma with galactorrhoea. The R_F of prolactin is 0·58, compared to an R_F of 0·47 for the major component of HGH at pH 4·3.

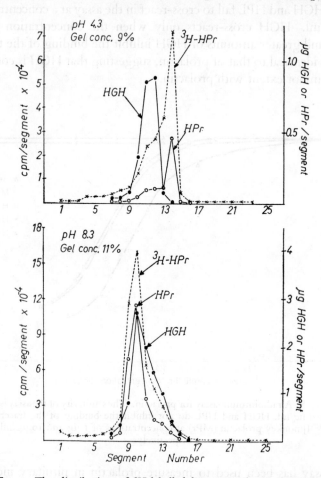

FIG. 9. The distribution of ^{3}H-labelled human prolactin (^{3}H-HPr), human prolactin (HPr) and HGH in eluates obtained after acrylamide gel electrophoresis of HGH and human prolactin at pH 4·3 and 8·3. There is a clear separation of HGH and human prolactin at pH 4·3 but almost no separation at pH 8·3, although the R_F of human prolactin and [^{3}H]human prolactin is slightly less than that of HGH at pH 8·3.

Table IX summarizes serum prolactin measurements made in patients and also in monkeys. Virtually all the subjects examined including children have detectable levels of prolactin. In patients with galactorrhoea, prolactin levels ranged from 12 to 1800 ng/ml whereas in acromegalics, prolactin levels were 'normal' with the exception of one young patient. One of the most surprising findings was that prolactin was detectable in many idio-

Table IX

SERUM PROLACTIN VALUES IN MONKEYS AND HUMANS MEASURED BY RADIOIMMUNOASSAY

	Number of subjects	Serum prolactin ng/ml	
		Mean	Range
Children	36	11	0–17
Adults (males)	42	10	0–28
Menstrual cycle	9		
Follicular		10	4·2–21
Luteal		11	4·9–42
Acromegaly	13	16	3–26
Galactorrhoea	21	100	12–1800
Idiopathic hypopituitary	14	10	0–32
Pregnancy			
First trimester	24	25	7–47
Second trimester	96	50	6–350
Third trimester	102	134	36–600
Term	19	207	44–600
Postpartum (6 weeks)	25	10	0–14
Newborn			
Cord blood	19	258	120–500
One week	7	192	40–400
Nursing mother			
1–3 weeks			
Before suckling	6	14	8–20
30 min after		259	175–400
60 min after		133	75–200
3–5 days			
Before suckling	2	130	120–140
30 min after		230	200–260
60 min after		185	180–190
Insulin (10 patients)	No response		
(2 patients)	Maximum increment of 20 ng/ml		
Arginine (6 patients)	No response		
Anaesthesia and surgery			
Human subjects	4		6–150
Pregnant monkeys	6		30–1000
Non-pregnant monkeys	4		2–60

pathic hypopituitary patients, whereas after hypophysectomy for diabetic retinopathy or mammary cancer, serum prolactin was undetectable.

In the normal menstrual cycle we have found no convincing evidence of any change in serum prolactin at any point in the cycle although the mean serum prolactin concentration was higher in the second half. During pregnancy there is a progressive increase in serum prolactin reaching a mean concentration at term of 200 ng/ml. The levels of prolactin in foetal serum were equal to those found in the mother's sera. In amniotic fluid they were ten-fold greater than those in the mother. These results are very different from HPL concentrations, which are high in the mother (mean 4 µg/ml) but low in the foetus (< 100 ng/ml) and amniotic fluid (< 0·5 µg/ml). In the neonatal period serum prolactin begins to decline immediately after birth, reaching concentrations which are usually less than 20 ng/ml by four to six weeks. In the mother, abrupt increases in serum prolactin occur 30 minutes after the onset of nursing (Hwang *et al.* 1971*a*). Finally, during surgery in pregnant and non-pregnant monkeys, sudden dramatic increases in serum prolactin have been observed (Friesen 1971) without any corresponding increase in serum placental lactogen. In man elevated prolactin levels have also been noted during surgery.

Table X shows that affinity chromatography can be applied to serum

TABLE X
AFFINITY CHROMATOGRAPHY OF SERA ON SEPHAROSE COLUMNS

Clinical status		Serum concentration (ng/ml)		
		Before chromatography	After Anti-HPL	After Anti-ovine prolactin
Pregnancy (30 wks)	HPL	6000	10	—
	Prolactin	120	129	7
Galactorrhoea (pituitary tumour)	HGH	1	1	<1
	Prolactin	160	—	5
Gigantism (pituitary tumour)	HGH	480	10	300
	Prolactin	190	140	0

samples in order to demonstrate unequivocally that growth hormone or placental lactogen, and prolactin can be adsorbed selectively from serum samples of patients by antibodies to placental lactogen and ovine prolactin respectively.

DISCUSSION

The studies which have been outlined on placental lactogen synthesis demonstrate that placental tissue can synthesize HPL at widely differing rates. The fundamental question, however, remains unanswered; namely, what factors determine the rate of HPL synthesis at any given time? Two radioactive proteins of widely differing molecular weight in placental tissue extracts are precipitated by antiserum to HPL. The smaller molecular weight species undoubtedly is HPL, but at this point we do not know the nature or significance of the large molecular weight component. After the incubation of pituitary glands, antiserum to HGH also precipitated radioactive proteins which emerged in the void volume (Friesen, Guyda and Hardy 1970). However, even with extended incubation periods the large molecular weight component was never released into the medium, whereas in the placenta the large molecular weight species predominated in the media in some experiments. As yet we have no good evidence for a precursor-product relationship between the two species, and therefore it is premature to refer to the large molecular weight species as a prohormone. Evidence has also been reported for a large molecular weight species of growth hormone in the rat pituitary (Frohman and Buret 1970).

The data we have presented can leave no doubt that the pituitary gland of primates, in the non-pregnant and pregnant states, synthesizes and secretes growth hormone and prolactin independently both *in vivo* and *in vitro*. There are three reasons why it has been possible to identify primate prolactin so readily *in vitro*. In the first place there is an immunological cross-reaction between sheep and human prolactin. Secondly, the pool size of prolactin and growth hormone in the pituitary differs by at least a factor of 100. Thirdly, the hypothalamic control of prolactin and growth hormone differ; in the former the predominant control is one of tonic inhibition, whereas in the latter a positive signal, GRF, is required for the release of growth hormone. Hence *in vitro*, growth hormone release is inhibited and prolactin release is enhanced. The last two factors therefore favour the rapid synthesis and release of prolactin, and this accounts for the fact that at an early period in the incubation, radioactive prolactin predominates and its specific activity is several orders of magnitude higher than that of growth hormone.

Now that prolactin has been identified as a separate human pituitary hormone, we are attempting to devise suitable methods to purify it from acetone-dried human pituitary powder. If we are successful it will be most interesting to compare the chemistry of pituitary and placental lactogens and to determine the biological effect of human prolactin. These studies should help to clarify the physiological role of prolactin in man.

SUMMARY

The biosynthesis of placental lactogen has been studied by incubating placental tissue fragments in Krebs-Ringer bicarbonate buffer containing L-[^{3}H]leucine. Immunoprecipitation studies were made on the tissue supernatant and incubation media using antiserum to HPL to precipitate [^{3}H]HPL. Two species of [^{3}H]HPL of widely differing molecular weight were noted after gel filtration of the tissue supernatant and the incubation media. In the tissue the large molecular weight species of [^{3}H]HPL predominated whereas in the media the smaller molecular weight species was the principal component. In the media [^{3}H]HPL formed between 10 and 50 per cent of the trichloroacetic acid precipitable proteins compared to 5 and 20 per cent in the tissue. These studies demonstrated that [^{3}H]-HPL is one of the principal proteins synthesized and secreted *in vitro*, and are consistent with the very high production rate of HPL *in vivo* (1 g per day).

Human and monkey pituitary glands were incubated *in vitro* in the same manner as placental tissue. The biosynthesis and secretion of growth hormone and prolactin were studied by immunoprecipitation of incubation media and tissue extracts with antiserum to HPL or HGH and antiserum to ovine prolactin to precipitate [^{3}H]growth hormone and [^{3}H]prolactin respectively. [^{3}H]growth hormone was not precipitated by anti-prolactin. In human pituitary glands [^{3}H]prolactin formed between 12 and 80 per cent of the ^{3}H-labelled proteins released into the medium and [^{3}H]prolactin was also the principal radioactive protein identified in the incubation medium of pituitaries from pregnant and lactating monkeys. As a result of these studies, we were able to identify and separate primate prolactin from growth hormone using affinity chromatography and to set up a specific radioimmunoassay for primate prolactin with which we have studied the synthesis and secretion of primate prolactin *in vitro* and *in vivo*. During pregnancy in humans, the amounts of serum HPL and human prolactin increased progressively to reach mean concentrations of 8 µg/ml and 200 ng/ml respectively. In monkeys no consistent increase in prolactin was observed during pregnancy, but in samples obtained during surgery the prolactin concentrations were remarkably high (> 500 ng/ml). The elevated concentrations appear to be related to a stress-induced release of prolactin.

In summary, our experiments on the biosynthesis of placental lactogen led to similar studies in the pituitary, which resulted in the eventual separation of primate growth hormone from prolactin and culminated in the development of a specific radioimmunoassay for primate prolactin which has been used to study the prolactin concentration in health and disease.

Acknowledgements

We wish to acknowledge the technical assistance of Miss Jean Henderson, Mrs Judy Halmagyi and Mrs Klara Holmwood. We are also indebted to Miss Francine Dupuis for the figures and to Mrs Inara Leimanis for typing the manuscript.

The research was supported by the Medical Research Council of Canada, MA 1862, and United States Public Health Service Child Health and Human Development grant HD 01727-06.

REFERENCES

FRANTZ, A. G. and KLEINBERG, D. L. (1970) *Science* **170**, 745–747.
FRIESEN, H. (1968) *Endocrinology* **83**, 744–753.
FRIESEN, H. (1971) In *Growth Hormone (Proceedings of the Second International Symposium),* ed. Pecile, A. and Müller, E. E. Amsterdam: Excerpta Medica Foundation. In press.
FRIESEN, H. and GUYDA, H. (1971) *Endocrinology* **88**, 1353–1362.
FRIESEN, H., GUYDA, H. and HARDY, J. (1970) *J. Clin. Endocrinol. & Metab.* **31**, 611–624.
FRIESEN, H., GUYDA H. and HWANG, P. (1971) *Nature New Biol.* **232**, 19–20.

Friesen, H., Suwa, S. and Pare, P. (1969) *Recent Prog. Horm. Res.* **25,** 161–205.
Friesen, H., Webster, B. R., Hwang, P., Guyda, H., Munro, R. E. and Read, L. (1971) *J. Clin. Endocrinol. & Metab.* in press.
Frohman, L. A. and Buret, C. L. (1970) *Program of the Endocrine Society, 52nd Meeting.*
Guyda, H. and Friesen, H. (1971) *Biochem. & Biophys. Res. Commun.* **42,** 1068–1075.
Guyda, H., Hwang, P. and Friesen, H. (1971) *J. Clin. & Endocrinol. Metab.* **32,** 120–123.
Hwang, P., Guyda, H. and Friesen, H. (1971) *Proc. Natl. Acad. Sci. (U.S.A.)* **68,** 1902–1906.
Hwang, P., Guyda, H., Friesen, H. and Tyson, J. (1971a) *Program of the Endocrine Society, 53rd Meeting,* p. 43 (abst. 2).
Hwang, P., Friesen, H., Hardy, J. and Wilansky, D. (1971b) *J. Clin. Endocrinol. & Metab.* **33,** 1–7.
Josimovich, J. B. (1969) In *Fetal Homeostasis,* vol. IV, pp. 109–121, ed. Wynn, R. M. New York: Appleton-Century-Crofts.
Suwa, S. and Friesen, H. (1969a) *Endocrinology* **85,** 1028–1036.
Suwa, S. and Friesen, H. (1969b) *Endocrinology* **85,** 1037–1045.
Tyson, J., Austin, K. L. and Farinholt, J. W. (1971) *Am. J. Obstet. & Gynec.* **109,** 1080–1082.

DISCUSSION

Spellacy: I was interested in the rise of HPL levels with prolonged incubation of human placental tissue. Have you looked at this in terms of the decreasing amount of glucose present in the medium and could you reverse this rise in HPL by adding glucose?

Friesen: We haven't done this systematically. I think Dr Swanson Beck had some evidence that HPL secretion is related to the glucose concentration of the medium.

Swanson Beck: We have not made systematic studies on the exact relationship between glucose concentration and HPL secretion, since we have not measured the hormonal output of many of our cultures (Hou, Ewen and Beck 1968), but it appears that the uptake of glucose rises over the first two or three days when the HPL output is falling rapidly: glucose consumption then reaches a peak and falls gradually, but HPL output continues at a very low level.

Could I go back to the question of the validity of tissue maintained in a culture medium? We have maintained normal human mid-pregnancy placental villi in organ culture with Medium 199 and we have noted remarkable changes in the morphology of this tissue. Within one to six hours mitotic figures become very numerous in the cytotrophoblast cells, which at this time actively incorporate [^{3}H]leucine into their cytoplasm. By two or three days, the cytotrophoblast forms an almost continuous layer underneath the syncytiotrophoblast, which at this time shows no uptake of [^{3}H]leucine (Ewen, Beck and Green 1971). Thus, it seems that in cultured tissue the roles of the cytotrophoblast and the syncytiotrophoblast are reversed, so I want to put back to you the question: is it possible that this cytopathology, if one could call it that, could be at least in part responsible for the high molecular weight HPL that you have detected?

Friesen: The large molecular weight HPL is present early on, within three or four hours of culturing, so I don't think its presence is due to incubation conditions. I think it's there in all circumstances.

Meites: I'd like to point out some differences between prolactin secretion in animals as compared to women. One difference is that during oestrous cycles in the rat, cow and sheep, a pronounced peak of prolactin is reached on the afternoon of pro-oestrus, prior to ovulation. I wonder whether you have tried collecting frequent blood samples to see whether you can detect such a peak before ovulation, during the menstrual cycle? Secondly, in regard to pregnancy, we reported in the rat that blood prolactin as measured by radioimmunoassay (Amenomori, Chen and Meites 1970) remains very low throughout pregnancy, and shows no increase until the very end of pregnancy, the last day or so. I don't know of any comparable studies of blood prolactin during pregnancy in other species. Pituitary studies in the rabbit and guinea pig (Meites and Turner 1948) also showed low prolactin levels until the end of pregnancy.

Friesen: The conclusions which I stated on the menstrual cycle were based on daily blood samples from ten women throughout the menstrual cycle, giving a total of 300 samples. There are some small minor peaks, but we found it impossible to correlate these minor variations in prolactin throughout the cycle with the mid-cycle peak of LH. We only measured it daily, and perhaps we ought to look more carefully around the LH peak.

I totally agree that there are species differences in prolactin elevation during pregnancy, even among primates, because in the rhesus monkey we find no systematic elevation of prolactin in pregnancy, as we do in the human.

Beck: Would you care to hypothesize why you think that is?

Friesen: I wonder whether it is related to the very striking differences in oestrogen metabolism at term in the two species. In the human the average excretion of oestrogen at term is 30 mg per day; the rhesus monkey excretes about 30 μg, so even allowing for differences in body size there is a very big difference in oestrogen metabolism, and something we would like to do is to see whether the administration of oestrogen to the rhesus monkey causes an elevation of serum prolactin. We have studied one male monkey to whom we gave very large amounts of oestrogen, 10 mg oestradiol polyphosphate per week for six weeks, and then removed the pituitary and incubated it in the presence of radioactive leucine. The pituitary from this male behaved the same as one from a pregnant monkey.

Approximately 80 per cent of the radioactive protein released into the medium in the first four hours of incubation was prolactin, compared with about 20 per cent in the normal male. So I think the basic difference is probably due to oestrogen.

Sherwood: The possible effect of oestrogen is interesting, particularly since Dr Frantz and Dr Rabkin (1965) described oestrogen-potentiated release of growth hormone in man. In lower forms, oestrogen stimulates prolactin primarily. Have you added oestrogen to incubated pituitary tumours or normal glands to see whether hormone production was affected?

Friesen: We tried only once and the results were equivocal. We did try to see whether, as Dr MacLeod has reported, noradrenaline would inhibit the release of prolactin *in vitro*. The changes we observed with noradrenaline in this regard were not very consistent.

Frantz: Have you studied the effect of glucose on prolactin secretion?

Friesen: We have only looked at human prolactin concentrations after giving glucose orally, but we observed no systematic effect on serum prolactin levels.

Greenwood: A point to Dr Meites. We haven't looked in the human menstrual cycle for a pattern in human prolactin. Certainly in the sheep, in collaboration with Dr J. R. Goding and Dr I. Cummins, sampling at half-hourly intervals and then five-minute intervals over the LH peak we found prolactin surges as the LH peak declined. Undoubtedly a fuller study in the human needs to be done on the same lines.

Li: Dr Friesen, I enjoyed your paper very much; you have convinced me there is a primate prolactin! How much work have you done so far on the normal pituitary?

Friesen: By now we have studied about twelve normal human pituitaries removed by transphenoidal hypophysectomy by Dr Jules Hardy of the Notre Dame Hospital, Montreal. These pituitaries were incubated as described in my paper. We have been able to demonstrate the synthesis of prolactin in each instance.

Li: Did I understand that the human growth hormone has no lactogenic activity?

Friesen: No. I would expect to find some intrinsic lactogenic activity in growth hormone. We want to take HGH that was absorbed by the anti-HPL–Sepharose column, elute it and compare the lactogenic activity present before and after absorption, to see whether there is a difference in the intrinsic prolactin activity.

Li: Have you any chemical data on your purified monkey prolactin?

Friesen: No.

Frantz: We find that highly purified human growth hormone preparations (from Dr Wilhelmi) all appear to have high intrinsic prolactin activity which can be completely neutralized with anti-growth hormone antiserum. The same antiserum is completely ineffective in neutralizing prolactin activity in the serum of patients with postpartum lactation or with galactorrhoea, which suggests that that activity is not due to growth hormone.

Friesen: What concentrations of HGH did you test?

Frantz: We have tested 2 μg/ml of preparation HS1103C. Since we get a maximal response at 50 ng, the abolition of the response at 2000 ng/ml seems to us significant.

Friesen: Our estimate is that there is about one part of human prolactin per 500 of HGH so you will probably have to test at a concentration 500 times that at which you obtain a minimal response in your bioassay.

Frantz: Yes; that would only give us about 4 ng, and that would perhaps be a little too low to detect.

Friesen: It's just at the limit of sensitivity of your assay. It would be very interesting and important to do this.

Forsyth: I think that the lactogenic activity of human growth hormone as shown in bioassays must be an intrinsic property. In the rabbit, the activity is so high (20–40 i.u./mg) that if it were due to some small prolactin contamination, the specific activity of human prolactin in the system would be fantastically high.

Pasteels: On species differences, I am not surprised by what Dr Friesen found in the monkey in comparison to the human pituitary. With Dr Nicoll we are studying pituitaries of rhesus monkeys and of humans using differential staining of the prolactin cells and immunofluorescence. In male monkeys or in adult females, not pregnant and not lactating, there were quite a lot of well-developed prolactin cells filled with secretory material, and in humans there were not, except during pregnancy and *post partum*. Perhaps this was the result of stress, because one never knows exactly how the monkeys were, whether they were pregnant before they were caught (in some we found evidence of a recent abortion), or if they were stressed by being kept in the laboratory. It might not be a true species difference, therefore.

Turkington: Following up Dr. Pasteels' point, in view of the fact that by a large number of criteria the prolactin-secreting cells in the normal human pituitary seem to be about one per cent of the number of acidophils which are secreting growth hormone, I wonder if this is consistent with Dr Friesen's findings that the serum concentration of prolactin is around 10 ng/ml in normal humans, male or female? This is very much the same concentration as growth hormone, and one would expect that the two hormones would have similar half-lives in the serum. The only way to account for this would be that prolactin is secreted at perhaps a hundred times the rate of growth hormone.

Friesen: These considerations concerned us too. There are three lines of evidence for thinking that the level of 10 ng/ml of prolactin is valid. The first is that if we pass the sera through the anti-ovine prolactin–Sepharose column, measuring prolactin before and after, the level is always zero afterwards. If we do the same experiment with an anti-HPL–Sepharose, it is unchanged. The second line of evidence is that when the sera of patients are examined before and after surgical hypophysectomy for diabetes or mammary cancer the blood prolactin after hypophysectomy is always zero. Thirdly, and this is the best evidence, when L-dopa is given to patients with Parkinson's disease, the levels initially are about 10 ng/ml of prolactin; within three hours, in eight out of ten patients, the levels had declined to zero. I think that dopamine or L-dopa in those individuals has shut off prolactin, just as Dr MacLeod found in the rat. I think your suggestion is quite right that the prolactin-secreting cells are synthesizing prolactin at a tremendous rate, which is why we find kinetic evidence for a rapid synthesis of prolactin, compared to the other hormones in the anterior pituitary.

Turkington: These hormones are defined operationally, in terms of immunological cross-reactivity. Have you any other criteria?

Friesen: In six studies where we did bioassays using the pigeon crop sac assay and immunoassay (Guyda and Friesen 1971) the difference in potency estimates usually was a factor of two; the immunoassay generally underestimates the biological activity by a factor of about two. So that is an independent way of looking at it.

Turkington: By bioassay (see p. 172) we have never encountered detectable prolactin in people whom we would call normal, and so I wonder whether this is a methodological discrepancy or not?

Friesen: I agree that this is an important issue that can only be resolved by assaying the same samples in two different laboratories. For the samples that were assayed by you from patients after pituitary stalk section and were also sent to us, our immunoassay results generally were considerably lower than yours, by a factor of five to ten; so I'm puzzled why you can't detect prolactin in normal people. At low levels our estimates are higher than yours, yet at high levels our estimates are lower. There appears to be an interesting difference here, which may be due to inherent differences in the assay.

On the question of stress, in a series of pregnant monkeys that we operated upon, the foetus being removed, the initial serum prolactin concentrations were all rather low, less than 20 ng/ml usually. Before surgery was begun but after Nembutal anaesthesia, there was an increase in prolactin. The highest concentration we found was 1000 ng/ml: a remarkable amount. During surgery the level fell, and during the next few days, samples taken in the unanaesthetized state were again low (less than 20 ng/ml). Two to six weeks later the same animals were re-operated upon and their levels after anaesthesia but before surgery were again elevated (200 ng/ml). In human subjects undergoing a variety of surgical procedures the initial serum prolactin levels are already high even before anaesthesia (30–70 ng/ml). One patient developed galactorrhoea after surgery, which is a rather rare occurrence, and her blood prolactin rose to 70 ng/ml from a preoperative level of 10 ng/ml.

Frantz: Do you interpret the results to indicate that removal of the foetus itself produced any change in circulating prolactin or do you interpret the results merely as stress?

Friesen: I think the data are too limited to conclude that removing the foetus had an effect, other than that due to stress.

Nicoll: Regarding the discrepancy between the small number and inactive appearance of human prolactin cells in hypophyses taken at autopsy and the apparently high synthetic rate *in vitro*, the patients from whom the glands were obtained were presumably emotionally stressed in preparation for surgery, and they also had some pathology, which may have had an influence on pituitary function.

Friesen: We must bear in mind that the rate of secretion as demonstrated *in vitro* is probably favoured. First of all, any hypothalamic inhibition has been removed, and secondly, there probably is a preferential release of prolactin. The chemical amount of prolactin released is really not very large—0·40 ng/mg/24 hours; the amount of growth hormone in our incubation media varies from twenty to a hundred times that of prolactin. The rapid synthesis of prolactin is shown by the rapid incorporation of labelled leucine into prolactin. If we had looked at chemical amounts of prolactin in the media, I don't think we would have been able to demonstrate so clearly the existence of a separate primate prolactin.

Greenwood: Certainly as far as we are concerned, human prolactin functions like sheep prolactin. Both hormones respond to the same three major stimuli of stress, phenothiazines and suckling. We have recently found a discrepancy in the half-lives of endogenous and exogenous sheep prolactin. The exogenous half-life is about 20 minutes (Bryant *et al.* 1968) whereas the endogenous half-life is not greater than seven minutes. The fall in plasma human prolactin is consistent with a half-life of about four or five minutes. Have you any data on the half-life of human prolactin in plasma?

Friesen: From some of the data on monkey prolactin we can infer a maximal estimate of half-life, and in some individuals it is as short as ten minutes, but it's hard to get a situation where you can shut off prolactin secretion.

Turkington: We made one observation on a patient who had hypophysectomy for a prolactin-secreting tumour and whose serum prolactin subsequently fell to zero. Her prolactin activity disappeared with a half-life between 12 and 15 minutes.

Greenwood: Dr Friesen, have you any evidence in your short-term pituitary incubations of degradation? Dr Stanley Ellis says that the pituitary contains a number of enzymes which will degrade peptide hormones down to amino acids. Have you evidence of partially synthesized prolactin? We have been trying to study this in cell-free systems and all we found was that ribosomes have a very high capacity for adsorbing prolactin non-specifically. Would ribosome-adsorbed material be readily pulled out by your extraction procedure?

Friesen: We haven't studied that point. The reason I doubt that there is much degradation is that in the pituitary from a patient with galactorrhoea, after 24 hours in culture half the total protein was prolactin, so that is a maximal estimate of degradation of 50 per cent.

Greenwood: In our foetal pituitary cultures degradation can't be very high otherwise we would not be able to find prolactin!

MacLeod: We looked at the question of degradation quite carefully and in the whole-tissue incubations we obtained 100 per cent recovery of added prolactin—there was no degradation at all.

Wilhelmi: If the cells are intact, proteolytic activity is nearly nil in pituitary cell cultures. If you freeze and thaw, rupturing the cells, disintegration begins. Prolactin is likely to go very quickly; Stanley Ellis, using frozen and thawed beef pituitary gland, found that the first hormonal activity to disappear was prolactin. Robert W. Bates also made the same observation.

Greenwood: Has Dr Friesen or anyone else an estimate of the prolactin content of fresh frozen or acetone-dried human pituitaries? I want to know whether it's worth proceeding with tissue culture as a source of isolation or whether somebody is going to isolate prolactin from ten thousand pituitaries.

Friesen: Of 10 g of acetone-dried powder, we estimate that $0 \cdot 09$ per cent was prolactin, or $0 \cdot 9$ mg per g. I think it's somewhat higher in frozen pituitary glands.

Wilhelmi: Ten g of acetone powder would be equivalent to about a hundred human pituitaries. Our figure would be close to Dr Friesen's figure, approximately 40 to 100 μg per g of fresh tissue.

Meites: Are these glands from children, men, pregnant women, lactating women, or which state? This must make a big difference.

Wilhelmi: With fresh glands from lactating women the concentration becomes five to ten times higher.

Friesen: I agree. The yields of prolactin that we obtained from pituitaries removed at surgery from cases of diabetic retinopathy or mammary cancer have been approximately 50 μg per gland. We haven't noticed any marked sex difference in prolactin content, but there is some variation from gland to gland (20–100 μg).

Sherwood: Were any differences noted in women before and after the menopause, since this may be relevant to the breast cancer problem? Most of the glands were presumably postmenopausal specimens.

Friesen: We haven't looked carefully, but we had at least a dozen samples from postmenopausal women and the levels are the same as those in premenopausal women. Patients with breast cancer (ten subjects) have the same concentrations as other adults.

REFERENCES

AMENOMORI, Y., CHEN, C. L. and MEITES, J. (1970) *Endocrinology* **87**, 506–510.
BRYANT, G. D., GREENWOOD, F. C. and LINZELL, J. L. (1968) *J. Endocrinol.* **40**, iv-v.
EWEN, S. W. B., BECK, J. S. and GREEN, J. (1971) *Br. J. Exp. Path.* submitted.
FRANTZ, A. G. and RABKIN, M. T. (1965) *J. Clin. Endocrinol. & Metab.* **25**, 1470–1476.
GUYDA, H. and FRIESEN, H. (1971) *Biochem. & Biophys. Res. Commun.* **42**, 1068–1075.
HOU, L. T., EWEN, S. W. B. and BECK, J. S. (1968) *Br. J. Exp. Path.* **49**, 648–657.
MEITES, J. and TURNER, C. W. (1948) *Res. Bull. Mo. Agric. Exp. Stn.* nos. 415, 416.

MOLECULAR BIOLOGICAL ASPECTS OF PROLACTIN

Roger W. Turkington

Department of Medicine, University of Wisconsin, Madison

It is the task of the mammary gland to develop and secrete an extracellular fluid of highly specialized composition at discrete periods in the life of the adult mammal and in response to specific physiological signals. The complex and relatively rapid processes of mammary growth and development depend upon responses to a variety of hormones which culminate in the formation of a new population of secretory cells, the mammary alveolar cells. Prolactin both participates in the regulation of alveolar cell differentiation and induces the secretory milk proteins after parturition. The mechanisms by which prolactin governs glandular development and exocrine function can only be understood in terms of specific molecular interactions which regulate gene activity and protein synthesis.

Studies in our laboratory on the mechanism of action of prolactin have utilized organ cultures of mouse mammary gland. Tissues from a highly inbred strain of mice have been studied at precise time points in pregnancy by culture on chemically defined medium. Specific hormones have been added to the medium at appropriate time points and at known concentrations. The effects of prolactin reported here can be obtained at physiological concentrations of prolactin, as will be shown later (pp. 169–184). Such a highly controlled system has permitted analysis of regulatory events throughout the cell cycle during hormonally induced cell proliferation, and has permitted detection of early perturbations in the pattern of cellular regulation in response to a specific hormonal signal.

The diagram shown in Fig. 1 summarizes the sequence of hormone action and cellular transition in the mammary organ culture system. Insulin can act to induce mammary epithelial stem cells to divide during the culture period (Turkington and Topper 1967; Stockdale and Topper 1966; Turkington 1968*a*). Although the action of insulin has been studied primarily, a number of other humoral factors can induce cell division (Turkington 1971*a*). Epithelial-epidermal growth factor (EGF) (Turkington 1968*c*) is a highly potent molecule in this respect, although its species distribution appears to be largely limited to rodents (Turkington, Males and

Cohen 1971). Preparations of growth hormone also stimulate mammary epithelial cells to divide (Turkington 1968a), as does a newly discovered protein in serum (Majumder and Turkington 1971a). Prolactin or human placental lactogen (Turkington and Topper 1966; Turkington 1968b) induce the milk proteins only in daughter cells formed *in vitro*. This

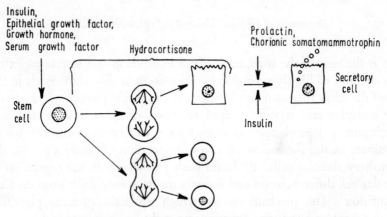

FIG. 1. General scheme of the sequence of hormone actions and cellular transitions during mammary epithelial cell differentiation *in vitro*. Chorionic somatomammotrophin ≡ placental lactogen.

process requires a period of pre-treatment with hydrocortisone or another active adrenal cortical hormone (Turkington, Juergens and Topper 1967). However, as will be mentioned below, a number of the early actions of prolactin can occur in cells not pre-treated with hydrocortisone. Stimulation of the rates of total nuclear and ribosomal RNA synthesis and the induction of the milk proteins requires the synergistic action of insulin. Although a number of other polypeptides can substitute for insulin to induce cell proliferation, none of these can replace the requirement for insulin's synergistic actions on the newly formed alveolar cells (Turkington 1969a). Despite the relatively large number of hormones to which the mammary cells respond, it has been possible to dissect out the actions of prolactin by using cultured cells.

INITIAL ACTION OF PROLACTIN: BINDING TO CELL MEMBRANE RECEPTORS

Our early interest in the molecular events regulated by prolactin was focused on the stimulation of milk protein synthesis (Turkington, Juergens and Topper 1965; Turkington *et al.* 1968). To discern the sequence of the large number of regulatory events which preceded this effect, however, it was essential to ask what molecular interaction occurred initially in the stimulation of the target cell. In attempts to identify this molecule it was found that specific prolactin receptor activity resides in preparations of mammary plasma membranes (Turkington 1971b). The binding reaction between prolactin and the receptor activity was assayed by using [^{125}I]-prolactin of very high specific activity and plasma membranes of high purity prepared by a modification of the method of Neville (1968) from lactational mammary gland. After a period of incubation of the membranes with [^{125}I]prolactin the mixture was sedimented by centrifugation and the radioactivity in the membrane fraction was counted. As shown in Fig. 2, approximately 30 per cent of the [^{125}I]prolactin was bound in a

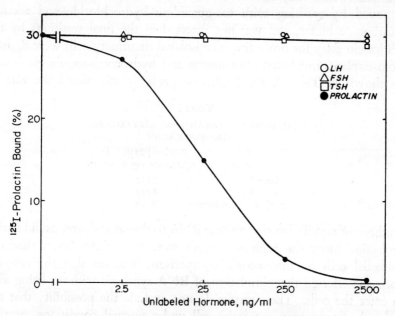

FIG. 2. Competitive displacement of [^{125}I]prolactin (ovine) from mouse mammary epithelial plasma membranes by prolactin (ovine). Bovine TSH, ovine LH and ovine FSH failed to compete with [^{125}I]prolactin for binding at these concentrations.

typical experiment. This binding was specific for prolactin since only prolactin, but not the other pituitary glycopeptides LH, FSH or TSH, could displace [^{125}I]prolactin from the binding sites on the membranes. This binding reaction was relatively specific for plasma membranes prepared from mammary gland. The rate of binding was relatively rapid, with nearly maximal values for binding observed after incubation at 0° or 30°C for 5 minutes. The [^{125}I]prolactin was found to retain 70–90 per cent of the biological activity of native prolactin as tested in the [^{32}P]casein assay (see pp. 169–184).

STIMULATION OF NUCLEAR TRANSCRIPTION BY ACTIVATION OF THE PLASMA MEMBRANE

To determine whether such interaction of prolactin with the superficial surface structures of the cell membrane could lead to a stimulation of intracellular activity, prolactin covalently bonded to beads of Sepharose was allowed to interact with isolated mammary epithelial cells. The reaction between ovine prolactin and cyanogen bromide-activated Sepharose beads was carried out at pH 6·0 to permit binding of a minimal number of amine groups in the prolactin peptide chain. Unreacted prolactin which may have been adsorbed to the Sepharose beads was removed by washing with 6M-guanidine hydrochloride and alkaline bicarbonate buffer, and was not detected in the final washings by the [^{32}P]casein assay for prolactin. The isolated mammary cells were derived from explants incubated with insulin and hydrocortisone for 72 hours. As shown in Table I, the Sepharose–prolactin stimulated the rate of

TABLE I

STIMULATION OF ISOLATED MAMMARY CELLS BY SEPHAROSE–PROLACTIN

System	Nuclear [^{3}H]RNA c.p.m./100 mg tissue
Control	2754
Prolactin	8539
Sepharose–prolactin	8505

synthesis of rapidly labelled nuclear RNA to the same degree as did native prolactin. Since the Sepharose beads were 5–10 times larger than the epithelial cells by microscopic comparison, it seems that the coupled hormone effected the stimulation of RNA synthesis without being able to enter the cells. These results do not exclude the possibility that the native hormone may enter the cell under normal conditions, or that proteolytic activity in the culture may have released biologically active peptides for cellular entry. However, no such soluble prolactin activity could be recovered from the culture medium after centrifugal separation of the cells and Sepharose beads (Turkington 1970c). Although these studies provide evidence for prolactin receptor activity in mammary cell membranes, little is known yet about the chemical nature of such activity or its organization in the membrane structure.

INDUCTION OF THE MILK PROTEINS: SEQUENCE OF EVENTS

As indicated in the diagram in Fig. 1 (p. 112), the epithelial cells both proliferate and differentiate during the period of culture. Only a single cell division occurs *in vitro*, and after 72 hours cell division ceases (Lockwood, Stockdale and Topper 1967; Turkington 1968a). The epithelial cell population then contains a large number of daughter cells in the G_1 phase of the cell cycle that can respond to prolactin. A characteristic sequence of biochemical changes occurs after the addition of prolactin to the culture medium of explants pre-treated for more than 72 hours with insulin and hydrocortisone. As shown in Fig. 3, an early effect is an in-

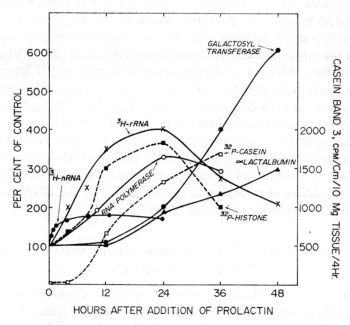

FIG. 3. Sequence of molecular events in the induction of milk proteins by prolactin. Mouse mammary explants were incubated with insulin and hydrocortisone for 96 hours prior to the addition of prolactin to the medium.

creased rate of synthesis of rapidly labelled nuclear RNA (Turkington 1970b). The activity of DNA-dependent RNA polymerase, as assayed in isolated cell nuclei, is also stimulated (Turkington and Ward 1969). The rate of phosphorylation of histones, primarily the F2a2 and F2b fractions, is increased, as is the rate of phosphorylation of certain non-histone nuclear proteins (Turkington and Riddle 1969). Following this period of increased biosynthetic activity at the chromosomal level the milk proteins are synthesized. During the first three days of culture the rate of synthesis of casein by the differentiated epithelial cells decreases as these cells lose their highly differentiated functions (Turkington, Lockwood and Topper 1967; Turkington 1968b). The synthesis of specific casein phosphoproteins (e.g. casein band 3 in Fig. 3) is undetectable until prolactin is added to the medium. Experiments with inhibitors of cell proliferation, such as colchicine, have demonstrated that the changes in macromolecular synthesis shown in Fig. 3 occur in the cells formed *in vitro*, and that casein synthesis is initiated in these cells from an apparently 'zero baseline'. Subsequently the enzymic activity of the lactose synthetase system is stimulated (Turkington *et al.* 1968; Turkington and Hill 1969). This system consists of a galactosyltransferase which can transfer galactose from UDP-galactose to N-acetylglucosamine or to glucose; and α-lactalbumin, which interacts with the galactosyltransferase to 'specify' transfer primarily to glucose for the formation of lactose (Brew, Vanaman and Hill 1968). The increased formation of [^{32}P]casein or [^{14}C]casein and the increased activity of lactose synthetase are prevented by actinomycin D, mitomycin C or cycloheximide, indicating that these effects of prolactin stimulation require concomitant synthesis of RNA and protein. The observations that the formation of the milk proteins was preceded by a stimulation of transcription (Fig. 3) and that synthesis of RNA was necessary to mediate the induction of the milk proteins led to a series of investigations on the regulation of transcription and the nature of the RNA products transcribed in response to prolactin stimulation.

INDUCTION OF CYCLIC AMP-ACTIVATED PROTEIN KINASE AND CYCLIC AMP-BINDING PROTEIN

Mammary epithelial cells contain two protein phosphokinase isoenzymes, which have been termed kinase I and kinase II. Kinase II requires cyclic AMP for maximal activity, while the activity of kinase I is not altered by cyclic nucleotides (Majumder and Turkington 1971*b*, *c*). These enzymes are found together in the cytosol with a protein that exhibits a high affinity for cyclic AMP. The specificity of the binding reaction for cyclic AMP is high, and other cyclic nucleotides compete for this binding site only at concentrations which are several orders of magnitude greater. As shown in Fig. 4B, the kinases and the cyclic AMP-binding protein are found to sediment together after partial purification from the cytosol fraction. After passage of these proteins through a DEAE-cellulose ion exchange column the binding activity is largely removed, and the kinase activities sediment with lower sedimentation constants. These observations suggest that in the cytosol these proteins exist together as a complex. Although the kinases exhibit the greatest substrate specificity toward histones, the proteins with which this protein complex interacts under physiological conditions are as yet unknown.

As shown in Fig. 5, increases in the intracellular concentrations of the cyclic AMP-activated protein kinase and the cyclic AMP-binding protein occur rapidly and coordinately in response to the addition of prolactin. These increases are prevented by actinomycin D and cycloheximide, a result consistent with the conclusion that prolactin induces the formation of

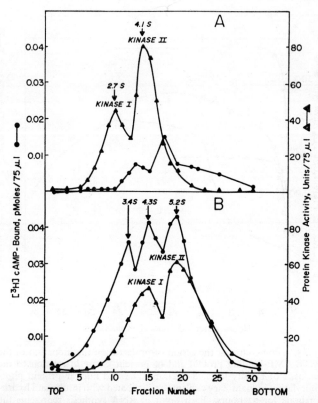

FIG. 4. Sedimentation profiles of protein kinase activity and cyclic AMP-binding protein in isokinetic sucrose density gradients: B, after partial purification from the cytosol by ammonium sulphate precipitation; A, after further purification of the kinases by DEAE-cellulose chromatography.

new protein molecules and does not merely activate the pre-existing proteins. This inductive process occurs as early as the increased rate of incorporation of [³H]uridine into rapidly labelled nuclear RNA. However, half-maximal induced levels of the kinase and cyclic AMP-binding protein are observed 30 minutes after addition of prolactin, and maximal levels are observed at four hours. Stimulation of rapidly labelled RNA continues beyond this time point. The phosphorylation of specific nuclear proteins is stimulated following the induction of the protein kinases, a result which is consistent with the possibility that phosphorylation of these proteins may be dependent in part upon the formation of the protein kinases. Prolactin, unlike several other pituitary polypeptide hormones, does not activate adenyl cyclase activity in the cell membranes of its target cell (Majumder and Turkington 1971c). Furthermore, cyclic AMP in the

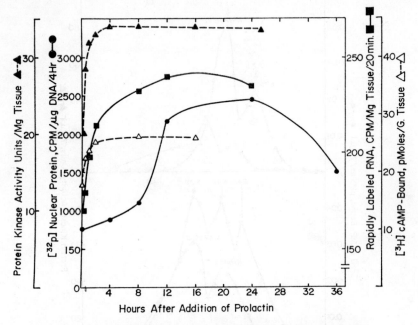

FIG. 5. Time course of the action of prolactin on the induction of the cyclic AMP-binding protein and of protein kinase in relation to the synthesis of rapidly labelled nuclear RNA and nuclear protein phosphorylation in mammary explants preincubated with insulin and hydrocortisone for 72 hours. Each point in RNA synthesis represents the incorporation of [^{3}H]uridine into RNA during the preceding 20-minute period. Each point in nuclear protein phosphorylation represents the incorporation of ^{32}P during the preceding 4-hour period.

culture medium cannot replace the action of prolactin or exert an additive effect with rate-limiting concentrations of prolactin for the induction of casein. These observations suggest that prolactin acts through a novel regulatory mechanism. The binding of prolactin to its membrane receptor(s) does not cause an activation of adenyl cyclase for the increased rate of formation of cyclic AMP. Rather, the proteins with which cyclic AMP interacts appear to be rate-limiting, since these are rapidly induced by the hormone.

THE PRODUCTS OF PROLACTIN-STIMULATED TRANSCRIPTION

Approximately 98 per cent of the RNA which becomes labelled by exposure of the explants for 20 minutes to medium containing [³H]uridine is found in the nucleus (Turkington 1970b). Analysis of this RNA by centrifugation in an isokinetic sucrose gradient indicates that it is composed of peaks of 45S and 32S RNA as well as some heavier, heterodisperse RNA which may extend to the region of 100S (Fig. 6A). The 45S and 32S peaks have a high guanine and cytosine content, while the heterodisperse RNA is DNA-like in its base composition. 'Chase' experiments of the

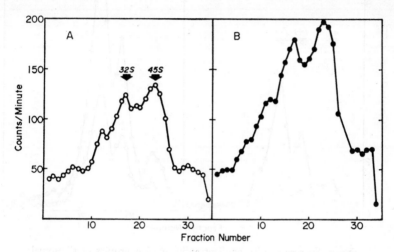

FIG. 6. Sucrose density gradient sedimentation profiles of rapidly labelled RNA in mammary explants incubated for 88 hours with insulin and hydrocortisone. A, control. In B the medium contained prolactin during the 72–88 hour period. The labelling time with [³H]uridine (8 μCi/ml) was 20 minutes in each case.

labelled RNA on non-isotopic medium containing actinomycin D during a 30-minute period subsequent to labelling indicate that the DNA-like RNA turns over nearly completely, and the 45S and 32S peaks are converted to 28S and 18S ribosomal RNA. Synthesis of these two classes of RNA, the heterodisperse DNA-like RNA and the pre-ribosomal RNA, is stimulated by prolactin (Fig. 6B). The increased rate of labelling of RNA does not reflect merely a change in the concentration of isotopic precursor available, since the specific activity of [³H]uridine triphosphate, the immediate precursor of RNA, is not significantly altered.

Figure 7 shows the sucrose gradient profiles of [³H]RNA labelled for four hours in mammary explants cultured in medium containing various added hormones. After culture for 24 hours with insulin the rate of synthesis in peaks of 28S, 18S and 4–5S ribosomal RNA is markedly stimulated. The combination of insulin and hydrocortisone does not further increase the rate of ribosomal RNA synthesis. After 72 hours of culture the addition

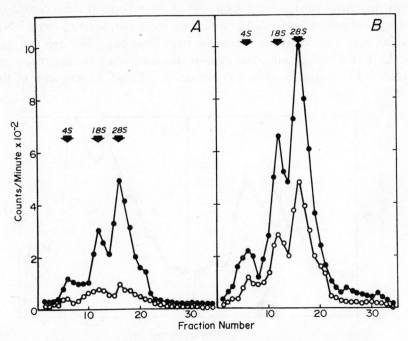

FIG. 7. Sucrose density gradient sedimentation profiles of ribosomal RNA in mouse mammary explants incubated with various hormones. A, 24-hour incubation. Control, o—o. Insulin, ●—●. B, 72-hour incubation with insulin and hydrocortisone, then 25-hour incubation with or without the addition of prolactin. Control, o—o. Prolactin, ●—●.

of prolactin for 20 hours results in a further marked increase in the rates of labelling of the ribosomal RNA's.

The rapidly labelled, heterodisperse, DNA-like nuclear RNA has been further characterized by DNA–RNA hybridization–competition experiments (Turkington 1970a). As shown in Fig. 8, rapidly labelled RNA recovered from highly purified nuclei of differentiated (lactational) tissue hybridized to homologous DNA, and the competitive displacement of label by unlabelled nuclear RNA from lactational tissue was complete, indicating the similarity of the labelled and unlabelled populations of RNA. The nuclear RNA prepared from undifferentiated (virginal) mammary epithelial cells competed incompletely with the lactational RNA, indicating the presence of species or sequences of RNA in the differentiated cells which were not detectable in the undifferentiated cells. These results were verified in reversed labelling experiments and with competitive hybridization using double labels. These results provide evidence to support the concept that new loci on the mouse genome are transcribed as mouse mammary cells differentiate. However, it is not known whether the

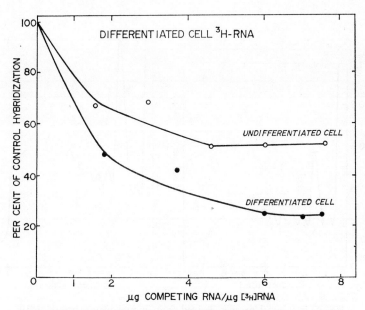

FIG. 8. The competition of unlabelled nuclear RNA from undifferentiated (virginal) mammary cells or differentiated (lactational) mammary cells in the hybridization of lactational mammary [³H]RNA with homologous DNA. Control hybridization (100 per cent) was 1100 c.p.m. above background.

hormones directly participate in the 'derepression' of these loci, or whether they merely permit expression of the derepressed genes by their actions at other sites for regulation of RNA and protein synthesis.

It is well accepted that polysomes are the site of protein synthesis. To determine how prolactin induces specific milk proteins it was of interest, therefore, to examine the effect of prolactin on the formation of polysomes. Fig. 9A shows the sucrose gradient profile of ribosomes isolated from mammary explants incubated with insulin and hydrocortisone for 72 hours. The largest peak near the top of the gradient is the monomer ribosomal peak with a sedimentation value of approximately 80S. The peaks which sediment nearer the bottom each represent a larger number of ribosomes aggregated in polysomes. The profile in Fig. 9B represents the polysomes derived from an amount of explant tissue equal to that used in Fig. 9A and obtained 16 hours after the addition of prolactin. Prolactin stimulates the formation of ribosomes and the aggregation of these ribosomes into polysomes. It was shown that the increase in area under the polysome profile occurred only with mammary epithelial cells formed *in vitro* and pre-treated with insulin and hydrocortisone. Experiments in

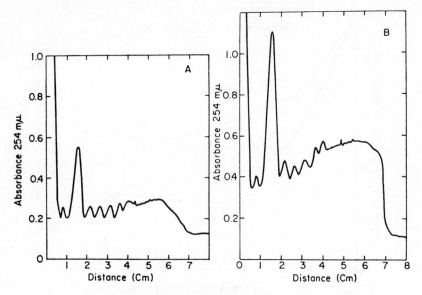

FIG. 9. Polysomal profiles derived from mouse mammary epithelial cell preparations after treatment with insulin and hydrocortisone for 72 hours (A). In B, prolactin was added and the explants were cultured for an additional 24 hours prior to assay.

which the ribosomes were labelled with [^{3}H]uridine before prolactin was added demonstrated that the prolactin-induced ribosomal RNA was processed into the ribosomes which then entered the pool of polysomes formed in response to the addition of prolactin (Turkington and Riddle 1970). To determine whether the population of polysomes induced by prolactin was qualitatively different from the pre-existing polysomes the time course of synthesis of a specific casein polypeptide was correlated with polysome formation. As shown in Fig. 10, the synthesis of casein band 3 is undetectable in the explant population after 96 hours of incubation. After the addition of prolactin there was a lag period of four hours, after which the synthesis of casein band 3 was initiated in association with the appearance of new polysomes in the cell. The rate of synthesis of this casein polypeptide continued to increase in parallel with the increase in total polysomes. This evidence indicates that the synthesis of casein band 3 is dependent upon the formation of prolactin-induced polysomes, and suggests that these polysomes represent new expression of a specific gene, since they presumably contain the messenger RNA for a polypeptide that was not synthesized at a detectable rate in the unstimulated post-mitotic cells.

In addition to messenger RNA and ribosomal RNA's a third general type of RNA is required for the assembly of polypeptides: transfer RNA's.

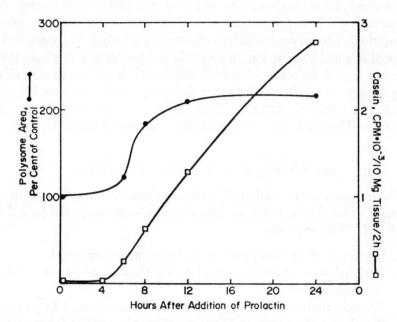

FIG. 10. Time course of the effect of prolactin on total polysomes and rate of incorporation of $^{32}P_i$ into casein. Mouse mammary explants were incubated on medium containing insulin and hydrocortisone for 90 hours, at which time prolactin was added to the medium.

Transfer RNA has been measured in these studies as material which absorbs at 260 nm and migrates in polyacrylamide gels as 4S RNA; and as RNA which has amino-acid acceptor activity in the presence of amino acyl synthetase enzymes. During development of the mammary gland in pregnancy the cellular content of transfer RNA increases more than 10-fold (Turkington and Riddle 1969). Although the content of low molecular weight RNA also increases in the epithelial cells, the increment which is transfer RNA is disproportionately larger at each developmental point. Studies on the hormonal regulation of transfer RNA in explants indicate that insulin, hydrocortisone and prolactin are required to stimulate the formation of transfer RNA in the epithelial cells formed *in vitro* (Turkington

1969*b*). The assembly of the transfer RNA's is completed by methylation of specific bases in the preformed polynucleotide chain. A number of transfer RNA methylases exhibiting specificity for particular bases and for transfer RNA from different biological sources have been described (Hurwitz, Gold and Anders 1964; Srinivasan and Borek 1964) and seven such enzymes have been identified in mouse mammary epithelial cells (Turkington 1969*b*). Synchronous increases in transfer RNA and transfer RNA-methylating enzymes are observed throughout the period of mammary gland development, and no further increase occurs after parturition. The enzymes which methylate transfer RNA are coordinately induced *in vitro* by the actions of prolactin and insulin, and this induction does not require the action of hydrocortisone. Since no new base-specific enzymes appear and the ratios of the methylating enzymes remain relatively constant, it is unlikely that altered patterns of transfer RNA methylation are required for the hormone-induced differentiation of mammary cells.

GENERAL SCHEME OF THE ACTION OF PROLACTIN

These studies on the action of prolactin illustrate several major steps through which this hormone's regulation of the mammary alveolar cell is mediated. These steps are:

1. Interaction of prolactin with the cell membrane receptor(s).
2. Translation of the altered state of the membrane into an intracellular signal to the nucleus.
3. Altered intranuclear transcription which is coordinated to provide the ribosomal, transfer and messenger RNA's required for synthesis of milk proteins and other mammary enzymes.
4. Synthesis of the milk proteins and other inducible proteins, and secretion of the milk products.

These steps are summarized in the diagram shown in Fig. 11. While evidence exists for highly specific binding activity for prolactin in the cell membrane, little information is available on the structural characteristics of the receptor and how it is organized into the membrane, possibly to affect other membrane functions. The mechanism by which a perturbation of the state of the plasma membrane could influence the nucleus is the weakest link in our knowledge of prolactin's action. It is likely that the cyclic AMP-binding protein and protein kinases participate in this process, and the way in which this could occur is currently under investigation. The mammary cell acts partly as an information-processing device. Its state at any one moment depends upon the spread of environmental signals through its regulatory mechanisms, and its responses to the processed information are those which are permitted by its state of cell differentiation. The single stimulus provided by prolactin in the environment could be processed in several ways. It is possible that a number of receptor types exist which could, by having differing affinities for prolactin, provide the cell with information about the concentration of the hormone in the extracellular fluid. The presence of this single chemical (prolactin) could

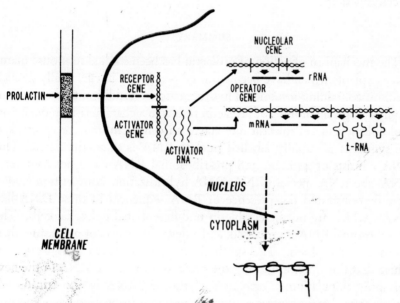

FIG. 11. Proposed scheme of the response of the mammary alveolar cell to prolactin, showing: hormone sensing, the processing of hormonal 'information' in the cytoplasm and nucleus; and the transcription of multiple classes of RNA for the synthesis of specific milk proteins and other induced proteins in the cytoplasm.

lead to alterations in the intracellular concentrations of several compounds through altered cell permeability and altered rates of metabolic conversions, thereby resulting in an amplification of the chemical signal. Translation of the transduced signal to the RNA sequences of nuclear RNA could then provide intranuclear signals required to activate transcription on the genes for ribosomal RNA, transfer RNA and messenger RNA. It is possible that the action of prolactin could be amplified further through the process of gene amplification (Brown and Dawid 1968) although there is no direct evidence for this mechanism in mammalian cells. A number of other factors regulate protein synthesis at the translational level. Although translational effects do not appear to be of major importance in these *in vitro* studies, this does not exclude the possibility that translation may be regulated by prolactin during maintained lactation *in vivo*. It is possible that the propagation of the prolactin signal in the cytoplasm of mammary cells may require macromolecular aggregates as highly organized as those which serve the passage of electrons in the mitochondrion. If so, it is possible that further studies on the effect of prolactin on mammary cells will lead to a new dimension in our concepts of cell function.

SUMMARY

The mechanism of action of prolactin has been studied in mouse mammary explants which have been cultured previously in chemically defined medium containing insulin and hydrocortisone. Addition to the medium of physiological amounts of prolactin results in a characteristic sequence of molecular events. Prolactin interacts with cell membrane receptors, and the synthesis of rapidly labelled nuclear RNA is then stimulated. This RNA consists of 45S and 32S pre-ribosomal RNA's and heterodisperse DNA-like RNA species. RNA–DNA hybridization–competition studies have demonstrated the presence of RNA sequences in these DNA-like RNA's which are not detectable in undifferentiated epithelial cells. The pre-ribosomal RNA is processed, and a new population of polysomes then enters the cytoplasm. The synthesis of specific casein polypeptides is initiated at this time, indicating the presence of casein mRNA in the new polysome population. Formation of transfer RNA is also stimulated. This increased transcriptive activity in response to prolactin is associated with an increase in chromosomal protein phosphorylation, and is preceded by the induction of specific cyclic AMP-activated protein kinases and a cyclic AMP-binding protein. The subsequent synthesis of milk proteins thus occurs as the result of a large number of steps which involve hormone sensing, information processing (initial responses to the hormonal signals) and transcriptional and translational read-outs (sequential inductive processes determined by cell differentiation).

REFERENCES

BREW, K., VANAMAN, T. C. and HILL, R. L. (1968) *Proc. Natl. Acad. Sci.* (*U.S.A.*) **50**, 491–498.
BROWN, D. D. and DAWID, I. B. (1968) *Science* **160**, 272.
HURWITZ, J., GOLD, M. and ANDERS, M. (1964) *J. Biol. Chem.* **239**, 3462–3472.
LOCKWOOD, D. H., STOCKDALE, F. E. and TOPPER, Y. J. (1967) *Science* **156**, 945–947.
MAJUMDER, G. C. and TURKINGTON, R. W. (1971a) *Endocrinology* **88**, 1506–1510.
MAJUMDER, G. C. and TURKINGTON, R. W. (1971b) *J. Biol. Chem.* **246**, 2650–2657.
MAJUMDER, G. C. and TURKINGTON, R. W. (1971c) *J. Biol. Chem.* **247**, in press.

NEVILLE, D. M., JR (1968) *Biochim. & Biophys. Acta* **154**, 540–552.
SRINIVASAN, P. R. and BOREK, E. (1964) *Biochemistry* **3**, 616–619.
STOCKDALE, F. E. and TOPPER, Y. J. (1966) *Proc. Natl. Acad. Sci. (U.S.A.)* **56**, 1283–1289.
TURKINGTON, R. W. (1968a) *Endocrinology* **82**, 540–546.
TURKINGTON, R. W. (1968b) *Endocrinology* **82**, 575–583.
TURKINGTON, R. W. (1968c) *Experientia* **243**, 226–228.
TURKINGTON, R. W. (1969a) *Exp. Cell Res.* **57**, 79–85.
TURKINGTON, R. W. (1969b) *J. Biol. Chem.* **244**, 5140–5148.
TURKINGTON, R. W. (1970a) *Biochim. & Biophys. Acta* **213**, 484–494.
TURKINGTON, R. W. (1970b) *J. Biol. Chem.* **245**, 6690–6697.
TURKINGTON, R. W. (1970c) *Biochem. & Biophys. Res. Commun.* **41**, 1362–1367.
TURKINGTON, R. W. (1971a) In *Developmental Aspects of the Cell Cycle*, pp. 315–355, ed. Cameron, I. L., Padilla, G. M. and Zimmerman, A. M. New York and London: Academic Press.
TURKINGTON, R. W. (1971b) *J. Clin. Invest.* **30**, 94a.
TURKINGTON, R. W., BREW, K., VANAMAN, T. C. and HILL, R. L. (1968) *J. Biol. Chem.* **243**, 3382–3387.
TURKINGTON, R. W. and HILL, R. L. (1969) *Science* **163**, 1458–1460.
TURKINGTON, R. W., JUERGENS, W. G. and TOPPER, Y. J. (1965) *Biochim. & Biophys. Acta* **111**, 573–576.
TURKINGTON, R. W., JUERGENS, W. G. and TOPPER, Y. J. (1967) *Endocrinology* **80**, 1139–1142.
TURKINGTON, R. W., LOCKWOOD, D. H. and TOPPER, Y. J. (1967) *Biochim. & Biophys. Acta* **148**, 475–480.
TURKINGTON, R. W., MALES, J. L. and COHEN, S. (1971) *Cancer Res.* **31**, 252–256.
TURKINGTON, R. W. and RIDDLE, M. (1969) *J. Biol. Chem.* **244**, 6040–6047.
TURKINGTON, R. W. and RIDDLE, M. (1970) *J. Biol. Chem.* **245**, 5145–5152.
TURKINGTON, R. W. and TOPPER, Y. J. (1966) *Endocrinology* **79**, 175–181.
TURKINGTON, R. W. and TOPPER, Y. J. (1967) *Endocrinology* **80**, 329–336.
TURKINGTON, R. W. and WARD, O. T. (1969) *Biochim. & Biophys. Acta* **174**, 291–301.

DISCUSSION

Sherwood: These are very elegant experiments. I was interested in the competitive displacement of ^{125}I-labelled prolactin from the membrane receptors by unlabelled prolactin. Following up my earlier discussion of common sequences in different proteins possibly activating a prolactin effect (p. 48), have you tried HGH or HPL in your system and do they displace prolactin from the cell membrane?

Turkington: We have done a few experiments with some human growth hormone preparations, and they seem to contain prolactin-like activity in terms of being able to compete with prolactin for binding. The relative activities seem to be the same as we find with these preparations in the organ culture system in terms of their relative abilities to induce synthesis of casein. But this doesn't resolve the question of whether biologically active homologous sequences in the polypeptide are responsible for that effect or contamination of the preparations with prolactin.

Sherwood: It does suggest, however, that they are binding to the same site?

Turkington: Yes, it suggests that possibility.

Beck: Does HPL do the same?

Turkington: HPL seems to compete very well for binding and to be highly active in the organ culture system as well.

Denamur: In 1967, 1968 and 1969 we published studies of the action of prolactin on mammary ribosomes from rabbits (Denamur and Gaye 1967; Gaye and Denamur 1968, 1969). Prolactin is able not only to increase the number of polyribosomes in each cell, and especially those bound to the membranes of the endoplasmic reticulum, but also to modify the sedimentation profile of the polyribosomes and their ability (per μg of polysomal RNA) to synthesize proteins. We recently showed that the bound polyribosomes synthesize the milk proteins, whereas the free polyribosomes are not able to synthesize these types of proteins (Gaye and Denamur 1970). So our results are comparable to yours, but we have shown that there is a very important difference between bound and free polyribosomes as regards their ability to synthesize milk proteins.

Turkington: These are very interesting studies and they go several steps beyond the evidence that I described. This represents a very important aspects of the hormone's action in terms of the binding of polyribosomes to the membranes, and allowing them to have a potentially increased level of biosynthetic activity.

Denamur: You showed several years ago (Turkington and Hill 1969) that progesterone inhibits α-lactalbumin synthesis. Do you know the mechanism of this action?

Turkington: We haven't studied that mechanism any further. I would assume that it probably involves binding to the progesterone receptor in the alveolar cells. Since we found that progesterone inhibited the rate of ribosomal RNA synthesis, and since you have also shown that progesterone can inhibit the formation of polysomes, it seems to have an inhibitory action at the transcriptional level. This is a very interesting effect, since the steroid–receptor complex apparently has an inhibitory effect in this cell type whereas in other cell types it has a stimulatory effect.

Greenwood: As a general comment, I don't think much of the specificity of the mammary gland receptor site if it can't tell the difference between growth hormone, prolactin and HPL! Would it be possible to show whether the stem cell might be able to distinguish between growth hormone and prolactin? In other words, could prolactin bind to the 'stem cell' and cause differentiation or induction of its own receptor?

Turkington: This is a very fundamental relationship, because when a cell differentiates and becomes one of the cells in the body which, perhaps uniquely, can respond to prolactin, it must utilize information encoded on its DNA to form a highly specific protein in its cell membrane. Whether the hormone can actually evoke its own receptor in a cell is a very interesting question. The only evidence which bears on this is in another system, the chick oviduct, in which oestradiol has been shown to induce the progesterone receptor. It may be that insulin or hydrocortisone is inducing the prolactin receptor in these cells. I think it unlikely that hydrocortisone is inducing it since we can observe prolactin-dependent changes in a cell which has not been treated with hydrocortisone. For example, the induction of the cyclic AMP-binding protein does not require pretreatment with hydrocortisone. But what the sequence of events is in terms of forming the receptor is probably quite important.

Greenwood: What I had in mind also is that we have somehow to produce a target tissue for prolactin other than the mammary gland, to explain the obvious role of prolactin in neutralizing a stress stimulus, since prolactin increases in the circulation in response to stress. Presumably it has a physiological role, as growth hormone does, in neutralizing the stress stimulus. I can't believe that the mammary gland is its only target tissue, although it has been shown that in a chronically stressed female rat, in which there is high prolactin, the mammary glands develop. In other words, mammary hyperplasia in females presumably is caused by chronic stress!

Frantz: We may be able to resolve this question of the metabolic action of prolactin on the body at large by further study on the distribution of receptor sites. For example, if we can identify some other tissue that has the prolactin receptor, that may be presumptive evidence that it may also respond to prolactin.

Beck: Dr Turkington, have you really responded to Dr Greenwood's challenge? He is very disturbed about a common receptor for three hormones!

Turkington: We find that in the mouse organ culture system, HPL is nearly as potent as ovine pituitary prolactin, and that human growth hormone has 10-20 per cent of the potency of ovine pituitary prolactin at physiological concentrations, and that these hormones seem to compete in the same way for the receptor site, in the sense that human growth hormone has 10-20 per cent of the affinity for the receptor site that HPL or ovine pituitary prolactin has. In that sense, the biological responses and the binding to the receptors seem to be parallel.

Friesen: Have you any information on the nature of kinases I and II; have any attempts been made to purify them, and have you used your Sepharose-prolactin to purify the receptor site? Is this a soluble system?

Turkington: It's not a soluble system. It is fragments of cell membrane, and we do the assay by sedimenting the membranes by centrifugation.

Friesen: Have you attempted to solubilize the membrane fragments?

Turkington: Yes, but without success. We have partly purified the kinases to about 1000-fold and characterized their substrate specificity and other properties. They have the highest substrate specificity for histones, but they also phosphorylate some of the non-histone nuclear proteins, casein and phosvitin and they are found in the cytosol in a complex together with the cyclic AMP-binding protein. And it appears that the cyclic AMP-binding protein is not a subunit of the enzymes because it can be dissociated from the enzymes and kinase II will retain activation by cyclic AMP. Cyclic AMP binds both to kinase II and to the cyclic AMP-binding protein.

MacLeod: Your system is a unique one, and you would have an opportunity to study the mode of action of many of the hormones that you have in your cultures, in particular their specific effects on the cyclic AMP system.

Turkington: Yes. Our evidence is that prolactin does not activate the adenyl cyclase of isolated cell membranes and if we incubate the cells with cyclic AMP or dibutyryl cyclic AMP, either by itself, or in combination with rate-limiting amounts of prolactin, we see no biological activity of cyclic AMP. We see a rather different kind of mechanism, in that prolactin does not increase the intracellular concentration of cyclic AMP as part of its early action. That apparently is not rate-limiting. What it does is to induce larger amounts of the protein with which cyclic AMP interacts.

MacLeod: Can this be accomplished in the absence of the other hormones, such as insulin or oestradiol?

Turkington: The kinases require insulin also for their induction, but not hydrocortisone, and the cyclic AMP-binding protein requires no hormones apart from prolactin.

Prop: In collaboration with Dr Th. M. Konijn, who has an elegant system for estimating cyclic AMP (Konijn 1970), we are investigating whether prolactin alone and/or in combination with other hormones that cause lobulo-alveolar growth or milk production *in vitro* causes an increase in cyclic AMP in organ cultures of mouse mammary glands. These glands are from seven-week-old virgin CBA mice and are cultured according to our method, using whole glands (Prop 1961). Our results so far indicate no considerable increase in cyclic AMP, which agrees with Dr Turkington's results.

Pasteels: To come back to the question of prolactin secretion during stress, it should be recalled that from the results of Neill (1970), prolactin discharge during stress occurs mainly in female rats. I have discussed that with Dr Nicoll, and he has a very satisfactory explanation.

Nicoll: The explanation is really speculation that perhaps one of the things prolactin does during prolonged stress is to have either an 'anti-gonadal' or an 'anti-gonadotropic' effect. In rats it would induce pseudopregnancy, and this would ensure that the female doesn't become pregnant while in a stressful situation. Another possibility concerns osmoregulation. Throughout the vertebrates prolactin seems to be involved in salt and water balance, and there is evidence that this may also be true in mammals. Perhaps in a stressful situation, the hormone is participating along with the adrenal cortical steroids in osmoregulation.

Short: I am surprised to hear you describe prolactin as 'anti-gonadal'!

Nicoll: I'm not denying that prolactin has gonadotropic effects, and I realize that 'anti-gonadal' or 'anti-gonadotropic' are rather poor terms. I am referring to the widespread effect that prolactin has among many vertebrates in bringing about physiological changes which block follicular development. For instance in the rat, prolactin can act as an anti-fertility agent by maintaining the animals in a pseudopregnant condition. This may also be true in other animals. There is evidence in some birds that prolactin inhibits gonadotropin secretion, possibly at the hypothalamic level.

Prop: One should be careful not to draw conclusions too rashly about the complete identity of the effects of prolactin and the prolactin-like effects of human growth hormone and placental lactogen. In experiments in which we stimulate lobulo-alveolar growth in virgin CBA mammary glands in organ culture (Prop 1961) we find that the maximal possible response to these latter hormones is less than the maximal possible response to sheep or mouse prolactin. Moreover, in experiments soon to be published we found that in mammary gland cultures derived from different mouse strains a completely prolactin-free bovine growth hormone gives a prolactin-like effect, but the maximally obtainable effect is much less than with prolactin and differs in different mouse strains. This indicates that the mechanism triggered by growth hormone differs from that triggered by prolactin. Looking for lobulo-alveolar growth of course is not the same as looking for stimulation of milk secretion, but both processes have in common that they require insulin and prolactin or prolactin-like substances, although the steroid hormones required are different. It is the steroids that determine whether lobulo-alveolar growth or milk production will occur.

Turkington: Our experience has been that our preparations of HPL are 80–90 per cent as effective as ovine prolactin over all concentrations of each hormone when compared on a mass basis. This might be somewhat less on a molar basis. This is my experience in terms of the morphology of the induced secretion, and of the induction of casein and lactose synthetase.

Denamur: Is there any similarity between the phosphokinases you described and those dependent on insulin and involved in the production of casein (Voytovich, Owens and Topper 1969)?

Turkington: I think the enzymic activity that Dr Topper postulated could be the same. The difference in our experiments is that we have isolated an enzyme whereas he was simply looking at the rate of phosphorylation of casein under a variety of culture conditions.

Sherwood: The interaction of steroid and prolactin in lactogenesis is of great interest. Dr Meites and others have shown that in the rabbit, for example, one can induce lactation during pregnancy by administering either prolactin (or related compounds) or cortisone. How do you understand this interaction? Are there critical concentrations of either hormone which when present prevent the other from triggering the lactogenic response?

Turkington: The *in vitro* explant model provides no adequate explanation at present for why lactogenesis is initiated *in vivo*, and so I have no good answer for that.

Meites: I'm not sure I understand it either, but it could be that in general lactation is not initiated during pregnancy because of a deficiency of prolactin or adrenal cortical hormones or both. In some species, such as rats and mice, you can give prolactin during pregnancy without obtaining initiation of lactation, but if you give adrenal cortical hormones, you initiate lactation. In the rabbit, either hormone will induce lactation during gestation, and the two together work best. We have also initiated lactation in cows in the first third or half of pregnancy by giving just a few injections of an adrenal cortical hormone (Meites 1966).

This also brings up the question of why there isn't enough circulating prolactin or adrenal cortical hormones to initiate lactation during pregnancy. The protein transcortin is present in plasma in large amounts and presumably binds to adrenal cortical hormones to prevent a sufficient amount of biologically active adrenal cortical hormones from acting on the mammary gland. There is also the idea that both Dr Folley and Dr Cowie and ourselves (Folley 1956; Meites and Turner 1948) postulated, that the combined action of progesterone and oestrogen during pregnancy inhibits the stimulating action of oestrogen alone on prolactin secretion.

There is the additional factor of the direct action of the steroid hormones on the mammary gland, which tends to prevent prolactin and adrenal cortical hormones from initiating lactation. For example, you can initiate lactation with either prolactin or cortisone in the pregnant rabbit, but lactation can be initiated in the non-pregnant rabbit with much smaller doses of either hormone, simply because there isn't the combined action of the two steroid hormones on the mammary gland. At parturition the two sex steroid hormones decline and prolactin and ACTH-adrenal cortical hormones increase, initiating lactation.

Cowie: Initiation of lactation has been examined carefully only in the pregnant rat. Dr N. J. Kuhn (1969a, b) showed that about 30 hours before parturition there is a change in the steroidogenic activity of the ovaries resulting in a fall in the level of progesterone in the blood which is associated with the initiation of milk secretion. But it must differ in different species, because in ruminants there is much secretory activity in the mammary gland at the beginning of the last third of pregnancy which is not associated with a fall in the level of blood progesterone and some quite different mechanism is evidently concerned in initiating milk secretion.

Greenwood: The beauty of Dr Turkington's story is that it brings together all areas and mechanisms of action of hormones except for one thing. There's no necessity, on this hypothesis, for prolactin ever to enter a cell. Is there any evidence, in fact, that prolactin localizes within, rather than on the outside of the target tissue? One thinks of the thymus, where cortisol is present in the thymus, but that cortisol is not active; it's the cortisol bound to the receptor which is delivered to the nucleus. It is suggested that the function of cortisol is to present a particular protein to the nucleus (cf. Munck 1971).

Turkington: I know of no convincing evidence that prolactin gets into the cell, but our studies don't exclude the possibility. Our evidence suggests that it does not have to get into the cell to be active. That pattern of polypeptide hormone action is consistent with the evidence for insulin and glucagon acting on specific receptor activities on liver cell membranes, and ACTH acting on adrenal cell membranes. This pattern seems to be different from the pattern for steroid hormones which have specific cytosol receptors, which require that the hormone enter the cell and react with the cytosol receptors.

Forsyth: I'm not entirely convinced by the experiments in which you bound insulin and prolactin to Sepharose. I'm not sure that the right control experiments have been done in this system. With Dr T. E. Barman at N.I.R.D. we tried to do this type of experiment and although we couldn't find any prolactin activity in the washings of columns of prolactin bound to Sepharose, we have so far been unable to convince ourselves that prolactin isn't coming off under the conditions of incubation.

Turkington: We have not been able to find free prolactin activity in the incubation medium, after the incubation period. We centrifuged down the cells and Sepharose beads and then tested the medium for prolactin activity, and have not detected any. One might argue here that the hormone went into the cell, or was degraded soon after either it came off the Sepharose bead to which it was adsorbed or when an enzyme released it from the Sepharose. Those are reservations about the conclusion of the experiment, although the evidence is against such possibilities.

Beck: Have you tried to dissociate the hormone from the Sepharose after it has induced activity, and recovered it?

Turkington: No, we haven't.

Friesen: When a hormone bound to Sepharose produces an effect, the conclusion which is usually drawn is that the hormone acts at the membrane and need not penetrate the cell to exert its action. However, a critical consideration is whether any hormone has been dissociated from the Sepharose and is exerting the biological effect. In the case of the mammary gland, have you attempted to see whether the minimal effective concentration of HPL coupled to Sepharose is the same as that of HPL in solution, and whether increasing concentrations of HPL and HPL-Sepharose exert comparable effects? If the same concentrations of HPL and HPL-Sepharose produced identical responses, one would think the effect would have to be exerted at the membrane level or else postulate that all the HPL-Sepharose had been dissociated.

Turkington: We haven't done that experiment. I think it would be interesting and could possibly further shore up our conclusions.

REFERENCES

DENAMUR, R. and GAYE, P. (1967) *Arch. Anat. Microsc. & Morphol. Exp.* **56** (suppl. 3-4), 596-615.
FOLLEY, S. J. (1956) *The Physiology and Biochemistry of Lactation.* Springfield, Ill.: Thomas.
GAYE, P. and DENAMUR, R. (1968) *Bull. Soc. Chim. Biol.* **50**, 1273-1289.
GAYE, P. and DENAMUR, R. (1969) *Biochim. & Biophys. Acta* **186**, 99-109.
GAYE, P. and DENAMUR, R. (1970) *Biochem. & Biophys. Res. Commun.* **41**, 266-272.
KONIJN, Th. M. (1970) *Experientia* **26**, 367-369.
KUHN, N. J. (1969a) *J. Endocrinol.* **44**, 39-54.
KUHN, N. J. (1969b) *J. Endocrinol.* **45**, 615-616.
MEITES, J. (1966) In *Neuroendocrinology*, vol. 1, pp. 669-707, ed. Martini, L. and Ganong, W. F. New York: Academic Press.
MEITES, J. and TURNER, C. W. (1948) *Res. Bull. Mo. Agric. Exp. Stn.* nos. 415, 416.
MUNCK, A. (1971) *Perspect. Biol. & Med.* **14**, 265-289.

Neill, J. D. (1970) *Endocrinology* **87**, 1192–1197.
Prop, F. J. A. (1961) *Path. & Biol. (Paris)* **9**, 640–645.
Turkington, R. W. and Hill, R. L. (1969) *Science* **163**, 1458–1460.
Voytovich, A. E., Owens, I. S. and Topper, Y. J. (1969) *Proc. Natl. Acad. Sci. (U.S.A.)* **63**, 212–217.

[For further discussion of the role of prolactin in lactation see pp. 184–196.]

PHYSIOLOGICAL AND PATHOLOGICAL SECRETION OF HUMAN PROLACTIN STUDIED BY *IN VITRO* BIOASSAY

Andrew G. Frantz, David L. Kleinberg and Gordon L. Noel

Department of Medicine, Columbia University College of Physicians and Surgeons and the Presbyterian Hospital, New York

Despite a considerable amount of physiological evidence favouring the existence of a separate prolactin in man, recognition of this hormone has been delayed in part because of the lack of a suitably sensitive and specific bioassay. While a great deal of valuable work has been done with the pigeon crop sac assay, it has been evident that a still more sensitive assay, applicable to plasma without preliminary extraction, and employing a mammalian end organ, would be desirable. Several years ago we became interested in using mid-pregnant mouse breast tissue in organ culture for the measurement of prolactin, and we found this system was capable of detecting the hormone at concentrations considerably below what had previously been reported (Kleinberg and Frantz 1969). The addition of normal human male plasma to the medium increased the sensitivity still further, and one year ago we reported a bioassay for prolactin based on this system, with a sensitivity of 5 ng/ml of ovine prolactin or somewhat better, which was capable of detecting the hormone in unextracted human plasma. Elevated levels of prolactin activity, with low immunoassayable growth hormone, were found in postpartum subjects, in patients with galactorrhoea, and in normal subjects after treatment with various tranquillizing drugs; in none of these could prolactin activity be neutralized with anti-human growth hormone antiserum. By contrast, elevated levels of prolactin activity with high growth hormone were found in acromegalic patients and in normal subjects after insulin-induced hypoglycaemia; the prolactin in these plasmas could be largely or completely neutralized with anti-human growth hormone antiserum (Frantz and Kleinberg 1970). In the present report we extend these observations and present additional physiological and immunological data.

ASSAY

The assay, which we have recently described in detail (Kleinberg and Frantz 1971), is based on the response of the breast tissue of Swiss albino mice of a local strain, bred in our own laboratory. Animals are sacrificed by ether anaesthesia on the eighth or ninth day of pregnancy. Breast tissues are cut into small pieces and incubated for four days under 95 per cent oxygen and 5 per cent carbon dioxide in synthetic Medium 199. This is a system which has been studied by a number of investigators for many years (Elias 1957; Topper 1968; Turkington 1968), but it had not previously been used for assay purposes or found capable of responding to such low concentrations of prolactin. An important finding has been the beneficial effects on this system of human plasma, which enhances the viability of the tissues and increases their sensitivity to prolactin. We routinely add plasma at 30 per cent concentration to all incubation dishes; in those containing prolactin standards or other purified substances, the plasma used is taken from a large pool of normal human male plasma which contains no detectable prolactin activity. When human plasma specimens are being tested they are usually run at 10 and 30 per cent concentrations, partly or completely replacing the pooled male plasma. The end-point is histological. A negative response, seen in the absence of prolactin (Fig. 1), is characterized by tubules which are empty and devoid of secretory material. A strongly positive response, shown in Fig. 2, has lumina nearly all of which are filled with dark red staining secretory material. Grading is done on an arbitrary scale of 0 to 4+, based on the amount of secretory material present. Each tissue fragment in a dish is assigned a grade and the resulting score for each dish is a mean of these grades. In practice grading is not hard, and the results of different observers are in excellent agreement. Although individual inspection and scoring of each sample is somewhat time-consuming, we feel that it affords a valuable control on the assay. In spite of the most careful technique, one occasionally encounters necrosis, which may affect a few or many of the dishes in a given assay. We have had experience with a very few human sera which, either because of something inherently toxic in the plasma or because of some toxic substance acquired in the process of collecting the specimen, regularly produced necrosis in every assay in which they were run. These effects are unusual but they would be missed if the microscopic appearance of the cultures were not visually inspected. Ovine prolactin standards are run in each assay at doses of 0, 1, 2, 5, 10, 20 and 50 ng/ml. When these results are plotted together a standard curve of the form shown in Fig. 3 is obtained, which represents the combined experience of 44 assays. Five ng/ml can be definitely distinguished from 0 in most assays, and this is what we usually consider the sensitivity of our assay, but in a certain number of incubations the effective sensitivity is somewhat better, or approximately 2 ng/ml. When human plasma to be tested is present in 30 per cent concen-

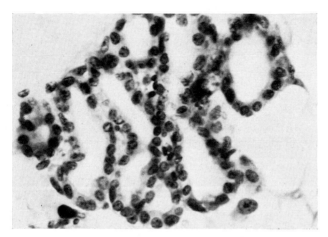

Fig. 1. Mouse breast tissue after incubation in absence of prolactin. ×215.

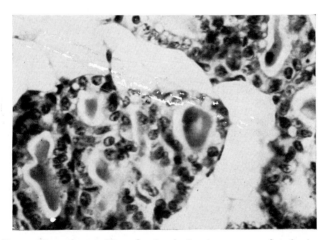

Fig. 2. Mouse breast tissue after incubation in presence of prolactin at 50 ng/ml. Note presence of darkly staining intraluminal secretory material. ×215.

tration the effective sensitivity of the assay therefore becomes approximately 15 ng/ml of ovine equivalents, or approximately 0·42 m i.u./ml.

The routine addition of normal human male plasma significantly improves the sensitivity of the assay, permitting the detection of prolactin at levels approximately one-half to one-quarter of those that can be

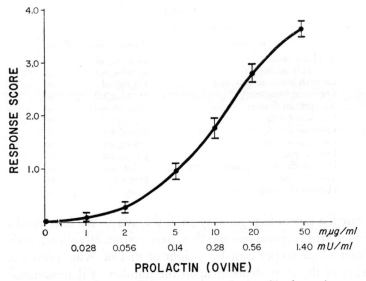

FIG. 3. Standard curve of bioassay, representing combined experience of 44 assays. Vertical bars indicate standard error of the mean.

detected in its absence. The precise reason for this improvement is not clear. It does not seem to us wholly explainable by a possibly significant, though undetectable, level of prolactin in the pooled male plasma which we add. The precise timing of the duration of pregnancy in these animals is important. We use tissues a little earlier in pregnancy than have been used by most other investigators and find that if the animals are more than nine days pregnant at the time the tissues are harvested, the control dish in the standard curve will generally have evidence of secretory activity and be graded as positive. This reduces the slope of the curve and degrades the sensitivity of the assay; we do not consider assays where this effect occurs suitable for use. On the other hand, if animals are used at times earlier than the eighth day of pregnancy, the tissues are much less sensitive to prolactin and positive responses may not be seen even at 50 ng/ml.

The specificity of the assay is very high. A number of pituitary and non-pituitary hormones have been tried in this system (Table I) with negative results. Oestrogen neither causes a positive response by itself nor alters the response to added prolactin. The only hormones which we have ever found to give positive responses have been ovine or other prolactins, human growth hormone, and human placental lactogen. We have not made extensive tests on the potency of placental lactogen, but a highly purified preparation supplied to us by Dr Hugh Niall had approximately one-third of the activity of ovine prolactin. Human growth

TABLE I
SPECIFICITY OF BIOASSAY

Hormone	Concentration	Response
TSH (bovine)	25 mi.u./ml	0
ACTH (porcine)	50 mi.u./ml	0
Growth hormone (bovine)	125 ng/ml	0
Oxytocin (synthetic)	0·25 i.u./ml	0
Vasopressin (lysine)	0·25 i.u./ml	0
Oestradiol-17β	2·5 μg/ml	0
Oestradiol-17β	5·0 ng/ml	0
Oestradiol-17β	0·5 ng/ml	0
Progesterone	5·0 μg/ml	0
Human chorionic gonadotropin	10 i.u./ml	0
Testosterone	5 μg/ml	0
Human placental lactogen	30 ng/ml	positive

hormone is a potent lactogenic agent in this assay. We have tried several highly purified preparations and they have consistently shown activity of the order of 50–80 per cent by weight of that of ovine prolactin. The precision of the assay depends upon the number of determinations one does. We consider it important to assay each sample at two different dose levels in at least two separate assays. Under these conditions, for samples not at the threshold of sensitivity of the assay, the mean of all determinations will show a coefficient of variation of 25–35 per cent. Single determinations at one dose level are inherently imprecise.

RESULTS ON HUMAN PLASMA

Fig. 4 gives an overall picture of the results we have achieved with human plasma measurements in a variety of conditions. In normal men and women, of whom we have studied over 50 to date, we have generally been unable to detect prolactin activity. This means that levels are less than 15 ng/ml ovine equivalents or 0·42 m i.u./ml. In a very few individuals we have obtained positive responses at or below this level in unusually sensitive assays. In some of these cases it has seemed that the levels of endogenous growth hormone, which were also measured, might have been sufficient to account for the prolactin effect observed. It seems to us that prolactin may very well circulate in normal plasma at levels not far below the sensitivity of our current assay, although all we are able to say is that we cannot detect it. It is worth pointing out that the routine use of pooled normal male plasma makes it impossible for us by definition to detect prolactin in large numbers of normal males. The accurate determination of normal levels of prolactin will probably be a task for radioimmunoassay, rather than a bioassay.

As shown in Fig. 4, we have examined plasmas from a number of nursing mothers. All have had detectable prolactin activity, ranging from 0·56 to

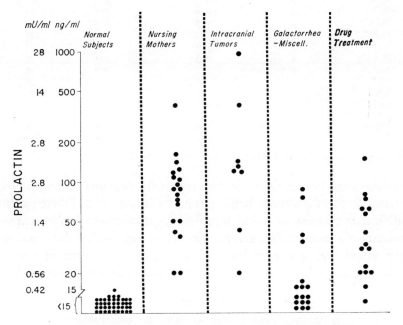

FIG. 4. Human plasma prolactin concentrations in various conditions. Values are expressed in terms of NIH-S8 ovine prolactin standard, for which 1 mg = 28 i.u.

10·8 m i.u./ml. Some of these patients are listed in Table II, together with their growth hormone levels as determined by radioimmunoassay (Frantz and Rabkin 1964). Prolactin is expressed both as m i.u./ml and ng/ml of ovine equivalents. As can be seen, the growth hormone levels are far too low to account for the prolactin activity measured in these subjects' plasmas. We also have evidence in a limited number of patients of a rise in prolactin due to suckling, with levels climbing two- to four-fold during the nursing period.

There are a number of conditions of abnormal lactation in which we have observed high levels of prolactin. Some of the highest, ranging up to 28 m i.u./ml, have been encountered in patients with pituitary tumours. Of these, chromophobe adenomas have yielded the highest levels, but we

Table II
PROLACTIN ACTIVITY IN PLASMAS OF NURSING MOTHERS

Subject	Days post partum	Human growth hormone ng/ml	Prolactin activity m i.u./ml	ng/ml*
A.R.	1	0·7	2·5	89
R.H.	1	3·6	2·3	82
U.G.	1	2·8	3·4	121
H.L.	3	1·0	10·8	386
G.W.	3	<0·3	4·5	160
S.R.	3	2·8	2·1	75
B.H.	3	1·8	2·7	96
G.B.	3	1·0	2·0	71
U.J.	3	2·1	1·2	43
B.W.	4	0·3	1·4	50
M.S.	5	2·1	2·7	96
M.P.	6	1·0	0·56	20
S.S.	6	0·3	2·5	89
M.H.	30	<0·3	1·4	50
Mean		1·41	2·86	102
Total range		(<0·3–3·6)	(0·56–10·8)	(20–386)

* In terms of ovine prolactin, NIH-S8 standard: 1 mg = 28 i.u.

have also seen elevated prolactin with craniopharyngiomas and with other lesions affecting the hypothalamic–pituitary axis. In some of these patients with intracranial lesions and high prolactin, galactorrhoea was present and prolactin was measured for this reason. As we have looked, however, we have found high prolactin levels in an increasing number of patients without galactorrhoea. In patients with galactorrhoea who do not have

Table III
PROLACTIN DURING CHRONIC TRANQUILLIZING DRUG THERAPY

Patient	Sex	Drug (time in weeks)	Dose mg/day	HGH ng/ml	Prolactin m i.u./ml
C.G.	F	Fluphenazine (2)	6	<0·3	0·42
A.E.	F	Perphenazine (78)	12	0·3	3·1
M.S.	M	Perphenazine (1)	24	3·3	0·42
A.M.	F	Perphenazine (12) + amitriptyline (12)	12 100	0·3	2·1
D.V.	M	Chlorpromazine (4)	2000	0·5	1·7
M.M.	M	Chlorpromazine (4)	1500	3·7	0·56
T.B.	M	Chlorpromazine (8)	2000	0·4	0·84
A.J.	F	Imipramine (4)	150	1·1	0·56
M.S.	M	Haloperidol (1)	4·5	0·3	1·7

known intracranial lesions, the highest prolactin levels have been found chiefly in those on tranquillizing drugs or various other medications.

We have also encountered a rather large number of patients with galactorrhoea in whom prolactin was unmeasurably low—that is, less than 0·42 m i.u./ml. In most of these patients the galactorrhoea has been of the idiopathic type, that is, not associated with intracranial pathology, drugs or other obvious causes, and most of these patients have continued to menstruate at regular intervals. It is our impression that galactorrhoea occurring in association with normal menses is far more common than is generally clinically recognized. The relatively low concentrations of prolactin we have observed in most of these patients are additional evidence that they have a lesser degree of hypothalamic or pituitary disturbance than those patients whose galactorrhoea is associated with amenorrhoea, in whom prolactin levels are generally higher.

In Table III are shown the results of prolactin determinations in a series of nine patients who were receiving tranquillizing drugs in high doses for psychiatric purposes. None of these patients had any endocrine abnormality prior to treatment. The drugs were mostly phenothiazine derivatives except for imipramine, which is closely related chemically, and haloperidol, a butyrophenone derivative. Growth hormone levels were much too low to account for the prolactin activity observed. Only two of these patients had galactorrhoea.

Acute as well as chronic administration of chlorpromazine will cause marked elevation of prolactin in humans. Table IV shows the results in seven normal volunteers, five men and two women, who were given intramuscular chlorpromazine in a single dose ranging from 12·5 to 50 mg. Within half an hour of injection, prolactin had risen to detectable levels in four of these subjects, and by two hours it was detectable in all. Levels ranged up to 3·0 m i.u./ml, and they tended to remain elevated for the whole six-hour period covered in Table IV. In three of four subjects

TABLE IV
CHLORPROMAZINE STIMULATION TESTS
Plasma prolactin m i.u./ml

Subject	Sex	0 hours	½ hour	1 hour	2 hours	3 hours	4 hours	5 hours	6 hours
1.	M	N.D.*	1·6	2·4	1·5	0·94	0·67	0·84	1·2
2.	M	N.D.	0·86	—	0·46	—	—	0·43	—
3.	M	N.D.	0·74	0·57	0·50	0·88	1·3	3·0	1·9
4.	M	N.D.	1·2	0·61	0·67	—	0·75	0·55	—
5.	M	N.D.	N.D.	N.D.	0·63	—	N.D.	—	0·52
6.	F	N.D.	—	N.D.	2·5	1·0	1·1	—	—
7.	F	N.D.	N.D.	1·9	1·3	1·0	N.D.	N.D.	0·76
Mean		N.D.	0·81	1·00	1·08	0·95	0·69	0·99	1·10

* N.D., not detected (<0·4 m.i.u./1 ml).

tested at 24 hours, prolactin was undetectable. These experiments indicate that despite the apparently small population of prolactin-secreting cells noted in the normal human pituitary, these cells can accomplish release and perhaps synthesis of prolactin within a very short period of time after appropriate stimulation, in men as well as in women.

The elevated prolactin levels so far discussed have all been associated with low or normal levels of immunoassayable growth hormone. As we have previously reported (Frantz and Kleinberg 1970), high levels of radioimmunoassayable activity are also associated with elevated prolactin activity. Fourteen normal subjects were tested for growth hormone and prolactin both before and after insulin-induced hypoglycaemia (Fig. 5).

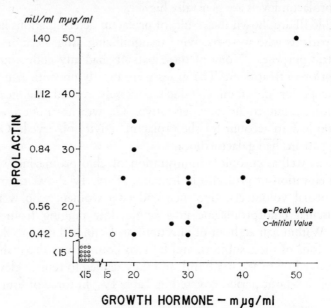

FIG. 5. Bioassayable plasma prolactin activity and immunoassayable growth hormone during insulin tolerance tests in normal individuals. Open circles represent levels at start of test; closed circles are levels one hour after intravenous insulin. Preincubation with anti-growth hormone serum abolished or greatly reduced the prolactin activity of all samples tested, indicating that the prolactin effect is largely attributable to growth hormone.

Prolactin was undetectable at the beginning of the test but rose concomitantly with growth hormone to become detectable in all subjects at levels ranging from 0·42 to 1·4 m i.u./ml (15 to 50 ng/ml ovine equivalents). Growth hormone in these subjects also rose, reaching peak levels between 16 and 50 ng/ml. The prolactin activity of these samples could be greatly reduced or abolished by preincubation with anti-human growth hormone serum, under conditions which did not lower the prolactin activity in plasma from nursing mothers, as we shall discuss below. These results indicated to us two things: first, that radioimmunoassayable growth hormone as it circulates in blood, like the material extracted from human pituitaries, possesses intrinsic lactogenic activity. Second, since in some post-hypoglycaemic plasma, prolactin activity could still be detected at low levels even after neutralization with anti-growth hormone, it seemed possible that in at least some patients hypoglycaemia had stimulated the release of prolactin as well as that of growth hormone. The magnitude of such prolactin release, however, in those cases where it occurred, would seem to be rather small.

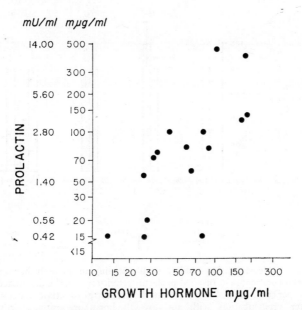

FIG. 6. Bioassayable plasma prolactin activity and immunoassayable growth hormone in patients with acromegaly.

We have also previously reported (Frantz and Kleinberg 1970) measurements of plasma prolactin activity in 16 acromegalic subjects (Fig. 6). Growth hormone ranged from 13 to 180 ng/ml, and prolactin activity was detectable in all these subjects with concentrations from 0·42 to 11 m i.u./ml (15 to 400 ng/ml ovine equivalents). Much of the bioassayable activity appeared to be due to growth hormone itself, but in some patients neutralization studies, which we mention below, indicated that prolactin was being hypersecreted together with growth hormone.

NEUTRALIZATION STUDIES WITH ANTISERA

The finding last year that seems to us to offer the best proof of the existence of a separate human prolactin was the complete inability of a potent anti-human growth hormone serum to effect any neutralization of the lactogenic activity of patients with high prolactin but low growth hormone in their plasma (Frantz and Kleinberg 1970; Kleinberg and Frantz 1971). This antiserum, prepared in a rabbit against a highly purified Wilhelmi preparation of human growth hormone (HS 612A), was capable of binding 60 per cent of trace amounts of ^{131}I-labelled growth hormone at a dilution of 1 in 150 000 and more than 95 per cent at a dilution of 1 in

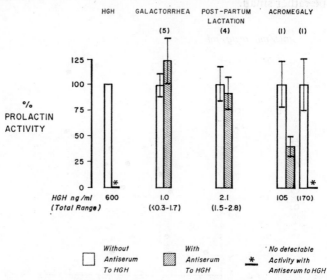

FIG. 7. Neutralization studies with anti-human growth hormone. Open bars represent original prolactin activity in sample, designated as 100 per cent. Shaded bars represent the percentage of activity remaining after preincubation with anti-growth hormone serum. Growth hormone concentrations are shown. The mean prolactin level for the galactorrhoea group was 2·4 m i.u./ml, and for postpartum subjects was 2·6 m i.u./ml. No significant neutralization was noted in any of the samples from these two groups. The prolactin activity of human growth hormone added to normal male plasma, in amounts up to 2000 ng/ml (subsequently bioassayed at 30 per cent concentration), was abolished by antiserum.

1000. In the experiments shown in Fig. 7 the antiserum was used at 1 in 10 concentration, 0·1 ml of antiserum being preincubated for 2 hours with 0·9 ml of plasma before the mixture was bioassayed in the usual manner. Parallel incubations of plasma without antiserum or with normal rabbit serum were always made and the two samples were always bioassayed together at several dilutions in the same assay. The prolactin effect of large amounts of human growth hormone, up to 2000 ng/ml, added to normal human male plasma was completely neutralized in this manner. Since this plasma like our others was bioassayed at 30 per cent concentration in the incubation medium with the mouse breast, the final concentration at the time of assay was 600 ng/ml. Under identical conditions of pre-incubation, the plasmas of five patients with galactorrhoea and four patients with postpartum lactation showed no significant neutralization of prolactin activity, as illustrated in Fig. 7. These were all samples containing high prolactin and low growth hormone. The results on two individual patients with acromegaly, shown at the right of Fig. 7, seem to us representative of a somewhat larger group of acromegalic subjects that we have studied. They indicate in one case complete, and in the other case partial, neutralization of prolactin activity after incubation with anti-human growth hormone antiserum. It appears therefore that in some but not all acromegalic patients excessive secretion of prolactin as well as growth hormone may occur. It should be noted in connection with the anti-human growth hormone studies that Peake and co-workers (1969) studied a human chromophobe adenoma extract which contained low growth hormone and high pigeon crop sac-stimulating material which could not be neutralized with anti-HGH antiserum. Dr Pasteels and his colleagues had previously obtained immunological, as well as biological, evidence, from cultures of human foetal pituitaries, pointing to the existence of a separate human prolactin (Pasteels, Brauman and Brauman 1965).

Turning to studies with ovine prolactin, it seemed desirable to confirm by biological neutralization what Herbert and Hayashida (1970) had first shown by immunofluorescent methods, namely that anti-ovine prolactin will react with primate prolactin. Table V shows that incubation with

TABLE V

NEUTRALIZATION OF PROLACTIN ACTIVITY IN HUMAN PLASMA WITH ANTI-OVINE PROLACTIN

Patient	Sex	Diagnosis	Prolactin mi.u./ml (Mean ± S.E.M.)	Growth hormone ng/ml	Prolactin after incubation with anti-ovine prolactin
B.J.	F	galactorrhoea: chromophobe adenoma	3·9±1·0	0·3	undetectable*
E.F.	F	galactorrhoea: hypothalamic glioma	3·6±0·7	0·8	undetectable
M.O.	F	galactorrhoea: chromophobe adenoma	3·5±1·0	0·3	undetectable
L.J.	M	galactorrhoea: possible oestrogen withdrawal	2·2±0·9	1·3	undetectable
A.C.	F	galactorrhoea: idiopathic	2·1±0·4	0·6	undetectable
J.W.	F	galactorrhoea: idiopathic	0·5	1·2	undetectable
H.L.	F	3 days *post partum*	10·8±3·1	1·0	undetectable
E.R.	F	2 days *post partum*	4·6±1·3	0·8	undetectable
L.F.	F	3 days *post partum*	4·0±0·5	0·6	undetectable
G.W.	F	2 days *post partum*	3·4±0·8	0·3	undetectable
C.P.	F	4 days *post partum*	2·8±1·2	0·5	undetectable
M.S.	F	3 days *post partum*	1·1±0·3	1·6	undetectable

* Less than 0·4 m i.u./ml.

anti-ovine prolactin, prepared against the highly purified NIH-S8 standard, under conditions similar to those we used with anti-growth hormone, completely abolished the prolactin activity in the plasma of six patients with galactorrhoea and six nursing mothers, all of whom had high prolactin and low growth hormone. This result was not unexpected, and it further confirms the close immunological similarity that must exist between ovine and human prolactin. More unexpected, however, were the findings shown in Table VI, depicting the results of testing a highly purified human growth hormone standard (HS 1103C) for prolactin activity before and after preincubation with anti-ovine prolactin. It can be seen that significant neutralization of the prolactin activity of this preparation occurred, with amounts up to 50 ng/ml showing no detectable activity after exposure to antiserum. One explanation of these findings, which seems to us unlikely although we do not have data to rule it out, would be that the ovine prolactin used for immunization contained some ovine growth hormone, against which some antibodies were also raised, and which exhibited cross-reaction with human growth hormone sufficient to cause some neutralization of the latter's biological prolactin activity. A more likely explanation of these findings seems to us that they tend to reflect in an immunological way the structural similarity between ovine prolactin and human growth hormone first noted by Bewley and

TABLE VI

EFFECT OF INCUBATION WITH ANTI-OVINE PROLACTIN ON PROLACTIN ACTIVITY OF HUMAN GROWTH HORMONE

Growth hormone ng/ml	Prolactin activity ng/ml*	Prolactin activity after incubation with anti-ovine prolactin ng/ml*
1000	> 50	> 50
200	> 50	8
50	33 ± 7 (S.E.M.)	< 5
20	16 ± 3	< 5
10	9 ± 4	< 5

* In terms of ovine prolactin, NIH-S8 standard: 1 mg = 28 i.u.

Li (1970), and further emphasized by Niall and his collaborators (1971) when they compared ovine prolactin to Niall's proposed revised structure for human growth hormone. In the converse experiment, that of assaying ovine prolactin before and after exposure to anti-human growth hormone, we have failed to see any neutralization of the prolactin by the anti-HGH. These latter results would indicate that the antigenic determinants against which this antiserum is directed are not present in the ovine prolactin molecule.

Finally, we too have been working on a radioimmunoassay, and have approached it by way of pituitary glands obtained from nursing baboons. Four of these animals obtained during the postpartum period have been further stimulated with daily injections of large doses of perphenazine for two weeks. They were then sacrificed and their pituitaries incubated in organ culture. Under these conditions very large quantities of prolactin, up to several hundred microgrammes of ovine equivalents/ml/day, are liberated into the incubation medium. Growth hormone in the medium is very much lower than this, one-tenth as much or less, and the media are therefore suitable sources of prolactin both for immunization and for radioiodination. Further purification of the radioiodinated material is necessary, by starch gel electrophoresis, affinity chromatography or other means, but is facilitated by the fact that very high losses can be accepted at this stage in the interest of obtaining a highly purified material. The antisera we have obtained show binding of ^{131}I-labelled ovine as well as baboon prolactin at concentrations in the range of 1 : 10 000, with specific displacement by ovine, baboon and human prolactin at varying levels of sensitivity. Correlation of radioimmunoassay with bioassay data, a project on which we have just begun, may well yield interesting additional information on the physiology of prolactin secretion.

Acknowledgements

This work was supported by grants from the National Institutes of Health and the American Heart Association. The authors are grateful to Miss Charity L. Young and Mr Robert E. Sundeen for excellent technical assistance.

SUMMARY

A sensitive *in vitro* bioassay for prolactin has been developed, based on the secretory response in organ culture of breast tissue from pregnant mice. Sensitivity is 5 ng/ml (0·14 m i.u./ml) or somewhat better for ovine prolactin. Unextracted human plasma can be satisfactorily assayed at 30–50 per cent concentrations in the incubation medium. Plasma prolactin activity is less than 0·42 m i.u./ml in normal men and women. Growth hormone as it circulates in blood is lactogenic, but this effect may be neutralized by anti-growth hormone antiserum added to the plasma under test. Elevated plasma prolactin activity, ranging from 0·56 to 10·8 m i.u./ml, has been found in all of 15 nursing mothers tested 1–30 days *post partum*. High prolactin activity (0·42 to greater than 28 m i.u./ml) has been found in the majority of over 30 patients with galactorrhoea of various causes, the highest levels being found in patients with pituitary tumours. Acute rises in prolactin activity, up to 2·5 m i.u./ml, occurred in all of seven normal subjects (five males, two females), $\frac{1}{2}$ to 2 hours after the intramuscular injection of chlorpromazine. Sustained oral administration of phenothiazine derivatives and certain other tranquillizers is associated with chronically elevated plasma prolactin activity, with or without galactorrhoea. Immunoassayable growth hormone in these subjects, as in the nursing mothers and in those with galactorrhoea, was not elevated. Antisera to human growth hormone, capable of completely neutralizing the prolactin activity of large amounts of growth hormone added to the incubation medium, produced no significant neutralization of prolactin activity in any of these subjects' plasma. Anti-ovine prolactin, on the other hand, under identical conditions of preincubation, completely abolished the prolactin activity in the plasma of all of these subjects so far tested.

REFERENCES

BEWLEY, T. A. and LI, C. H. (1970) *Science* **168**, 1361–1362.
ELIAS, J. J. (1957) *Science* **126**, 842–844.
FRANTZ, A. G. and RABKIN, M. T. (1964) *New Engl. J. Med.* **271**, 1375–1381.
FRANTZ, A. G. and KLEINBERG, D. L. (1970) *Science* **170**, 745–747.
HERBERT, D. C. and HAYASHIDA, T. (1970) *Science* **169**, 378–379.
KLEINBERG, D. L. and FRANTZ, A. G. (1969) *Program of the Endocrine Society, 51st Meeting*, p. 46 (abst. 32).
KLEINBERG, D. L. and FRANTZ, A. G. (1971) *J. Clin. Invest.* **50**, 1557–1568.
NIALL, H. D., HOGAN, M. L., SAUER, R., ROSENBLUM, I. Y. and GREENWOOD, F. C. (1971) *Proc. Natl. Acad. Sci. (U.S.A.)* **68**, 866–869.
PASTEELS, J. L., BRAUMAN, H. and BRAUMAN, J. (1965) *C. R. Hebd. Séance Acad. Sci., Sér. D* **261**, 1746–1748.
PEAKE, G. T., MCKEEL, D. W., JARETT, L. and DAUGHADAY, W. H. (1969) *J. Clin. Endocrinol. & Metab.* **29**, 1383–1393.
TOPPER, Y. J. (1968) *Trans. N.Y. Acad. Sci.* **30**, 869–874.
TURKINGTON, R. W. (1968) *Endocrinology* **82**, 575–583.

[For discussion, see pp. 184–196].

USE OF A RABBIT MAMMARY GLAND ORGAN CULTURE SYSTEM TO DETECT LACTOGENIC ACTIVITY IN BLOOD

Isabel A. Forsyth

National Institute for Research in Dairying, Shinfield, Reading

The development of radioimmunoassay methods has made possible the measurement of pituitary prolactin levels in the plasma of a number of species, in various physiological states. However, since the sites on the prolactin molecule responsible for biological and immunological activity may differ, it is desirable that where possible some measurements are made by both radioimmunoassay and bioassay on the same plasma samples. None of the classical methods for the assay of prolactin activity has proved entirely suitable for detection of the hormone in plasma. All are insufficiently sensitive or are subject to interference from non-specific factors. These considerations led us, early in 1968, to use a rabbit mammary gland organ culture system for the detection of lactogenic activity in the plasma of goats at parturition (Brumby and Forsyth 1969), following reports of high levels of circulating pituitary prolactin, measured by radioimmunoassay, at this time (Bryant, Greenwood and Linzell 1968). A direct comparison of the bioassay method with radioimmunoassay for prolactin has now been undertaken in lactating goats. In addition the method has led to most interesting findings about the nature of human prolactin and of the lactogenically active hormones circulating in the pregnant goat; it provides evidence of a human prolactin immunologically distinct from human growth hormone (HGH) (Forsyth 1970; Forsyth and Myres 1971) and of a placental lactogenic hormone circulating in the goat from mid-pregnancy to term.

The conclusions depend very largely on the opportunity to compare the bioassay results with those of a radioimmunoassay, for HGH in human plasma and for pituitary prolactin in goat plasma. I am therefore indebted to Dr A. D. Wright and Professor Russell Fraser and to Dr G. M. Besser and Professor J. Landon by whose courtesy the radioimmunoassays for HGH were performed and to Dr H. L. Buttle for permission to quote his unpublished work on prolactin levels in the goat during pregnancy and lactation.

THE BIOASSAY METHOD

A series of papers from Professor H. A. Bern's laboratory established the ability of the mouse mammary gland to show a lactogenic response to prolactin when maintained under suitable conditions on a synthetic medium *in vitro* (Elias 1957, 1959; Rivera and Bern 1961; Rivera 1964). Similar responses were later reported for a number of other species, including the rabbit, which showed a marked sensitivity to the hormone (Barnawell 1965). It is this *in vitro* lactogenic response which forms the basis of the assay, and a similar method using mouse mammary gland has recently been described by Frantz and Kleinberg (1970). Forsyth and Myres (1971) give full details of the *in vitro* rabbit assay. In brief, pseudopregnancy is induced in virgin female Dutch rabbits, about six months old, by a single intravenous injection of human chorionic gonadotropin. On the 11th day of pseudopregnancy, when development of lobulo-alveolar mammary tissue has occurred, the animal is killed and explants about 1 mm^3 are prepared from its mammary glands. The explants are maintained at 37°C in an atmosphere of 95 per cent O_2:5 per cent CO_2 on synthetic Medium 199 containing insulin (5 μg/ml) and corticosterone (1 μg/ml) in all cultures. The standards used are ovine prolactin (NIH-P-S-6, 25 i.u./mg) or HGH (NIH-GH-HS612A). Plasma samples are added to the medium, usually at a concentration of 10 per cent, and an assay is possible using as little as 0·5 ml of plasma which requires no pretreatment. The medium is changed on day 3 and cultures terminated on day 5. Explants are then prepared for histological examination and a blind-test assessment of secretory response is made on serial sections, stained with haematoxylin and eosin. A 1-4 + grading system is used, based on that of Barnawell (1965). In any one assay, eight explants, contained in two culture dishes, receive the same treatment and grades shown are based on the mean response of eight explants. When comparisons are made between samples in a series, for example from one goat at intervals during pregnancy, or during a suckling episode, all samples are examined together in the same bioassay (see Figs. 2 and 3). Tests of significance are carried out using the Mann-Witney U-test (Siegel 1956).

Specificity of the lactogenic response

Tests of specificity have been made using the following hormones: human luteinizing hormone, human follicle-stimulating hormone, human thyroid-stimulating hormone, human adrenocorticotropic hormone, human chorionic gonadotropin, oxytocin, vasopressin, oestradiol and progesterone. None of these hormones, alone or in various combinations, produced any lactogenic response, nor did they modify the response to prolactin. The only hormones found able to stimulate the rabbit mammary gland to secrete *in vitro* have been prolactin itself and the human hormones, HGH and human placental lactogen, both of which show prolactin-like activity in other assay systems (see Forsyth and Folley 1970). The sensitivity to these hormones varies from assay to assay, but a positive response is often obtained to a dose of 10 ng/ml while 50 ng/ml regularly has significant lactogenic effect.

The results of Barnawell (1965) indicate that rabbit mammary gland does not give a secretory response to bovine GH *in vitro*, the responses obtained to high doses being due to prolactin contamination of the bovine GH preparation. All these findings are consistent with observations on the rabbit *in vivo* (Bradley and Clarke 1956; Chadwick 1963; Cowie, Hartmann and Turvey 1969).

ASSAY OF HUMAN PLASMA SAMPLES

Fig. 1 illustrates the results of an assay in which the lactogenic activity of various doses of HGH in the presence of 10 per cent pooled male plasma was assessed and compared with that produced by six plasma samples added to the medium at 10 per cent. Two of the plasma samples were from premenopausal, non-pregnant women, two from women breast-feeding their babies on the fifth and seventh postpartum days and two from women with amenorrhoea and galactorrhoea, one of whom (E.A.) had a pituitary tumour. It will be seen that positive lactogenic responses were obtained to HGH at concentrations of 50 and 100 ng/ml and to the plasma samples from women lactating, either normally *post partum* or in abnormal lactation. The male plasma and the samples from premenopausal, non-pregnant women produced no response. The levels of HGH in the plasma samples have been measured for me by radioimmunoassay techniques and these values are also shown (*), the amount of HGH in the culture medium resulting from addition of plasma being 10 per cent of these values. Thus, the HGH level in the medium was in all cases less than 1 per cent of that required to give the positive lactogenic responses observed. Moreover, the higher HGH concentrations do not necessarily correspond with positive responses.

From evidence of this kind we have concluded (Forsyth and Myres 1971) that during lactation human plasma contains a hormone which does not cross-react immunologically with HGH and has the biological characteristics of a prolactin. Using mouse mammary gland *in vitro*, Frantz and Kleinberg (1970) have obtained similar results.

The conclusion is supported by the results of an experiment in which rabbit antiserum to HGH was also included in the medium. It abolished the response to 50 ng/ml HGH but did not significantly affect the response to plasma from a woman with galactorrhoea (Forsyth and Myres 1971).

Plasma samples from a total of 84 people have now been examined

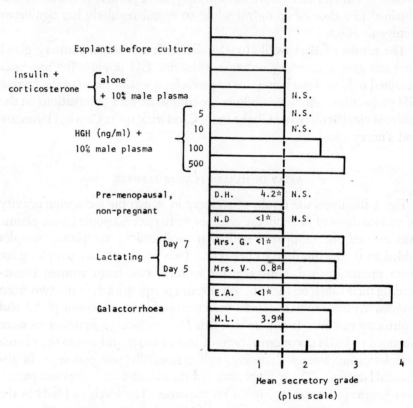

FIG. 1. Lactogenic activity of HGH compared with that of 10 per cent plasma from six women and a pool of male plasma. Mean responses of eight explants.
*Plasma HGH (ng/ml) measured by radioimmunoassay. N.S., not significantly different from response to culture with insulin + corticosterone alone, $P > 0.1$. All other responses were significant, $P < 0.001$.

(Forsyth and Myres 1971; Forsyth et al. 1971; Edwards, Forsyth and Besser 1971). In 26 control subjects, not known to have any endocrinopathy, no plasma lactogenic activity could be detected. This group of control subjects consisted of one girl of seven years, five premenopausal, non-pregnant women in the age range 20–25 years, 19 women past the menopause, aged 55–81 years, and one male of 67 years. All had HGH levels within the normal range (less than 11 ng/ml).

Of 18 women breast-feeding their babies in early lactation, 12 showed detectable lactogenic activity in their plasma and all had HGH levels below 7 ng/ml. Nine patients have been examined who showed abnormal breast development without evidence of secretion. These were one girl of 19 with massive breast hypertrophy, and a boy of five years and seven adult men with gynaecomastia. None showed detectable plasma lactogenic activity. By contrast one male and 14 of 19 female patients with galactorrhoea showed significant lactogenic activity. The three female patients known to have pituitary tumours all showed very marked plasma lactogenic activity. In all these patients HGH levels were within the normal range.

Lactogenic activity has been detected in the plasma of six patients with acromegaly. In this case their high plasma concentrations of HGH could be responsible for the activity observed. However, in at least one case, a girl of 20 with the added symptoms of amenorrhoea and galactorrhoea, it does seem that lactogenically active material not cross-reacting with HGH may also be present.

One of the female patients with galactorrhoea was found to be hypothyroid (Edwards, Forsyth and Besser 1971). A syndrome of galactorrhoea associated with primary hypothyroidism has been described (see Kinch, Plunkett and Devlin 1969) and it is of particular interest that assays of plasma from five further patients with primary hypothyroidism, but without evidence of galactorrhoea, have shown two of them to have detectable lactogenic activity in their plasma.

ASSAY OF PLASMA FROM PREGNANT AND LACTATING GOATS

Plasma samples have been taken from the jugular vein of five primiparous goats in mid-October, during the anoestrum, then at first oestrus 8–52 days later, on the following morning after mating, and at three-week intervals throughout pregnancy to term. Samples were also taken from the same goats during a single suckling episode after overnight separation of the kids, between the 11th and 15th days of lactation. Six multiparous goats, five of which were simultaneously lactating, were also sampled at three-week intervals during the second half of pregnancy and one non-pregnant lactating goat was similarly sampled over the same time period. Bioassays for lactogenic activity have been carried out on many of these samples and in all cases the standards contained 10 per cent plasma from a castrated male goat. Information is also available on the plasma prolactin levels, as measured by a double antibody radioimmunoassay method using ovine prolactin as antigen (H. L. Buttle, unpublished), for all the samples from primiparous goats and for samples from three multiparous animals.

The effect of suckling on plasma prolactin levels in the goat

Fig. 2 illustrates the results of a bioassay of nine samples taken from goat 404 during suckling. Lactogenic activity was detected in all samples, the highest activity being present in the last sample, taken ten minutes after the

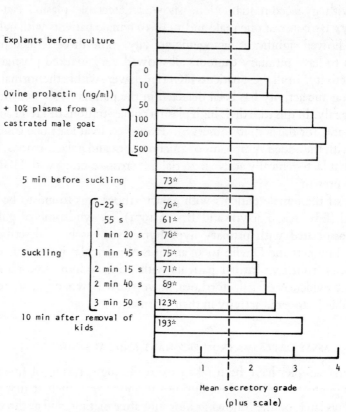

FIG. 2. Bioassay of plasma from goat 404 during suckling on day 15 of lactation. Mean responses of eight explants.

*Plasma prolactin (ng/ml) measured by radioimmunoassay.

All responses were significantly different from response to culture in the absence of prolactin, $P < 0.001$.

removal of the kids. Comparing these results with the values for pituitary prolactin indicated by radioimmunoassay, it will be seen that the bioassay indicates a higher prolactin level in the last sample but that in general there is a reasonable agreement between the two methods. In three other goats rather similar results were obtained. The pituitary prolactin levels from five minutes before suckling to the removal of the kids were below or close to the limits of the bioassay, in the range 34–250 ng/ml plasma

(3·4–25 ng/ml medium). The highest concentrations (260, 520 and 523 ng/ml) were again found in the final sample ten minutes after removal of the kids and were all detected by the bioassay. In one goat (399) high and rather constant levels of prolactin (370–520 ng/ml) were detected throughout the suckling episode by both methods. Stimulation of pro-

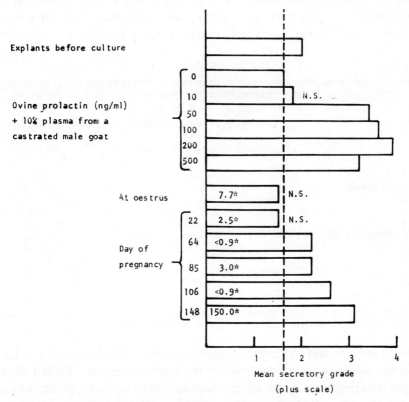

FIG. 3. Bioassay of plasma from goat 404 during first pregnancy. Mean responses of eight explants.

*Plasma prolactin (ng/ml) measured by radioimmunoassay.

N.S., not significantly different from response to culture in the absence of prolactin, $P > 0·1$. All other responses were significant, $P < 0·001$.

lactin secretion by suckling has been reported by other workers (for references see Cowie and Tindal 1971).

Plasma lactogenic activity and prolactin levels in primiparous goats

Fig. 3 gives details of the bioassay and radioimmunoassay of plasma from goat 404 at intervals during pregnancy and the results on all five primiparous goats studied are summarized in Table I. In goat 404, plasma prolactin levels as measured by radioimmunoassay remained extremely low until shortly before term. The sample taken at 148 days, on the morning of parturition, contained 150 ng/ml prolactin, but the previous sample, taken at 127 days of pregnancy, contained only 1·2 ng/ml; the highest level detected earlier in pregnancy was only 15 ng/ml at 43 days. By contrast, the lactogenic activity in plasma was already marked in the sample taken at 64 days of pregnancy when pituitary prolactin was less than 0·9 ng/ml (Fig. 3). The lactogenic activity detected in the samples

TABLE I

BIOASSAY AND RADIOIMMUNOASSAY OF JUGULAR VEIN SAMPLES FROM FIVE GOATS AT INTERVALS DURING THEIR FIRST PREGNANCY

Plasma sample		Lactogenic activity in plasma of individual goats measured by bioassay Goat no.					Plasma prolactin level (ng/ml, measured by radioimmunoassay)
		380	390	397	399	404	Mean ± S.D.
1. Anoestrus		—	0	+	—	—	16·4 ± 18·97
2. At oestrus		0	0	—	+	0	40·1 ± 52·59
3.	1	—	0	0	—	—	2·4 ± 1·70
4.	22	0	—	—	0	0	1·4 ± 0·74
5.	43	—	0	0	—	—	4·6 ± 5·97
6. Pregnancy, day	64	+	—	—	0	+	3·3 ± 3·44
7.	85	—	+	0	+	+	2·9 ± 2·58
8.	106	+	—	—	+	+	12·1 ± 10·69
9.	127	—	+	+	—	—	38·2 ± 40·69
10.	148	+	—	—	+	+	137·8 ± 115·69
11. Post partum		—	+	+	—	—	216·2 ± 94·40

Lactogenic activity: —, not examined; 0, no activity detected, < 100 ng/ml ovine prolactin equivalents; +, activity detected, ⩾ 100 ng/ml ovine prolactin equivalents.

taken at 64, 85 and 106 days of pregnancy corresponded to a prolactin level in the plasma somewhere between 100 and 500 ng/ml. Table I shows that significant lactogenic activity had been detected in the plasma of four of the five goats by 85 days of pregnancy.

In goat 397, slight lactogenic activity was detected by bioassay in sample 1, taken 52 days before oestrus, while in goat 399, sample 2, taken at oestrus, was lactogenic. These responses correspond with the highest prolactin levels detected by radioimmunoassay at these times (44 ng/ml and 130 ng/ml respectively). In all goats prolactin levels were higher at oestrus than on the following morning, day 1 of pregnancy.

Plasma lactogenic activity in multiparous goats

Six goats in their second or subsequent pregnancy were sampled at three-week intervals, the first sample being taken between days 69 and 87 of pregnancy and the last sample within 24 hours *post partum*, in most cases within one hour. In five of the six goats all samples were positive. In one goat the sample taken at 111 days was negative though the samples taken both before, at 69 and 90 days, and the samples taken later, were positive.

Unfortunately, the majority of the samples taken from multiparous goats had been used for other purposes before it was possible to obtain radioimmunoassay results on them. However, this has been possible for two multiparous pregnant goats and for one goat (116) which proved not to be pregnant. The results are presented in Table II. It will be seen that

TABLE II

BIOASSAY AND RADIOIMMUNOASSAY OF JUGULAR VEIN SAMPLES FROM ONE NON-PREGNANT AND TWO MULTIPAROUS GOATS

Goat no.	Sampling date	Day of pregnancy	Total milk yield (kg)	Plasma prolactin level* (ng/ml)	Mean secretory grade	P†
116	13.1.69	—	4·9	44	1·6	0·368 N.S.
	3.2.69	—	3·6	330	3·1	<0·001
	24.2.69	—	4·3	75	1·6	0·368 N.S.
	17.3.69	—	6·2	320	2·6	<0·001
	21.3.69	—	2·3	127	2·2	<0·001
	3.4.69	—	0	500	3·2	<0·001
277	16.1.69	84	4·4	82	2·0	<0·001
	6.2.69	105	4·6	95	2·1	<0·001
	28.2.69	127	4·6	100	2·4	<0·001
	23.3.69	35 min *post partum*	3·1	300	2·1	<0·001
351	15.1.69	84	0	19	2·6	<0·001
	5.2.69	105	0	<10	2·7	<0·001
	26.2.69	126	0	<10	2·8	<0·001
	21.3.69	149	0	320	3·0	<0·001
	23.3.69	23 h *post partum*	—	260	2·9	<0·001

* Measured by radioimmunoassay.
† Significance of responses compared with explants cultured in the absence of prolactin.
N.S., not significant.

for the non-pregnant goat there is reasonable consistency between the immunoassay and bioassay results. In goat 351 which had ceased to give milk before sampling began, marked lactogenic activity is evident from the first sampling day, day 84 of pregnancy, although the level of prolactin measured by radioimmunoassay was low until the 149th day, the day before parturition. Goat 277 was lactating throughout pregnancy and shows much higher levels of radioimmunoassayable prolactin than any of the other goats studied, levels which could in this case be just high enough to account for the lactogenic activity observed. However, the mean secretory grade is lower *post partum* than on day 127 of pregnancy, despite an increase in plasma prolactin level from 100 to 300 ng/ml, and this may be of significance. The final sample was taken only 35 minutes *post partum* but the half-life of protein hormones in plasma is generally quite short. For example, estimates ranging from 15 to 30 minutes have been obtained for HPL (see Forsyth 1967).

Co-culture of goat foetal and placental tissues with mouse mammary gland

Comparison of the data obtained by bioassay and radioimmunoassay on plasma samples from pregnant and lactating goats suggests a fair measure of agreement between the two methods in the lactating goat, but reveals a far greater biological activity during the second half of pregnancy than can be wholly accounted for by material cross-reacting with antisera

TABLE III

CO-CULTURE OF MID-PREGNANT MOUSE MAMMARY GLAND WITH FOETAL AND PLACENTAL TISSUE OF THE GOAT

Tissues tested	Number of cultures*	Percentage of cultures in which lactogenic activity was detected
Foetal skin	6	0
Foetal small intestine	2	0
Amnion	4	0
Allanto-chorion	6	0
Uterine endometrium	12	0
'Maternal' cotyledon	7	29
'Foetal' cotyledon	7	86
Cotyledon (maternal + foetal)	17	100

* In each culture the tissue tested was combined with four mammary gland explants.

prepared against purified ovine pituitary prolactin. It thus seems possible that the goat placenta produces a lactogenically active hormone. Foetal and placental tissues have, therefore, been examined by co-culturing them, on Medium 199 containing insulin and corticosterone, with explants of mammary gland from mid-pregnant strain A mice. This strain of mice is known to show a lactogenic response to prolactin but not to non-primate growth hormone *in vitro* (see Rivera, Forsyth and Folley 1967). Cultures were continued for four days and at the end of this time all tissues were examined histologically.

The foetal and placental tissues were taken from seven goats killed on days 80, 90, 111 (2), 120, 126 and 140 of pregnancy respectively. Four of the goats were in their first pregnancy. Essentially similar results were obtained in all cases and these are summarized in Table III. The only tissue examined in the presence of which mouse mammary gland explants were induced to secrete was the cotyledon. Activity was unaffected by two medium changes, at approximately six hours and 21 hours after the cultures had been set up. This, together with the good survival of all the tissues in culture, suggests that the active material may have been continuously secreted.

It is not yet possible to say which component of the cotyledon is responsible. Complete separation of maternal and foetal cotyledon is not feasible, but some explants could be identified which contained a preponderance of maternal or foetal tissue and these are designated as 'maternal' cotyledon and 'foetal' cotyledon in Table III. The detection of lactogenic activity in a high proportion of cultures with mainly foetal tissue may indicate this as the site of synthesis.

DISCUSSION

Human prolactin

Purified preparations of HGH have frequently been found to show the biological activities of a prolactin in addition to their growth-promoting abilities (Lyons, Li and Johnson 1960; Chadwick, Folley and Gemzell 1961; Peckham et al. 1968; Forsyth and Folley 1970). The synthesis recently achieved by Li and Yamashiro (1970) adds certainty to the view that prolactin-like activity is intrinsic to the HGH molecule. Nevertheless, radioimmunoassay of HGH in normal lactation and galactorrhoea has shown that its concentration in plasma in the fasting state and its response to stimuli such as arginine or insulin infusion is not significantly different from that in non-lactating women (Board 1968; Spellacy, Carlson and Schade 1968; Benjamin, Casper and Kolodny 1969; Spellacy, Buhi and Birk 1970). The sexual ateliotic dwarfs studied by Rimoin and co-workers (1968) show normal mammary development and lactation although HGH cannot be detected in their plasma by a radioimmunoassay method.

Several other lines of evidence have suggested that a human prolactin distinct from HGH is secreted (for reviews see Forsyth 1969; Pasteels 1969; Boot 1970; Forsyth and Folley 1970). In pregnancy and lactation changes in the acidophil cell population of the human pituitary have been noted (Pasteels 1963; Herlant and Pasteels 1967; Goluboff and Ezrin 1969). Cells which are thought to be prolactin-secreting become numerous at these times but are otherwise scarce, though they have also been detected in a pituitary adenoma associated with galactorrhoea (Herlant and Pasteels 1967; Linquette et al. 1967). Similarly, studies by Peake and co-workers (1969) of a pituitary tumour associated with galactorrhoea have led them to conclude that the tumour was secreting prolactin but very little HGH. Herbert and Hayashida (1970) have used antibody to sheep prolactin in an indirect immunofluorescence method to detect prolactin-containing cells in the normal monkey adenohypophysis.

The experiments of Pasteels and co-workers (see Pasteels 1963, 1969) involving culture of human pituitaries have shown that HGH production, as measured immunologically, declined quite rapidly *in vitro*, while pigeon crop-stimulating activity detected in the culture medium rose. The pattern could be reversed by addition of hypothalamic extracts. Essentially similar results have been obtained by Nicoll and co-workers, using monkey pituitaries (Nicoll *et al.* 1970; Channing *et al.* 1970).

The pigeon has frequently been used in attempts to measure prolactin activity in extracts of human blood, but in general with rather equivocal results (see Meites and Nicoll 1966; Forsyth 1967). Interference by non-specific factors has always been a great problem in the identification and measurement of the response, and the efficacy of the extraction method (Sulman 1956) adopted by several authors (Simkin and Goodart 1960; Pasteels 1969) has been questioned (Kurcz *et al.* 1969).

An alternative method devised by Bates has enabled elevated prolactin activity to be demonstrated in the plasma of six of seven patients with galactorrhoea (Canfield and Bates 1965) and in eight women *post partum*, but not in acromegalics, unless their HGH levels exceeded 400 ng/ml (Roth, Gorden and Bates 1968).

Use of mammary gland organ culture, as described by ourselves (Forsyth 1970; Forsyth and Myres 1971) and Frantz and Kleinberg (1970), enables prolactin activity to be detected in unextracted human plasma and distinguished from activity due to HGH, both by simultaneous measurement of HGH levels in the samples and by incorporation of antisera to HGH into the culture medium. Using this method we have been unable to detect prolactin activity in 26 control subjects and in nine subjects with abnormal breast development without secretion, but have detected it in 12 of 18 women breast-feeding their babies in the early puerperium and in 15 of 20 subjects with abnormal lactation of diverse aetiology.

It is, for a number of reasons, difficult to put values on the prolactin levels observed. In most cases samples have been assayed at only one dilution so no test of regression or parallelism with standards is possible. The subjective scale used to score the responses cannot be considered linear, and the use of parametric statistics is inappropriate. However, if comparison with sheep prolactin standards has any validity, it is clear that human prolactin levels are very high in some cases. In a female patient with galactorrhoea associated with a pituitary tumour, a level of 2 μg/ml plasma was estimated in a 6-point assay. Similarly, in a woman breast-feeding on the seventh day of lactation the level estimated was 0·5 μg/ml.

Isolation of primate prolactin has proved an extremely difficult problem. It may be that the prolactin content of the pituitary is very low, except during pregnancy and lactation, but polyacrylamide gel electrophoresis has failed to identify a separate prolactin band even in pituitaries from women dying at this time (Nicholson 1970). By contrast, gel electrophoresis of a prolactin-secreting pituitary tumour revealed a band with the same electrophoretic mobility as haemoglobin which was tentatively identified as prolactin (Peake et al. 1969). However, no prolactin activity could be found in the haemoglobin band of a pituitary from a woman 24 weeks pregnant (Nicholson 1970).

Recent experiments by Friesen, Guyda and Hardy (1970) provide evidence of very rapid synthesis of small amounts of prolactin by human pituitaries *in vitro*. This material is very similar to HGH in molecular weight, isoelectric point and solubility. It does not cross-react with HGH and much of the other evidence outlined above points to this as a required feature of a primate prolactin molecule. Exploitation of this characteristic may soon lead to the isolation of primate prolactin (Guyda and Friesen 1971).

Placental lactogen in the goat

The human placenta secretes a material (human placental lactogen, HPL) which partially cross-reacts with antiserum to HGH and which combines prolactin-like activity with some of the biological characteristics of a growth hormone (see Josimovich and MacLaren 1962; Kaplan and Grumbach 1964; Forsyth 1967; Selenkow et al. 1969). Similar material is present in the monkey placenta (Grant, Kaplan and Grumbach 1970). Prolactin-like activity has also been demonstrated in the placenta of the rat (Ray et al. 1955; Matthies 1967, 1968; Cohen and Gala 1969; Shani et al. 1970) and the mouse (Cerruti and Lyons 1960; Kohmoto and Bern 1970). A haemagglutination-inhibition method has been used to demonstrate material cross-reacting with HPL in placental fractions of the monkey, rat, dog, pig, horse, sheep, rabbit and cow (Gusdon et al. 1970).

The present experiments suggest that the circulation of the pregnant goat from mid-pregnancy to term contains appreciable quantities of a lactogenically active material which is not detected by a radioimmunoassay for pituitary prolactin. Co-culture experiments indicate that this material is secreted by the cotyledon. It is not yet possible to say which cells are responsible but there is some suggestion that the foetal component is involved.

This material was detected in the circulation of two goats at 64 days of pregnancy. The earliest co-culture experiments were carried out with tissue from a goat on the 80th day of pregnancy, so I do not yet know at what stage of pregnancy this material is first secreted. In humans, HPL has been detected in the foetus as early as 12 days (Beck 1970). It is, however, interesting to note that in the goat, the mid-pregnancy appearance of appreciable quantities of lactogenic activity in the circulation coincides with the time when there is an acceleration of mammary proliferation (Cowie 1971). Hypophysectomy results in abortion in the goat at all stages of pregnancy (Cowie et al. 1963). This suggests that the placental material does not have any appreciable luteotropic effect. Any possible metabolic role will require further investigation, as will the relationship between the placental lactogen and pituitary prolactin. The very low levels of the latter in primiparous goats may be noted and it is possible that the circulating placental lactogen exerts some central feedback effect depressing pituitary prolactin production, in the same way that HPL may be involved in suppressing HGH release during pregnancy (e.g. Spellacy and Buhi 1969). Any such relationship may be modified in goats simultaneously pregnant and lactating.

Attempts to isolate the goat placental lactogen are now in progress.

CONCLUSIONS

Further evidence has been presented to support the view that the human pituitary does secrete a prolactin distinct from HGH and that this material is present in the circulation in quite large quantities in lactation. It has also been shown that during the second half of pregnancy the circulation of the goat contains appreciable quantities of a lactogenically active material apparently originating from the placenta.

Gel filtration studies have indicated that both these materials have molecular weights in the region of 20 000 (Forsyth and Myres 1971; Andrews and Forsyth, unpublished). This approximate molecular weight seems to be common to all the pituitary prolactins and pituitary growth hormones which have been examined (e.g. Andrews 1966) and to human placental lactogen (Andrews 1969). Studies of hormone structure are revealing interesting similarities. Thus, for example, some common amino acid sequences have been noted between ovine prolactin and HGH (Bewley and Li 1970), between HGH and bovine GH (Santomé, Dellacha and Paladini 1968; Fellows and Rogol 1969) and between HGH and HPL (Sherwood 1969; see also Niall 1971). Overlap in biological activity among these different hormones, which has frequently been noted (see Bern and Nicoll 1968; Forsyth and Folley 1970), becomes readily explicable on this basis. This, together with the recognition of placental lactogens in an increasing number of species, leads to the concept that the pituitary and placenta secrete a family of related hormones with important effects both on the mammary gland and on the general metabolism of the mammal.

SUMMARY

The secretory response of rabbit lobulo-alveolar mammary tissue *in vitro* has been used to bioassay lactogenic activity in the plasma of man and goats. (1) In man, prolactin activity distinct from radioimmunoactive growth hormone has been demonstrated in the plasma of 12 of 18 women breast-feeding in the early puerperium, in one male and 14 females of 20 patients with galactorrhoea of diverse aetiology, but not in 26 control subjects without known endocrinopathy or in nine subjects with simple gynaecomastia. (2) In primiparous goats, prolactin bioassay and radioimmunoassay gave similar results for plasma samples taken during a suckling episode *post partum*. However, during the second half of pregnancy, plasma lactogenic activity, unaccounted for by low levels of immuno-reactive pituitary prolactin, was demonstrated by bioassay. Co-cultures of goat foetal and placental tissues with mammary gland of mid-pregnant mice suggest that this activity is due to a lactogenic hormone secreted by some part of the cotyledon.

Acknowledgements

It is a great pleasure to acknowledge the assistance I have received from my collaborators and colleagues Dr G. M. Besser (St. Bartholomew's Hospital, London) and Dr H. L. Buttle, Miss Lynn Francis and Mrs Rita P. Myres (N.I.R.D.). I am also most grateful to the following who have helped me to obtain the human plasma samples studied: Dr I. D. Cooke, Professor A. P. M. Forrest, Professor Russell Fraser, Dr J. F. Hale, Dr M. Hartog, Dr D. P. M. Howells, Professor P. J. Huntingford, Dr J. S. Jenkins, Professor F. T. G. Prunty and the Pathology Laboratories, Royal Berkshire Hospital. Mrs Maureen Barrett gave skilled technical help. The ovine prolactin and HGH used as standards were gifts of the Endocrine Study Section, U.S. National Institutes of Health.

The studies on human prolactin were in part supported by a grant from the Cancer Research Campaign to the late Professor S. J. Folley, F.R.S. in honour of whom this meeting is being held and whose help, unfailing interest, advice and constructive criticism were an important stimulus at the initiation of this work.

REFERENCES

ANDREWS, P. (1966) *Nature (Lond.)* **209**, 155–157.
ANDREWS, P. (1969) *Biochem. J.* **111**, 799–800.
BARNAWELL, E. B. (1965) *J. Exp. Zool.* **160**, 189–206.
BECK, J. S. (1970) *New Engl. J. Med.* **283**, 189–190.
BENJAMIN, F., CASPER, D. J. and KOLODNY, H. H. (1969) *Obstet. & Gynecol. (N.Y.)* **34**, 34–39.
BERN, H. A. and NICOLL, C. S. (1968) *Recent Prog. Horm. Res.* **24**, 681–713.
BEWLEY, T. A. and LI, C. H. (1970) *Science* **168**, 1361–1362.
BOARD, J. A. (1968) *Am. J. Obstet. & Gynecol.* **100**, 1106–1109.

Boot, L. M. (1970) *Int. J. Cancer* **5**, 167-175.
Bradley, T. R. and Clarke, P. M. (1956) *J. Endocrinol.* **14**, 28-36.
Brumby, H. I. and Forsyth, I. A. (1969) *J. Endocrinol.* **43**, xxiii-xxiv.
Bryant, G. D., Greenwood, F. C. and Linzell, J. L. (1968) *J. Endocrinol.* **40**, iv-v.
Canfield, C. J. and Bates, R. W. (1965) *New Engl. J. Med.* **273**, 897-902.
Cerruti, R. A. and Lyons, W. R. (1960) *Endocrinology* **67**, 884-887.
Chadwick, A. (1963) *J. Endocrinol.* **27**, 253-263.
Chadwick, A., Folley, S. J. and Gemzell, C. A. (1961) *Lancet* **2**, 241-243.
Channing, C. P., Taylor, M., Knobil, E., Nicoll, C. S. and Nichols, C. W. (1970) *Proc. Soc. Exp. Biol. & Med.* **135**, 540-542.
Cohen, R. M. and Gala, R. R. (1969) *Proc. Soc. Exp. Biol. & Med.* **132**, 683-685.
Cowie, A. T. (1971) In *Lactation*, pp. 123-140, ed. Falconer, I. R. London: Butterworth.
Cowie, A. T., Daniel, P. M., Prichard, M. M. L. and Tindal, J. S. (1963) *J. Endocrinol.* **28**, 93-102.
Cowie, A. T., Hartmann, P. E. and Turvey, A. (1969) *J. Endocrinol.* **43**, 651-662.
Cowie, A. T. and Tindal, J. S. (1971) *The Physiology of Lactation*. London: Arnold.
Edwards, C. R. W., Forsyth, I. A. and Besser, G. M. (1971) *Br. Med. J.* **3**, 462-464.
Elias, J. J. (1957) *Science* **126**, 842-844.
Elias, J. J. (1959) *Proc. Soc. Exp. Biol. & Med.* **101**, 500-502.
Fellows, R. E. and Rogol, A. D. (1969) *J. Biol. Chem.* **244**, 1567-1575.
Forsyth, I. A. (1967) In *Hormones in Blood*, vol. 1, 2nd edn, pp. 233-272, ed. Gray, C. H. and Bacharach, A. L. London: Academic Press.
Forsyth, I. A. (1969) In *Lactogenesis*, pp. 195-205, ed. Reynolds, M. and Folley, S. J. Philadelphia: University of Pennsylvania Press.
Forsyth, I. A. (1970) *J. Endocrinol.* **46**, iv-v.
Forsyth, I. A., Besser, G. M., Edwards, C. R. W., Francis, L. and Myres, R. P. (1971) *Br. Med. J.* **3**, 225-227.
Forsyth, I. A. and Folley, S. J. (1970) In *Ovo-implantation: Human Gonadotropins and Prolactin*, pp. 266-278, ed. Hubinont, P. O., Leroy, F., Robyn, C. and Leleux, P. Basel: Karger.
Forsyth, I. A. and Myres, R. P. (1971) *J. Endocrinol.* **51**, 157-168.
Frantz, A. G. and Kleinberg, D. L. (1970) *Science* **170**, 745-747.
Friesen, H., Guyda, H. and Hardy, J. (1970) *J. Clin. Endocrinol. & Metab.* **31**, 611-624.
Goluboff, L. G. and Ezrin, C. (1969) *J. Clin. Endocrinol. & Metab.* **29**, 1533-1538.
Grant, D. B., Kaplan, S. L. and Grumbach, M. M. (1970) *Acta Endocrinol. (Copenh.)* **63**, 736-746.
Gusdon, J. P., Leake, N. H., van Dyke, A. H. and Atkins, W. (1970) *Am. J. Obstet. & Gynec.* **107**, 441-444.
Guyda, H. J. and Friesen, H. G. (1971) *Biochem. & Biophys. Res. Commun.* **42**, 1068-1075.
Herbert, D. C. and Hayashida, T. (1970) *Science* **169**, 378-379.
Herlant, M. and Pasteels, J. L. (1967) In *Methods and Achievements in Experimental Pathology*, vol. 3, pp. 250-305, ed. Bajusz, E. and Jasmin, G. Basel: Karger.
Josimovich, J. B. and MacLaren, J. A. (1962) *Endocrinology* **71**, 209-220.
Kaplan, S. L. and Grumbach, M. M. (1964) *J. Clin. Endocrinol. & Metab.* **24**, 80-100.
Kinch, R. A. H., Plunkett, E. R. and Devlin, M. C. (1969) *Am. J. Obstet. & Gynec.* **105**, 766-773.
Kohmoto, K. and Bern, H. A. (1970) *J. Endocrinol.* **48**, 99-107.
Kurcz, M., Nagy, I., Kiss, C. and Halmy, L. (1969) *Acta Physiol. Hung.* **35**, 153-166.
Li, C. H. and Yamashiro, D. (1970) *J. Am. Chem. Soc.* **92**, 7608-7609.
Linquette, M., Herlant, M., Laine, E., Fossati, P. and Dupont-Lecompte, J. (1967) *Ann. Endocrinol.* **28**, 773-780.
Lyons, W. R., Li, C. H. and Johnson, R. E. (1960) *Acta Endocrinol. (Copenh.)* suppl. 51, 1145.

MATTHIES, D. L. (1967) *Anat. Rec.* **159,** 55–67.
MATTHIES, D. L. (1968) *Proc. Soc. Exp. Biol. & Med.* **127,** 1126–1129.
MEITES, J. and NICOLL, C. S. (1966) *A. Rev. Physiol.* **28,** 57–88.
NIALL, H. D. (1971) *Nature New Biol.* **230,** 90–91.
NICHOLSON, P. M. (1970) *J. Endocrinol.* **48,** 639–647.
NICOLL, C. S., PARSONS, J. A., FIORINDO, R. P., NICHOLS, C. W. and SAKUMA, M. (1970) *J. Clin. Endocrinol. & Metab.* **30,** 512–519.
PASTEELS, J. L. (1963) *Archs. Biol. (Liège)* **74,** 439–553.
PASTEELS, J. L. (1969) In *Lactogenesis*, pp. 207–216, ed. Reynolds, M. and Folley, S. J. Philadelphia: University of Pennsylvania Press.
PEAKE, G. T., MCKEEL, D. W., JARETT, L. and DAUGHADAY, W. H. (1969) *J. Clin. Endocrinol. & Metab.* **29,** 1383–1393.
PECKHAM, W. D., HOTCHKISS, J., KNOBIL, E. and NICOLL, C. S. (1968) *Endocrinology* **82,** 1247–1248.
RAY, E. W., AVERILL, S. C., LYONS, W. R. and JOHNSON, R. E. (1955) *Endocrinology* **56,** 359–373.
RIMOIN, D. L., HOLZMAN, G. B., MERIMEE, T. J., RABINOWITZ, D., BARNES, A. C., TYSON, J. E. A. and MCKUSICK, V. A. (1968) *J. Clin. Endocrinol. & Metab.* **28,** 1183–1188.
RIVERA, E. M. (1964) *Endocrinology* **74,** 853–864.
RIVERA, E. M. and BERN, H. A. (1961) *Endocrinology* **69,** 340–353.
RIVERA, E. M., FORSYTH, I. A. and FOLLEY, S. J. (1967) *Proc. Soc. Exp. Biol. & Med.* **124,** 859–865.
ROTH, J., GORDEN, P. and BATES, R. W. (1968) In *Growth Hormone*, pp. 124–128, ed. Pecile, A. and Müller, E. E. Amsterdam: Excerpta Medica Foundation.
SANTOMÉ, J. A., DELLACHA, J. M. and PALADINI, C. (1968) In *Growth Hormone*, pp. 29–37, ed. Pecile, A. and Müller, E. E. Amsterdam: Excerpta Medica Foundation.
SELENKOW, H. A., SAXENA, B. N., DANA, C. L. and EMERSON, K. (1969) In *The Foeto-Placental Unit*, pp. 340–362, ed. Pecile, A. and Finzi, C. Amsterdam: Excerpta Medica Foundation.
SHANI, J., ZANBELMAN, L., KHAZEN, K. and SULMAN, F. G. (1970) *J. Endocrinol.* **46,** 15–20.
SHERWOOD, L. M. (1969) In *Progress in Endocrinology*, pp. 394–401, ed. Gual, C. Amsterdam: Excerpta Medica Foundation.
SIEGEL, S. (1956) *Nonparametric Statistics: for the Behavioral Sciences*, pp. 116–127. New York: McGraw-Hill.
SIMKIN, B. and GOODART, D. (1960) *J. Clin. Endocrinol. & Metab.* **20,** 1095–1106.
SPELLACY, W. N. and BUHI, W. C. (1969) *Am. J. Obstet. & Gynec.* **105,** 888–896.
SPELLACY, W. N., BUHI, W. C. and BIRK, S. A. (1970) *Am. J. Obstet. & Gynec.* **107,** 244–249.
SPELLACY, W. N., CARLSON, K. L. and SCHADE, S. L. (1968) *Am. J. Obstet. & Gynec.* **100,** 84–89.
SULMAN, F. G. (1956) *J. Clin. Endocrinol. & Metab.* **16,** 755–774.

[For discussion, see pp. 184–196.]

MEASUREMENT OF PROLACTIN ACTIVITY IN HUMAN SERUM BY THE INDUCTION OF SPECIFIC MILK PROTEINS *IN VITRO:* RESULTS IN VARIOUS CLINICAL DISORDERS

ROGER W. TURKINGTON

Department of Medicine, University of Wisconsin, Madison

THE studies reviewed earlier (pp. 111–127) have characterized a sequence of molecular reactions which occur in cultured mammary epithelial cells after stimulation with prolactin. Multiple classes of RNA molecules are transcribed and subsequently participate in the synthesis of casein and lactose synthetase, and the formation of these proteins depends upon prolactin with a high degree of hormonal specificity. The synthesis of the milk proteins in the induced state represents 90 per cent of the protein synthesis in these cells (Lockwood, Turkington and Topper 1966) so that the induced rate of milk protein biosynthesis represents a substantial amplification of the original hormonal signal. It has been possible, therefore, to utilize the induction of specific milk proteins in organ cultures of mouse mammary gland as a sensitive and specific bioassay for prolactin activity in human serum. Several features of this assay and the results it has yielded in clinical studies are reviewed here.

CHARACTERISTICS OF THE MOUSE MAMMARY GLAND *IN VITRO* BIOASSAY: INDUCTION OF MILK PROTEINS

The stimulatory effect of prolactin on the induction of [^{32}P]casein or lactose synthetase is not proportional to prolactin concentration in the 5–500 ng/ml range when the prolactin is added to chemically defined Medium 199. As shown in Fig. 1, human serum albumin at a concentration of 1 per cent acts to correct this effect, probably by preventing adsorption of the hormone to the culture dishes during the 48-hour period of incubation. Serum (final concentration, 50 per cent) has an additional effect of making the response linear with respect to the log concentration of prolactin, perhaps by some beneficial effect on the explants. A similar dose–response relationship has been obtained with the induction of N-acetyl-lactosamine synthetase (the galactosyl transferase of lactose synthetase)

(Fig. 2). Lowenstein and co-workers (1971) have also used this reaction to measure prolactin activity in human serum. In our studies the induction of α-lactalbumin as measured enzymatically in the lactose synthetase reaction

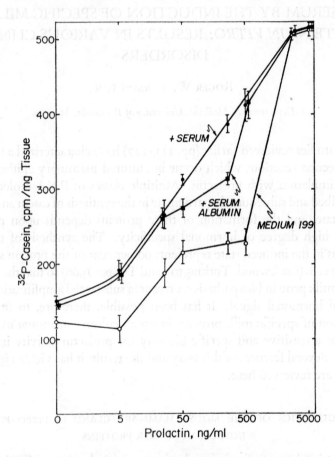

FIG. 1. Synthesis of [^{32}P]casein by mouse mammary explants incubated for 48 hours in various media containing various concentrations of ovine prolactin. Each point represents the incorporation of ^{32}P$_i$ (25 μCi/ml) during the 44–48 hour period. (Reproduced by permission of the publishers of the *Journal of Clinical Endocrinology and Metabolism*.)

has yielded erratic results, perhaps because this small polypeptide readily leaves the extracellular spaces of the explants and cannot be recovered from the medium. The [^{32}P]casein exists in micellar aggregates of 8–9 million daltons (Baldwin 1970) and does not leave the explants during the time periods studied in our experiments. We have chosen the [^{32}P]casein response in preference to the prolactin-induced increase in *N*-acetyl-lactosamine synthetase activity. This is primarily because the [^{32}P]casein assay offers greater precision, as indicated by the data plotted in Fig. 2 (Turkington 1971a).

The hormonal specificity of the induction of [^{32}P]casein is high for prolactin. At concentrations in the physiological or clinical ranges, only human placental lactogen (HPL) and growth hormone (HGH) gave positive

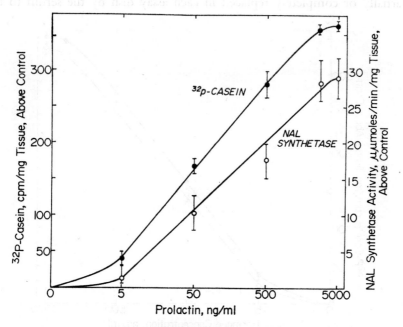

FIG. 2. Rates of [^{32}P]casein synthesis and activity of N-acetyl-lactosamine synthetase in explants incubated for 48 hours in the standard incubation medium containing various concentrations of ovine prolactin (30·3 i.u./mg). (Reproduced by permission of the publishers of the *Journal of Clinical Endocrinology and Metabolism*.)

responses, and no hormones were found to be inhibitory to the action of prolactin (Turkington 1971a). As shown in Fig. 3, HPL was nearly as potent as prolactin when compared by mass. Thus, in the absence of pituitary prolactin the mammary gland *in vitro* assay can serve as a sensitive assay for HPL. It has also been possible to prepare antisera to HPL to determine its contribution to the observed 'prolactin' activity of a given serum sample (Turkington 1971c). The 'prolactin' activity of human growth hormone preparations becomes significant at concentrations of growth hormone above 10–20 ng/ml, but is negligible at concentrations below this level. This complication has been avoided by selecting sera known to contain growth hormone at 10 ng/ml or less, or by treating the sera with anti-human growth hormone prior to assay. Such antisera have been found not to reduce the values for prolactin activity observed in serum having high prolactin activity and low or undetectable serum growth hormone as determined by radioimmunoassay. In our laboratory the culture medium has been made to contain 50 per cent human serum derived from normal blood donors. This serum component of the medium is partially or completely replaced in each assay dish by the serum to be

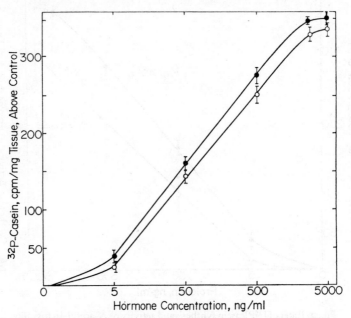

FIG. 3. Rates of [^{32}P]casein synthesis in mammary explants incubated for 48 hours with various concentrations of ovine prolactin (●) or human placental lactogen (o). (Reproduced by permission of the publishers of the *Journal of Clinical Endocrinology and Metabolism*.)

assayed. The activity which is measured, therefore, is that which exceeds the activity present in pooled donor serum. In assays in which serum samples from a large number of apparently healthy controls have been individually tested we have not detected prolactin activity by this technique. Prolactin activity has been found in all patients with clinically significant galactorrhoea and in all patients treated with the drugs shown in Figs. 7 and 8. The sensitivity of the assay (lowest concentration of prolactin which can be reliably detected) has been determined by serial dilution of human and ovine standards to be 2 ng (0·06 m i.u.) per ml of serum.

The variation within assays and between assays has been quite satisfactory in our laboratory. In serum samples with elevated prolactin levels the standard deviation has been less than ±10 per cent, and quite often has been ±6–7 per cent. This bioassay derives its high degree of precision from the fact that all determinations are performed with tissue from a single animal, the hormonal treatment is carried out under carefully controlled conditions *in vitro*, and the end-point is an isotopically labelled protein which can be assayed with a high degree of precision. It does require expertise in organ culture and at present would not appear to yield itself easily to automation or the performance of several hundred assays, as can be achieved with radioimmunoassays. However, it can be completed in three days as opposed to the 7–10 days required for most radioimmunoassays, and it yields values which represent biological activity rather than merely immunoreactivity. It is possible that polypeptide hormones in serum may be inactivated but yet retain their immunological reactivities. It may therefore be more relevant to determine the concentration of biologically active material, if one wants to relate its actions to the patterns of regulation observed under physiological or pathological conditions.

IDENTIFICATION OF THE HUMAN PROLACTIN MOLECULE

In the course of these studies it became evident that the prolactin activity in human sera varied independently of the growth hormone concentration as determined by radioimmunoassay. It was also found that the prolactin activity in the serum of patients with the Forbes-Albright syndrome stimulated [^{32}P]casein synthesis to yield a dose–response curve which was parallel to that of ovine prolactin standards (Turkington 1971*a*), and such sera often contained little or no immunoreactive growth hormone. This result did not constitute proof of the existence of a human prolactin separate from the growth hormone molecule. It did, however, support the proposition that human serum prolactin activity could be assayed using ovine prolactin as a standard. Direct identification of the human prolactin molecule was achieved by physical separation of the biologically active molecule from immunoreactive growth hormone by polyacrylamide gel electrophoresis (Chrambach, Bridson and Turkington 1971). This technique demonstrated that the prolactin activity in the serum of a patient with the Forbes-Albright syndrome was a single, homogeneous protein which was distinct in its molecular weight and net charge from the dominant HGH-B. Fig. 4 is the Ferguson plot showing the logarithm of the relative mobilities of prolactin activity from a Forbes-Albright patient and from a woman lactating *post partum*. The values are plotted against the gel concentration (% T) (Rodbard and Chrambach 1971). The statistical analysis of the results obtained indicated that the prolactin molecule from the patient with a pituitary tumour was identical by polyacrylamide gel electrophoresis to that from the normal, lactating woman. The Ferguson plots for human prolactin and HGH-B were not parallel. The molecular weight at pH 7·8 of human prolactin was found to be 17 200 as compared to 20 800 for HGH-B. The molecular net charges (valence)

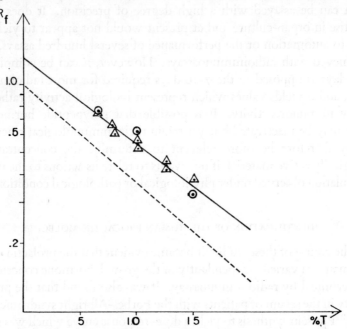

FIG. 4. Ferguson plots of human prolactin in serum from a postpartum lactating woman (o) or from a male patient with a pituitary tumour and galactorrhoea (△) and of human growth hormone (-----). The relative mobilities of human prolactin and human growth hormone are plotted at different gel concentrations (% T) and are shown to be distinctly different with >99 per cent confidence limits.

of the two molecules under the experimental conditions used were $-2\cdot 82$ net protons per molecule of prolactin and $-1\cdot 85$ net protons for HGH-B (pH 7·8). These studies provided the first direct identification of human prolactin as a homogeneous protein distinct from HGH, and formed a firm basis for the conclusion that the biologically active material in human serum which was detected by the mammary gland *in vitro* assay was in fact a distinct prolactin molecule.

EFFECT OF PITUITARY STALK SECTION ON PROLACTIN SECRETION IN MAN

A large number of experimental studies indicate that the synthesis and secretion of prolactin in mammals is chronically inhibited by the hypothalamus, and that prolactin secretion can be initiated when the influence of the hypothalamus is removed. This regulatory relationship of the hypothalamus to the pituitary has been demonstrated by the placement of specific hypothalamic lesions, pituitary transplantation, organ culture of pituitary tissue, administration of certain central nervous system depressant drugs, and pituitary stalk section (Meites and Nicoll 1966). Although the effects of drugs on the secretion of prolactin in the human probably depend upon an effect on the hypothalamus, these agents do not afford as direct an opportunity to study hypothalamic regulation of prolactin secretion as does surgical section of the pituitary stalk. Such a procedure permits the study of prolactin secretion in the same patient before and after a specific experimental variable has been introduced, and can be contrasted with the effects produced by surgical hypophysectomy. In man this procedure is judged to be indicated in selected patients with metastatic breast carcinoma or severe diabetic retinopathy to obtain the therapeutic benefits reported by a number of clinics (Dugger, Van Wyk and Newsome 1958; Ehni and Eckles 1959; Antony et al. 1969; Field et al. 1961).

In the series of 13 patients shown in Table I serum prolactin activity was

TABLE I

EFFECT OF PITUITARY STALK SECTION ON PROLACTIN SECRETION IN 13 PATIENTS

Number of patients	Serum prolactin ng/ml	
	Preoperative	Postoperative
3	0	110–246
2	0	0
8	—	42–508

measured at various intervals after pituitary stalk section for metastatic breast carcinoma or diabetic retinopathy (Turkington, Underwood and Van Wyk 1971). In five of these patients we had the opportunity to measure serum prolactin activity before stalk section also. In none of the sera obtained preoperatively from these patients or from unoperated patients with breast carcinoma or diabetic retinopathy, or from a patient with a chromophobe adenoma, was there detectable prolactin activity. Following pituitary stalk section there was markedly elevated prolactin activity in the sera of a majority of these patients. This secretion was evident by at least five days after stalk section, and was known to persist in one patient for at least 12·5 years. In one of the patients who did not secrete prolactin after stalk section it was concluded that the vascular insult of the surgical procedure resulted in the infarction of all pituitary cells, since this patient had no detectable secretion of other pituitary hormones either. In the second patient who did not secrete detectable prolactin after the operation, the secretion of growth hormone after the induction of hypoglycaemia suggested that some vascular connection of the pituitary to the hypothalamus may have been re-established after the surgical procedure. Secretion of prolactin was not an essential feature of either remission or progression of the activity of breast carcinoma following pituitary stalk section in these patients. Their clinical courses appeared to represent primarily the growth patterns intrinsic to their individual carcinoma cell phenotype, rather than a common response to the presence or absence of prolactin. All four patients with diabetic retinopathy experienced objective arrested progression of their retinopathy and showed increased prolactin secretion. However, additional studies will be required to evaluate any functional role which prolactin may play in the activities of these disease processes. These studies demonstrate that pituitary stalk section in man, as in other mammals studied, usually converts the pituitary gland into a primarily prolactin-secreting organ. The results are consistent with the belief that the human hypothalamus regulates pituitary prolactin secretion primarily through an inhibitory mechanism.

SECRETION OF PROLACTIN BY PATIENTS WITH HYPOTHALAMIC AND PITUITARY TUMOURS

We have also assayed the prolactin activities in the sera of a limited number of patients with tumours of the hypothalamus or pituitary. Elevated prolactin levels have been found in patients with craniopharyngioma, ectopic pinealoma, chromophobic pituitary adenoma and basophilic-chromophobic pituitary adenoma, and in some patients with an eosinophilic pituitary adenoma (Turkington 1971b). Fig. 5 shows the values obtained in patients with the Forbes-Albright syndrome secondary to a chromophobic adenoma. The values are shown before and after surgical removal of the tumour. In three of these nine patients serum prolactin declined to undetectable levels as determined by the *in vitro* [^{32}P]casein assay, but in the majority of the patients in this series significant amounts of prolactin continued to be secreted following the operation. This residual secretory function may represent autonomously secreting residual tumour cells, or the activity of non-neoplastic pituitary cells which were removed from or continued to be separated from the inhibitory influence of the hypothalamus. In one patient in whom the secretion of prolactin was completely terminated by surgical removal of a chromophobe adenoma we measured the rate of disappearance of the biologically active form of prolactin from the serum. As shown in Fig. 6 a serum prolactin concentration of 210 ng/ml was observed in this patient before her tumour was removed. After removal of the tumour the prolactin concentration declined exponentially with a half-life of approximately 15 minutes. The rate of disappearance of prolactin activity then decreased

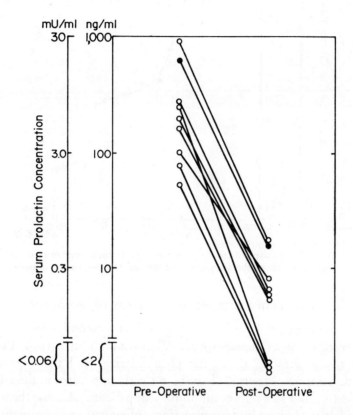

FIG. 5. Serum prolactin levels in patients with a pituitary chromophobe adenoma and galactorrhoea before and after surgical removal of the tumour. Females, o. Male, ●.

markedly. This second phase of disappearance of the hormone may represent an effect of the entry of prolactin from extravascular spaces into the intravascular compartment, the release of prolactin from infarcted pituitary cells, as well as depressed hypothalamic inhibitory function during the post-operative recovery period. The apparent half-life of biologically active prolactin during the period of rapid decline is similar to the half-lives of other polypeptide hormones as determined by their residual serum immunoreactivities (Berson 1968).

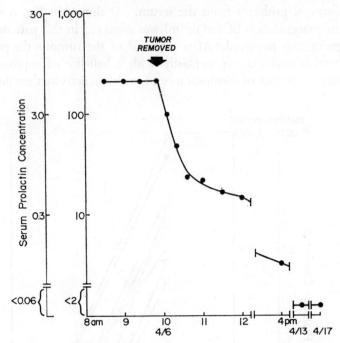

FIG. 6. Time course of the change in serum prolactin activity in a patient with a chromophobe adenoma after surgical removal of the tumour.

EFFECTS OF DRUGS ON THE SECRETION OF PROLACTIN

Galactorrhoea has long been recognized as a complication of therapy with certain drugs (Sulman 1970a; Winnik and Sulman 1956; Platt and Sears 1956). Among these, the phenothiazines and their derivatives, α-methyl dopa and reserpine have been clearly implicated in the aetiology of galactorrhoea in a significant number of patients. Because these agents are widely used in medical practice, it was of interest to measure the serum prolactin levels in groups of patients treated with these drugs. Fig. 7 summarizes the values obtained in 60 psychiatric patients treated with a single phenothiazine or a phenothiazine congener for 2–6 weeks. The patients were selected consecutively to include ten patients in each group. All treated patients in this series were found to have markedly elevated serum prolactin levels. No attempt was made to compare the potencies of these agents in stimulating prolactin secretion on a dose–response basis. The highest serum prolactin levels were observed in those patients treated with perphenazine and fluphenazine, while lower mean values were observed in patients treated with the tricyclic compounds imipramine and amitriptyline. The untreated psychiatric patients and those treated with a variety of other drugs were found to have serum prolactin levels of < 2 ng/ml. A striking feature of the phenothiazine-induced secretion of prolactin was the prolonged duration of secretion which was observed after the cessation of therapy. Serum prolactin levels declined slowly and did not

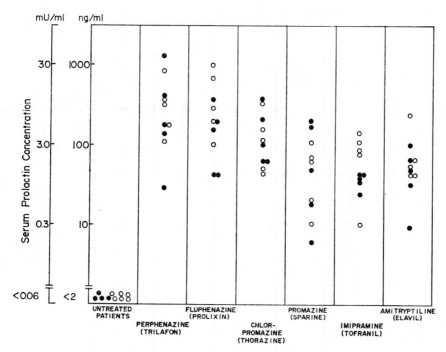

FIG. 7. Serum prolactin levels in psychiatric patients treated chronically with various phenothiazine or tricyclic drugs. Females, o. Males, ●.

reach normal values for 10–15 days in many patients (Turkington 1971c). The possibility that a patient may have ingested tranquillizing medications even one to two weeks previously must therefore be carefully considered in evaluating an elevated value for serum prolactin.

The effect of chronic treatment with α-methyl dopa or reserpine on prolactin secretion is shown in Fig. 8. Each of these agents induced markedly elevated serum prolactin levels in all patients studied.

PROLACTIN LEVELS IN PATIENTS WITH GYNAECOMASTIA

Gynaecomastia has traditionally been considered a sign of endocrine imbalance. Although disorders of the adrenal cortex or testis and of oestrogen metabolism have been clearly implicated in its aetiology in some patients, in a large number of patients the pathogenesis of gynaecomastia remains unclear. Since the normal mammary tissue can respond to the pituitary mammotropic hormone it has been postulated by several investigators that increased prolactin stimulation may lead to growth of the male breast (Sulman 1970b; Paulsen 1968). It was of interest, therefore, to

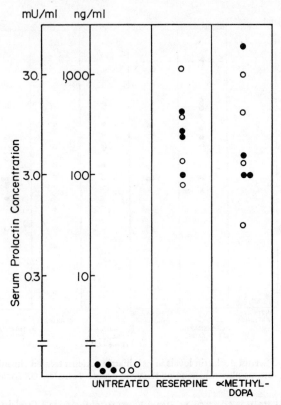

FIG. 8. Serum prolactin levels in hypertensive patients treated chronically with reserpine or α-methyl dopa. Females, o. Males, ●.

measure serum prolactin in patients with gynaecomastia associated with various clinical conditions. Table II lists 27 such patients. Although each category contains a relatively small number of patients, it is clear that an elevated serum level of prolactin is not a general requirement for the development of gynaecomastia. Increased serum prolactin activity has been observed in patients with gynaecomastia who were also treated with phenothiazines or reserpine (Turkington 1971d). However, the elevated prolactin levels in these patients may be regarded as an associated finding and unrelated to the aetiology of the gynaecomastia. These conclusions are consistent with previous experimental results on cultured mouse mammary cells, which indicated that prolactin does not affect the rate of mammary epithelial cell proliferation *in vitro* (Turkington 1968). Thus it would appear that the clinical investigation of the origin of gynaecomastia in patients presenting with this condition should be directed primarily toward the detection of altered patterns of oestrogen metabolism or

TABLE II

PROLACTIN ACTIVITIES IN SERA FROM 25 PATIENTS WITH GYNAECOMASTIA ASSOCIATED WITH VARIOUS CONDITIONS

Associated condition	Number of patients	Serum prolactin activity m i.u./ml
Adolescence	4	<0·06
Bronchogenic carcinoma	4	<0·06
Hepatic cirrhosis	4	<0·06
Traumatic paraplegia	1	<0·06
Idiopathic hypergonadotropuric hypogonadism	2	<0·06
Klinefelter's syndrome	1	<0·06
Chronic azotaemia, maintenance haemodialysis	3	<0·06
Digitalis therapy	1	<0·06
Thyrotoxicosis	2	<0·06
Hypothyroidism	1	<0·06
Chlorpromazine therapy	4	1·70–5·80

other factors which may condition the sensitivity of the mammary tissue to oestrogens.

BIOASSAY OF HUMAN PLACENTAL LACTOGEN

Fig. 9 shows the serum levels of human placental lactogen measured in 25 patients at various stages of pregnancy. The culture media containing the sera were pretreated with sufficient anti-ovine prolactin antiserum to neutralize the pituitary prolactin activity of a human serum standard over the range 0–1000 ng/ml. A relatively small proportion of the initial activity of the samples in the latter half of gestation was removed by this procedure, suggesting that low amounts of pituitary prolactin may be present in the serum during pregnancy. The values for human placental lactogen were determined by comparison with the activity of a purified standard preparation (see Fig. 3). During gestation the maternal venous serum level increases steadily to a maximum value of approximately 5 μg/ml. These values for biologically active hormone are somewhat less than the serum concentrations reported for immunologically reactive HPL in pregnancy serum (Beck, Parker and Daughaday 1965; Saxena, Emerson and Selenkow 1969; Spellacy 1972). Although these results suggest that some biologically inactive material may be assayed by the radioimmunoassay, evaluation of this possibility will require that HPL is assayed by both techniques on the same individual serum samples.

Our experience with the *in vitro* [^{32}P]casein bioassay for prolactin activity indicates that it has a number of features which would appear to make it useful for future clinical studies. It has been illuminating for us to observe how it has united two diverse areas of interest, molecular biology (see

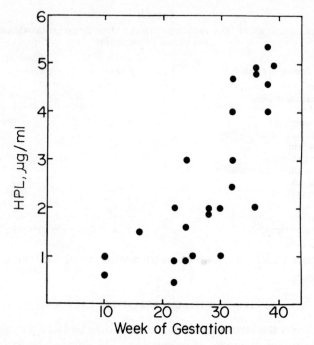

FIG. 9. Serum concentrations of human placental lactogen at various stages of pregnancy as determined by the [^{32}P]casein *in vitro* bioassay.

pp. 111–127) and clinical neuroendocrinology. The mouse mammary gland has pointed out to us that prolactin is an interesting material which can serve to relay changes in the state of the central nervous system to a target cell for the regulation of multi-molecular cell membrane aggregates and the activities of specific genes.

SUMMARY

The induction of [^{32}P]casein or N-acetyl-lactosamine synthetase in mouse mammary explants has been utilized as a highly sensitive and specific bioassay system for prolactin activity in human serum. The sensitivity of the [^{32}P]casein assay was 2 ng/ml for prolactin or human placental lactogen. The rate of induction of specific milk proteins was proportional to the log concentration of ovine prolactin over the range 2–1000 ng/ml, and dilutions of human serum containing prolactin activity paralleled this activity over the same activity range. The within assay and between assay precision of the [^{32}P]casein determination was represented by a standard deviation of less than ±10 per cent. This assay was used to identify, for the first time, the human prolactin molecule, and to separate it from growth hormone and characterize the physical properties of human prolactin by polyacrylamide gel electrophoresis. Serum prolactin levels were found to be elevated in patients with galactorrhoea, pituitary stalk section, pituitary tumours, and during treatment with certain drugs, but not generally in patients with gynaecomastia. The half-life of human prolactin in serum after the removal of a prolactin-secreting tumour was found to be approximately 15 minutes. This assay appears to yield meaningful clinical data on the serum concentration of biologically active human prolactin and should be a useful tool for future clinical research.

REFERENCES

ANTONY, G. J., VAN WYK, J. J., FRENCH, F. S., WEAVER, R. P., DUGGER, G. S., TIMMONS, R. L. and NEWSOME, J. F. (1969) *J. Clin. Endocrinol. & Metab.* **29,** 1238.
BALDWIN, R. L. (1970) *The Milk Proteins.* New York: Academic Press.
BECK, P., PARKER, M. L. and DAUGHADAY, W. H. (1965) *J. Clin. Endocrinol. & Metab.* **25,** 1457.
BERSON, S. (1968) *New Engl. J. Med.* **278,** 751.
CHRAMBACH, A., BRIDSON, W. E. and TURKINGTON, R. W. (1971) *Biochem. & Biophys. Res. Commun.* **43,** 1296.
DUGGER, G. S., VAN WYK, J. J. and NEWSOME, J. F. (1958) *Am. Surg.* **24,** 603.
EHNI, G. and ECKLES, N. E. (1959) *J. Neurosurg.* **16,** 628.
FIELD, R. A., HALL, W. Z., CONTRERAS, J. S. and SWEET, W. H. (1961) *New Engl. J. Med.* **264,** 689.
LOCKWOOD, D. H., TURKINGTON, R. W. and TOPPER, Y. J. (1966) *Biochim. & Biophys. Acta* **130,** 493.
LOWENSTEIN, J. E., MARIZ, I. K., PEAKE, G. T. and DAUGHADAY, W. H. (1971) *J. Clin. Endocrinol. & Metab.* in press (Aug.).
MEITES, J. and NICOLL, C. S. (1966) *Ann. Rev. Physiol.* **28,** 57.
PAULSEN, C. A. (1968) In *Textbook of Endocrinology,* p. 442, ed. Williams, R. H. Philadelphia and London: Saunders.
PLATT, R. and SEARS, H. T. N. (1956) *Lancet* **1,** 401.
RODBARD, D. and CHRAMBACH, A. (1971) *Anal. Biochem.* **40,** 95.
SAXENA, B. N., EMERSON, K., JR and SELENKOW, H. A. (1969) *New Engl. J. Med.* **281,** 225.
SPELLACY, W. N. (1972) This volume, pp. 223-235.
SULMAN, F. G. (1970a) *Hypothalamic Control of Lactation,* p. 67. (Monographs on Endocrinology, vol. 3.) New York: Springer.
SULMAN, F. G. (1970b) *Hypothalamic Control of Lactation,* p. 194. (Monographs on Endocrinology, vol. 3.) New York: Springer.
TURKINGTON, R. W. (1968) *Endocrinology* **82,** 540.
TURKINGTON, R. W. (1971a) *J. Clin. Endocrinol. & Metab.* in press (Aug.).
TURKINGTON, R. W. (1971b) *J. Clin. Endocrinol. & Metab.* in press.
TURKINGTON, R. W. (1971c) *Arch. Intern. Med. (Chic.)* in press.

TURKINGTON, R. W. (1971d) *J. Clin. Endocrinol. & Metab.* in press.
TURKINGTON, R. W., UNDERWOOD, L. and VAN WYK, J. J. (1971) *New Engl. J. Med.* **285**, in press.
WINNIK, H. Z. and SULMAN, F. G. (1956) *Nature (Lond.)* **178**, 365.

DISCUSSION

Pasteels: I would like to ask the three speakers who have described their bioassay systems about specificity control. Dr Frantz has shown that progesterone alone is not active in his mammary gland system, but have any of you tried the simultaneous administration of progesterone and prolactin? I wonder whether the presence of some progesterone in the plasma samples might interfere with the lactogenic activity.

Frantz: That's a very good question. We have not been aware of such an effect of progesterone, but we haven't done enough experiments to give a definite answer. We have studied the question more carefully with regard to oestrogens. Oestradiol, when added to the medium in concentrations ranging from $0 \cdot 5$ ng/ml to $2 \cdot 5$ µg/ml, neither inhibited nor enhanced the effect of ovine prolactin at multiple dose levels; it similarly had no effect on the prolactin activity of plasma samples from any of ten patients.

Turkington: We have added up to 1 µg/ml progesterone simultaneously with prolactin and this does not interfere with the known amount of prolactin in the medium when measured with the casein assay. It does interfere with the amount of α-lactalbumin activity in the tissue but we have not used that as a standard response. We have tested all the known hormones at concentrations that would be encountered in the clinical range (say two to five times above physiological concentrations) and have not found that any of these interfere with the known amount of prolactin in the medium.

Forsyth: We have only looked at one progesterone concentration $(0 \cdot 5$ µg/ml) and at this concentration there was no effect on the response to prolactin observed histologically (Forsyth and Myres 1971).

Prop: We found (unpublished observation) that the addition of only 2 ng/ml of progesterone completely blocks the production of milk proteins in mammary glands from 6–7 week-old CBA virgin mice cultured with the hormones insulin, prolactin and hydrocortisone. The milk proteins were determined immunologically. It should be kept in mind that at the start of these experiments the mammary glands are not lactating nor preparing for lactation, so that they start secretory activity from the resting state. Without the addition of progesterone, however, these immature glands which have not yet developed an alveolar system show histologically vacuolization in the cells of the end buds and a secretion in the ducts that contains milk proteins.

Greenwood: Do you still see milk being produced, when progesterone is present?

Prop: No! Histologically we don't see any milk if progesterone is added.

Meites: I have wondered about the effects of adrenal cortical steroids. None of the three speakers described the effect of holding prolactin levels constant in the culture system and determining the effects of different doses of adrenal cortical steroids on initiation of lactation. Presumably adrenal cortical steroids are present in the blood.

Frantz: The amount of adrenal steroids that we routinely use in the incubation medium is far in excess of what would be encountered even in patients with the most severe Cushing's disease. We have increased the concentrations of adrenal steroids over four-fold ranges to see whether this increased the sensitivity of the assay, and we found no effect of increasing the cortisol concentrations from 20 to 80 ng/ml in the assay on the response to prolactin.

Forsyth: We regularly use corticosterone at 1 µg/ml in the medium. We tried adding cortisol as well at various concentrations, and found no effect.

Turkington: We studied all the structural permutations of the adrenal cortical steroid ring to determine which structures are biologically active, and cortisol is one of those with most activity. We studied these steroids over wide concentration ranges and cortisol at 'physiological' concentrations has nearly 'maximal' activity. We have also added cortisol to all assays at very high concentrations, a thousand times above physiological concentrations, so that any change in the amount of endogenous cortisol in the serum could not influence the result. So I don't think that steroids do influence the activity of prolactin in the assay.

Friesen: In the pituitary stalk-sectioned patients, did you notice any difference in the clinical response of the mammary cancer in patients with high or low prolactin levels? Secondly, did any of those with high concentrations of prolactin have galactorrhoea?

Turkington: There was no correlation of the clinical course of these patients with the presence or absence of serum prolactin. One patient with no prolactin activity had a very good remission; another patient with no prolactin activity showed no response to surgery. In the patients with very elevated prolactin activity, there was also a variable clinical response; some had a good response and others had not. Our limited evidence seemed to indicate that the subsequent clinical course is not primarily a function of whether prolactin is present but more a function of the phenotype of the particular carcinoma. I can't make any general statements about dependence or independence of prolactin in this particular cancer.

None of our patients had galactorrhoea, but all those with mammary cancer had previously been oophorectomized and we know that the response of the mammary gland is conditioned by oestrogenic hormones. A number had had unilateral or bilateral mastectomy which would preclude any galactorrhoea. Of the two women with diabetic retinopathy, neither had galactorrhoea, even though their serum prolactin concentrations were very high. I should add that in all patients with clinically significant galactorrhoea we have never failed to detect elevated serum prolactin. The two diabetic patients therefore did not show a correlation between the response of the mammary gland in the patient and the bioassay result. I don't know why these patients did not have galactorrhoea, except that they had severe diabetes.

Frantz: We encountered high concentrations of prolactin in the absence of galactorrhoea in the patients receiving tranquillizing drugs. Only two out of nine patients had galactorrhoea although all had markedly elevated amounts of prolactin.

Turkington: We also found high prolactin in the phenothiazine-treated group without galactorrhoea.

MacLeod: Could Dr Turkington or Dr Frantz give us some perspective on the importance of oestrogens in the regulation of prolactin secretion? I am very interested in Dr Frantz's observations that chlorpromazine given to *male* patients increased the serum concentration of prolactin and I wonder if he thinks that chlorpromazine is acting through another mechanism, as compared to the direct oestrogenic action, to affect the synthesis and release of prolactin? And I wonder if any of Dr Turkington's patients who underwent pituitary-stalk section for diabetic retinopathy were males and if their prolactin levels increased after surgery?

Turkington: Two were males, and their serum prolactin levels increased after surgery.

Frantz: The question of oestrogens as agents causing mammary hyperplasia is a very interesting one. Some years ago Lyons, Li and Johnson (1958) observed in an extended series of studies that oestrogens were completely ineffective in producing mammary development in rats in the absence of the pituitary or without the simultaneous administration of growth hormone and prolactin. To my knowledge this study has never been done in man. We have studied two hypopituitary men who were shown to have no immunoassayable growth hormone after stimulation, to whom we gave oestrogens in large amounts, 5 mg of ethinyl oestradiol per day, and did serial breast biopsies. We had evidence of some degree of stimulation of ductal growth with oestrogens alone in these patients. However, this was before we were able to measure prolactin and we don't know what the state of endogenous prolactin in these patients may have been. I would assume that chlorpromazine is probably acting directly on the hypothalamic–pituitary axis causing release of prolactin, and it is this which causes the occasional gynaecomastia and lactation in some patients who are being treated with this drug.

Beck: Did your patients have the idiopathic type of hypopituitarism?

Frantz: No. Both had pituitary tumours, removed by surgery.

Turkington: I want to modify this picture a little. We have tested many patients with gynaecomastia from different causes and associated clinical conditions and we have never detected elevated prolactin concentrations in any of them. Furthermore, we have never seen any effect of prolactin *in vitro* on the rate of DNA synthesis in normal mouse mammary tissue, so I would conclude that prolactin is not a growth-promoting substance for the normal mammary gland, nor for a number of experimental mammary cancers, *in vitro* at least. In addition to the requirement you point out, that the pituitary must be present for oestrogens to have any effect, we also have to keep in mind that normal mammary cells have cytosol receptor proteins which bind oestradiol and therefore that presumably these cells are targets for oestradiol. We have also found that oestradiol profoundly regulates the rate of growth of mammary cells in organ culture; at very low concentrations (10^{-12}M) oestradiol can be very inhibitory, and at the concentrations found in the adult (10^{-10}M) it can permit maximal rates of growth.

Frantz: There are also studies by Talwalker and Meites (1961) and Meites (1965) which indicated that in hypophysectomized, gonadectomized and adrenalectomized rats, prolactin and growth hormone alone were capable of causing lobulo-alveolar development.

Cowie: From the experiments of Lyons, Li and Johnson (1958) and of Nandi (1959) there is no question but that prolactin and growth hormone are involved in mammary growth *in vivo*, in both rats and mice, growth hormone being required for duct growth while both hormones are essential for the extensive formation of alveolar structures.

Meites: It has also been shown by Dilley and Nandi (1968) that in the presence of insulin, prolactin alone can induce lobulo-alveolar growth *in vitro* in mammary tissue from the rat. I think there's no question about the direct action of prolactin on the mammary gland.

Turkington: 'Lobulo-alveolar growth' is in fact a contradiction in terms, because lobulo-alveolar development depends on two things: cell proliferation and cell differentiation. In our studies prolactin is required for lobulo-alveolar development, in order to cause differentiation of the cells, but it has no effect on cell division. The insulin in the medium causes the cells to divide.

Meites: I must disagree, because one can take a gland containing only a few ducts and branches, and obtain growth of alveoli and lobules *in vitro* with prolactin (provided that insulin is present).

Turkington: That's primarily cell differentiation, not merely 'growth'.

Nicoll: Dr Turkington, have you tested prolactin in combination with oestrogen and progesterone *in vitro*? Do you find that prolactin is not a proliferative agent in these conditions?

Turkington: Oestrogen profoundly modifies the rate of cell division. By itself it has no effect but in the presence of a growth-stimulating substance, such as insulin, oestradiol modifies the rate at which the cells grow, and depending on its concentration it can permit maximal rates of growth or be profoundly inhibitory.

Nicoll: However, your conclusion of the ineffectiveness of prolactin as a growth-promoting agent appears to assume that the cells have not been exposed to prolactin. When they are taken from intact animals they may have been exposed to some pituitary prolactin, and certainly to some placental prolactin if tissue from pregnant mice is used.

Turkington: That's an inevitable condition of the *in vitro* studies. We can't exclude the possibility that some prolactin was necessary at some time, but the main regulation of the rate of growth of the epithelial cells is through the rate at which they can initiate DNA synthesis and subsequently divide, and exogenous prolactin has no influence on that process *in vitro*.

Friesen: Dr Turkington, wouldn't you like to restrict your view of the role of prolactin to the *in vitro* situation, or do you think it is a general phenomenon? Certainly in the rabbit there is no question that prolactin can cause proliferation of breast tissue *in vivo*.

Forsyth: In Dr Meites' *in vivo* study (Talwalker and Meites 1961), larger doses of prolactin and growth hormone were needed to give lobulo-alveolar development in rats in the absence of steroids than in their presence, so it would seem that dosage must be considered in deciding whether or not a hormone can produce a particular effect.

Turkington: All I can say from my studies is that in the rat and mouse, prolactin has no effect *in vitro* on the rate of cell division. There are reports of a slight increase in the rate of incorporation of tritiated thymidine into DNA in the rat mammary gland in organ cultures, but those studies did not allow for the increase of about 10 per cent in the intracellular concentration of tritiated thymidine in the presence of prolactin. The 1 or 2 per cent increase in the rate of DNA labelling which was observed is explained by the increased intracellular concentration of the precursor and not by any change in the mitotic index.

Prop: Ten years ago I published experiments (Prop 1961a) on organ cultures of whole mammary glands from virgin B.CBA (F_1 hybrids, $C_{57}BL \times CBA$) mice, in which the stimulation of [^{3}H]thymidine incorporation by prolactin was demonstrated. In a medium containing insulin, progesterone and hydrocortisone the [^{3}H]thymidine incorporation on the fourth day was negligible (< 1 per cent), whereas addition of prolactin to this medium together with the above hormones caused lobulo-alveolar growth with a histologically determined increase in cell number and very intense [^{3}H]thymidine incorporation in the nuclei (60–90 per cent of all nuclei in the end buds were labelled after 18 hours of exposure to labelled thymidine). The incorporation of thymidine starts before the fourth day but most experiments were done adding [^{3}H]thymidine on the fourth day and fixing the cultures on the fifth.

Turkington: I have done the same study, measuring the rate of incorporation of [^{3}H]thymidine into DNA, and also looking at the mitotic indices and the labelling index, with and without insulin. We made these observations every four to eight hours for two weeks and found no difference in the presence of prolactin. The point is that we can now examine the interaction of a hormone with a cell more critically and define growth in molecular terms, rather than relying on the ambiguous criterion of increase in tissue mass. Our *in vitro* studies have certainly given us the insight that prolactin acts on the alveolar cell to induce secretory activity in it, and to cause the cell to hypertrophy. We have also found that this is a cell which does not divide.

Cowie: How did you get an alveolar cell in the first instance? Prolactin is necessary to produce alveoli!

Turkington: No; prolactin is not a sufficient stimulus *in vitro* to form the alveolar cell.

Cowie: Then you probably need prolactin plus growth hormone, depending on the species.

Turkington: We tested all these hormone combinations. We start with a simple duct, taking a mammary gland in any physiological state, either during pregnancy, or before pregnancy, or in a virgin preadolescent or adolescent mouse, and the changes are the same. The alveolar cells are formed in response to the hormones insulin and hydrocortisone. Once the alveolar cell is formed, it responds to prolactin by making milk, and we can't detect any activity of prolactin in the culture until that particular cell type is formed. Various factors cause the cells to multiply and to form new prolactin-responsive cells. This type of analysis allows us only to look at one cell cycle, and to look at a cell population where there are probably maximal rates of cell division, so it is a very limited point of view; we don't see the kind of effects one might see in an animal over many days. We can only speak about what happens to mouse mammary cells in organ culture over a few days in chemically defined medium, and that allows us to dissect out some of the actions of prolactin that normally occur in the animal, but they cannot be extrapolated to say that that is the only thing prolactin does in the whole animal. It's conceivable that it acts on the liver to induce a factor which causes the mammary gland to grow, or acts indirectly in some other way, but we can't consider those factors in our *in vitro* situation.

The most critical factor determining growth is the rate at which a cell initiates DNA synthesis; that is, how long the G1 phase of the cell cycle is. We find that all mitogenic agents shorten this phase and that the stem cell very rapidly initiates DNA synthesis, replicates its chromosomes and subsequently divides. We detect no effect of prolactin on that process. It might have an effect in the intact animal.

Meites: But you are excluding all the earlier evidence that prolactin increases DNA in mammary tissue. Dr Denamur has done a lot of this work.

Denamur: Using biochemical methods, we measured the DNA content of whole mammary glands during pseudopregnancy or pregnancy in the rabbit (Denamur 1963). This parameter increases slowly up to the twentieth day of pregnancy and then the rate of increase becomes rapid. The daily injection of small doses of prolactin (12·5 i.u. twice daily) induces lactogenesis (assessed histologically and by lactose production) even at the beginning of pregnancy or pseudopregnancy, but also immediately causes a very important increase in the DNA content. After only five days of prolactin injections it is possible to obtain the same amount of DNA on day 19 as that normally found on day 30 of pregnancy. The results are qualitatively identical in hypophysectomized, adrenalectomized and ovariectomized rabbits. Therefore, it seems clear that prolactin can induce DNA synthesis. Recently Norgren (1967a, b, 1968) also found that large doses of insulin were not able to provoke mammogenesis in rabbits.

Turkington: DNA synthesis occurs in our organ cultures with insulin, but I think it is unlikely to be the physiological mediator of this kind of growth during pregnancy. I believe your data and I think they are very interesting, and they certainly provide convincing evidence that the secretion of prolactin is associated with mammary gland growth, but I don't think they show that prolactin interacts with the mammary cells themselves, and is the mediator for cell growth; it could interact with some other organ.

Cowie: Returning to *in vitro* studies, Rivera (1964), Ichinose and Nandi (1966) and Singh, DeOme and Bern (1970) all agree that prolactin and growth hormone are necessary in the medium for the cellular proliferation and histological organization required for the formation of lobules of alveoli in mammary explants. Indeed most *in vitro* studies on the requirements for mammary growth are in good agreement with the *in vivo* studies on triply-operated rats and mice (Lyons, Li and Johnson 1958; Nandi 1959). During lobulo-alveolar development, moreover, cellular proliferation is intense (Bresciani 1971).

Sherwood: When Dr Turkington adds prolactin alone he gets no effect *in vitro*.

Frantz: How about direct intramammary injections of prolactin? Don't you see greater effects than from systemic injections?

Greenwood: There's no difference between a localized effect and a systemic effect, is there?

Cowie: There is an effect of local (intraductal) injection of prolactin in the mammary gland of the pseudopregnant rabbit, which disproves a systemic effect. The responses, both mammogenic and lactogenic, are limited to the injected sector of the mammary gland. Early signs of the responses may be detected after a couple of days.

MacLeod: This is different from what Dr Turkington described after a matter of hours.

Turkington: We have cultured these glands for as long as two weeks.

Meites: This effect of local injection is one of the most striking demonstrations of the ability of prolactin to increase cell growth, and was demonstrated first by Dr Lyons in 1941 (see Lyons 1958). He showed that an injection of a small amount of prolactin into one of the ducts in a nipple of the rabbit caused localized lactation only in the segment of the gland drained by that duct. We (Meites and Turner 1948) and the Reading laboratory (Folley 1956) confirmed this. There was a large increase in the number of cells in this sector, as first described by Lyons in 1941; this was not only an increase in differentiation. This local growth and lactation demonstrates that prolactin acts locally without intercession of the liver or kidney. If prolactin were acting through a systemic route, you would get generalized lactation and growth throughout the mammary gland.

Greenwood: Dr Turkington could argue that in that situation you already have insulin, which he is regarding as a growth hormone. You are adding prolactin, in local, non-physiological excess, at one spot in an animal which has already been exposed to prolactin; it also has its own insulin and adrenal steroids. I think that's the same situation as he has *in vitro*.

Meites: Insulin is not a growth hormone for the mammary gland. It's necessary for the maintenance of the structural integrity of the gland but it doesn't induce growth.

Greenwood: But a complex of hormones is required; they are all necessary, and when you do the intraductal injection you are just adding an excess of the last one.

Meites: You can take a similar preparation to that used by Dr Turkington, namely mammary tissue from rats, containing only ducts, culture it, add insulin to maintain the integrity of the gland, and there is no growth. When you also add prolactin there is good proliferation of alveoli and lobules (Dilley and Nandi 1968). How can you avoid concluding that prolactin is a growth hormone for the mammary gland, on the basis of the large increase in cell number, and increase in cell differentiation?

Turkington: We now have more objective criteria for determining numbers of cells than counting them under the microscope, cutting through one plane. I've done this enough to realize that it can be treacherous.

Meites: Apparently you are unwilling to accept the earlier data of Dr Denamur, C. W. Turner, H. Tucker and others on the stimulatory effect of prolactin on DNA synthesis.

Turkington: There are two relevant points here. Firstly, the prolactin concentration is very high during lactation, and secondly, there's really no DNA synthesis during lactation at all.

Meites: I doubt that anyone would agree with you who has reported DNA synthesis during lactation.

Wilhelmi: All the experiments with intraductal injection of prolactin are interesting in that they are being done in an intact animal, with an endocrine status that is essentially complete, and this means, if you admit the possibility of the permissive actions and preparatory effects of all the other hormones, that a local increase in prolactin concentration might initiate changes that Dr Turkington wouldn't necessarily see in a much simpler system endocrinologically—a much more meagre system if you like. So you are both in quite a clear sense correct, but one must be conscious of the interpretation.

Turkington: Yes. I am not convinced that these workers have demonstrated *growth* from the local injection, but there's no question that the induction of secretion is a consequence of the local injection.

Prop: In drawing general conclusions from tissue culture work one should always be aware that slight, seemingly insignificant differences in technique can modify the results considerably. This must be the reason for the results that led Dr Turkington to conclude that prolactin has nothing to do with stimulation of DNA synthesis, whereas my results described earlier (p. 189) show the contrary. Since prolactin was the sole variable in my experiments the only conclusion from them can be that under suitable conditions prolactin *does* stimulate DNA synthesis in organ culture. This conclusion is amplified by the experiments mentioned by Dr Meites. Failure to reproduce this effect using a different culture system only means that this system is inadequate for the purpose of showing prolactin stimulation of DNA synthesis. In my experience, Medium 199 is a reasonably good medium for obtaining milk production in organ cultures, but it is a rather bad medium for lobulo-alveolar development, at the basis of which lies DNA synthesis. Our medium contains 5 per cent human serum from young men and 0·5 per cent lactalbumin hydrolysate in Hank's balanced saline, and this makes an excellent medium for the purpose. The use of different media may be one reason for the discrepancies between Dr Turkington's results and mine.

There is a further point. Our experiments were done with whole mammary glands (Prop 1961b). Thus wounding of the duct system of the glands is avoided. I have found wounding to be a stimulus for DNA synthesis as potent as the addition of prolactin, and the cells under this wounding stimulus generally cease to react to hormones. And thirdly, the reactivity of the mammary gland to hormones *in vitro* changes with the age of the donor (Prop 1966) and also with the hormonal status of the donor just before explantation (Ichinose and Nandi 1964). Thus glands with a different history may react differently.

Turkington: We have recently partially purified a protein from mouse and human serum which causes mouse mammary cells to divide. This factor is not insulin or prolactin, or any of the other known polypeptide hormones. I would certainly agree that if you have serum present, of this kind, at least, you would get cell division as a result of this factor. I am surprised that culturing these explants with insulin gives less than 1 per cent of the cells labelled. We routinely find at least 70 per cent.

Prop: This high percentage may be due to a wound-healing reaction. Miss S. E. A. M. Hendrix and I (Prop and Hendrix 1965) have found that insulin causes up to 6 per cent of mitoses arrested by colchicine over 18 hours. In the presence of hydrocortisone or corticosterone the number of [^{3}H]thymidine-labelled cells falls to extremely low levels ($\ll 1$ per cent; Prop 1961a). Wounding, however, is an extremely potent trigger for DNA synthesis; so at the inevitably wounded surface of the cultures we find intensive labelling with [^{3}H]thymidine that is not influenced by hormones.

Turkington: This percentage of labelling occurs with or without all concentrations of progesterone and hydrocortisone in our cultures.

Bryant: I understand that prolactin increases sulphation factor. Does sulphation factor affect any of the assay systems that have been described?

Beck: Ovine prolactin added *in vitro* has no effect on the sulphation factor assay, in our experience.

Greenwood: Human growth hormone and maybe human prolactin seem to 'target' the liver to produce sulphation factor, which not only acts on cartilage but has other targets. It seems to me that the hormones may act directly on the breast, or certainly human prolactin does, but the other metabolic effects of prolactin and growth hormone may be mediated through the sulphation factors.

Friesen: What is the evidence that ovine prolactin enhances sulphation factor?

Greenwood: Dr W. H. Daughaday has a case in whom growth hormone was low and prolactin was high after the removal of a craniopharyngioma. The patient grew rapidly after operation.

Sherwood: Dr Daughaday suggested that prolactin was stimulating growth but he didn't assay sulphation factor in such patients.

Friesen: He didn't have preoperative levels of sulphation factor either. We have seen patients with high prolactin levels and no growth hormone, measured by immunoassay, who are not growing.

Greenwood: I thought he had a situation of high sulphation factor activity in hypopituitary dwarfs with demonstrable plasma prolactin but low growth hormone.

Friesen: This was one case. To make the deduction with confidence one has to show that there is no sulphation factor preoperatively when prolactin levels are low and that sulphation factor is present postoperatively when prolactin levels are high. I believe Dr Daughaday didn't have this information.

Greenwood: What interests me by extrapolation is whether sulphation factor, induced by growth hormone or prolactin, targets the breast as well as other tissues. Because if so, we are going to have a lot of receptors on the breast—for sulphation factors, growth hormone, and prolactin.

Turkington: We have observed some receptor-like activity for prolactin in liver cell membranes. I don't know what its specificity is or whether it represents some cross-reactivity to the intrinsic growth hormone receptor, but it is possible that prolactin could interact with the liver cells.

Frantz: Have you looked at other tissues besides the liver for receptor site activity, such as spleen and kidney?

Turkington: Yes, and it seems to be relatively specific for mammary cells, though as I say we have run into this problem in the liver cell.

Spellacy: Dr Forsyth, you obtained a curious positive prolactin assay at oestrus in one of your goats. Since in women both the basal and stimulated growth hormone concentration in blood is increased in the preovulatory period and since you find a cross-reaction of your assay with growth hormone, did you look specifically at growth hormone levels?

Forsyth: Barnawell (1965) has looked at the effect of bovine growth hormone on the rabbit mammary gland *in vitro* and found that the responses he obtained at high doses (about 5 µg/ml) could be accounted for by prolactin contamination of the bovine growth hormone preparation. Our limited experience is similar. Also, bovine growth hormone is not lactogenic when injected intraductally into the rabbit mammary gland (Bradley and Clarke 1956), and we have confirmed this many times.

Li: Has anyone any information on prolactin activity in the plasma of breast cancer patients? Secondly, Dr Frantz, what was the prolactin activity in HCS (HPL) in your system? And finally, has anyone correlated these *in vitro* assays with the pigeon crop sac assay?

Frantz: Dr Forsyth has made correlations on the rabbit intraductal mammary assay with the pigeon crop sac assay on preparations from Dr Wilhelmi. Our activity for HPL is about 30–50 per cent of that of ovine prolactin by weight in our assay. There is a slight discrepancy here, and also with regard to the prolactin potency of human growth hormone, between Dr Turkington's results and our own. In the few plasmas from breast cancer patients that we have looked at, we have detected no prolactin activity.

Li: Does this suggest that prolactin is not important in breast cancer?

Frantz: No, because a more sensitive assay might possibly reveal significant differences between normal individuals and those with breast cancer.

Forsyth: We have detected elevated prolactin levels in one of six patients with advanced breast cancer but we don't yet know what the significance of this finding is.

Friesen: We have examined about a dozen patients and prolactin levels were within normal limits.

Greenwood: Dr Forsyth, we have detected 'ovine placental lactogen' as a cross-reaction in our anti-ovine prolactin assay. In the later stages of pregnancy, were you picking up a 'caprine placental lactogen'? In your radioimmunoassay, does goat pregnancy plasma dilute like standard prolactin?

Forsyth: We haven't detected any cross-reaction between ovine prolactin and the goat placental lactogen.

Greenwood: So as far as you are concerned the plasma prolactin in sheep pregnancy is authentic ovine prolactin?

Forsyth: Yes.

Bryant: But this was done in the goat; we found the cross-reaction in the sheep. It is also important which antiserum you are using.

Friesen: Were you able to neutralize the biological activity with either anti-HPL or anti-ovine prolactin?

Forsyth: We haven't tried this yet.

Herlant: Dr Frantz, I think your adenoma with high levels of prolactin was not a true chromophobe adenoma. It is very difficult to distinguish between chromophobe adenoma and functional adenoma, because in the latter the granules are very sparse.

Frantz: This tissue was subjected to routine histological studies only, and not to any special stains which might have detected more clearly the presence of prolactin-secreting cells.

REFERENCES

BARNAWELL, E. B. (1965) *J. Exp. Zool.* **160,** 189–206.
BRADLEY, T. R. and CLARKE, P. M. (1956) *J. Endocrinol.* **14,** 28–36.
BRESCIANI, F. (1971) In *Basic Actions of Sex Steroids on Target Organs,* pp. 130–159, ed. Hubinont, P. O., Leroy, F. and Galand, P. Basel and London: Karger.
DENAMUR, R. (1963) *C. R. Hebd. Séance Acad. Sci., Sér. D* **256,** 4748–4750.
DILLEY, W. G. and NANDI, S. (1968) *Science* **161,** 59–60.
FOLLEY, S. J. (1956) *The Physiology and Biochemistry of Lactation.* Springfield, Ill.: Thomas.
FORSYTH, I. A. and MYRES, R. P. (1971) *J. Endocrinol.* **51,** 157–168.
ICHINOSE, R. R. and NANDI, S. (1964) *Science* **145,** 496–497.
ICHINOSE, R. R. and NANDI, S. (1966) *J. Endocrinol.* **35,** 331–340.
LYONS, W. (1958) *Proc. R. Soc. B* **149,** 303–325.
LYONS, W. R., LI, C. H. and JOHNSON, R. E. (1958) *Recent Prog. Horm. Res.* **14,** 219–248.
MEITES, J. (1965) *Endocrinology* **76,** 1220–1223.
MEITES, J. and TURNER, C. W. (1948) *Res. Bull. Mo. Agric. Exp. Stn.* nos. 415, 416.
NANDI, S. (1959) *Univ. Calif. Publs. Zool.* **65,** 1–128.
NORGREN, A. (1967a) *Acta Univ. Lund.* **34,** 1–17.
NORGREN, A. (1967b) *Acta Univ. Lund.* **37,** 1–15.
NORGREN, A. (1968) *Acta Univ. Lund.* **40,** 1–41.
PROP, F. J. A. (1961a) *Exp. Cell Res.* **24,** 629–631.
PROP, F. J. A. (1961b) *Path. & Biol. (Paris)* **9,** 640–645.
PROP, F. J. A. (1966) *Exp. Cell Res.* **42,** 386–389.
PROP, F. J. A. and HENDRIX, S. E. A. M. (1965) *Exp. Cell Res.* **40,** 277–281.
RIVERA, E. M. (1964) *J. Endocrinol.* **30,** 33–39.
SINGH, D. V., DEOME, K. B. and BERN, H. A. (1970) *J. Natl. Cancer Inst.* **45,** 657–675.
TALWALKER, P. K. and MEITES, J. (1961) *Proc. Soc. Exp. Biol. & Med.* **107,** 880–883.

THE CONCENTRATIONS OF HUMAN PROLACTIN IN PLASMA MEASURED BY RADIOIMMUNOASSAY: EXPERIMENTAL AND PHYSIOLOGICAL MODIFICATIONS

GILLIAN D. BRYANT AND FREDERICK C. GREENWOOD

Department of Biochemistry and Biophysics, University of Hawaii, Honolulu

OUR interest in human prolactin stemmed from its possible role, with growth hormone, in the aetiology and maintenance of human breast cancer. This led to two attempts, which we judged unsuccessful, to develop radioimmunoassays for human prolactin in plasma (Stephenson and Greenwood 1967; Stephenson and Greenwood 1969).

The successful outcome of the third attempt is inherent in the title and is due to the generous gift by Professor J. L. Pasteels of a sample of human prolactin isolated from the media of foetal pituitaries in culture. Preliminary (Bryant *et al.* 1971*a*) and more detailed reports (Bryant *et al.* 1971*b*) have been made on the development of a radioimmunoassay, albeit imprecise, for Pasteels human prolactin and its application to plasma. These papers presented evidence, in the guarded jargon of radioimmunoassay, that human plasma contained material which was immunologically distinguishable from human growth hormone but indistinguishable from a human prolactin preparation isolated by Professor Pasteels and his colleagues from the media of human foetal pituitaries in culture.

Evidence for the specificity of the radioimmunoassay was obtained from radioimmunological studies using the Pasteels antigen, labelled and unlabelled, antiserum to the human prolactin sample, labelled human growth hormone (HGH) and human placental lactogen (HPL), and antisera to these unlabelled peptides. Specificity of plasma measurements was sought by applying the radioimmunoassay to plasmas obtained from subjects treated with phenothiazines and from lactating women during suckling. These latter stimuli had been previously shown to cause a marked release of prolactin in the sheep and goat (Bryant, Connan and Greenwood 1968; Bryant, Linzell and Greenwood 1970).

This paper extends the preliminary measurements in plasma to further adduce specificity of the radioimmunoassay for human prolactin. An interdependent study of the release of human prolactin, and a number of other peptide hormones, into the culture media of human foetal pituitary and human pituitary tumour tissues is reported in this volume (Siler, Morgenstern and Greenwood 1972). It may be noted that developments in the biological, immunological and radioimmunological identification of a human and monkey prolactin have been sufficiently rapid and continuous to outdate reviews not yet published (Bryant and Greenwood 1971; Greenwood et al. 1971).

MATERIALS AND METHODS

Details of the radioimmunoassay for human prolactin, growth hormone and placental lactogen are given elsewhere (Bryant et al. 1971b). Results for plasma prolactin given here are expressed in terms of Pasteels human prolactin in ng/ml plasma. It may be noted that a stock plasma from a patient with a pituitary tumour with a high biological activity has been distributed by Dr Griff Ross, National Cancer Institute, Bethesda, to ourselves and others.* We shall subsequently report values in terms of this stock plasma until we obtain a well-characterized sample of human prolactin from our tissue culture studies. Supplementary information on the radioimmunoassay for human prolactin:

Chromatoelectrophoretic and incubation damage. Chromatoelectrophoresis of the labelled human prolactin obtained on routine radioiodination has invariably yielded a value for 'damaged' peptide of 20–30 per cent. Since this does not increase further during incubation and is not reflected in the maximum binding of the label to antiserum it seems likely that some part of the apparent damaged material is an unreactive impurity. It may be noted that immunoreactive growth hormone is detectable as a 0·1 per cent contaminant of the unlabelled human prolactin.

Storage of unlabelled hormone. Aliquots of antigen stored at $-20°C$ in 0·01 ml of phosphate buffer (0·05 M, pH 7·4) at a concentration of 1 mg/ml were used for radioiodinations. Over a six-week period similar yields of reaction and damage were obtained but maximum binding to antiserum progressively decreased.

Quality of antiserum. On the usual criterion the limited antiserum available would be considered virtually unusable for radioimmunoassay, because of its poor maximum binding and slope. The imprecision and lack of sensitivity of the assay is somewhat offset by the high concentration of prolactin in plasma. Immunization to prolactin produced in our own foetal pituitary tissue cultures is in progress.

* See also pp. 396–400.

RESULTS

Plasma prolactin levels in lactation

Plasma samples were taken by repeated venepuncture or by an indwelling catheter at six weeks *post partum* in 13 subjects in established lactation. These subjects were all volunteers with laboratory or nursing experience. A resting sample was taken before breast-feeding. The infants were not

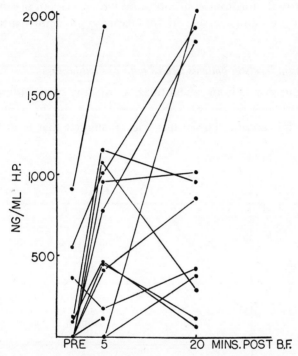

FIG. 1. Prolactin levels in lactation. A sample was taken by repeated venepuncture or by an indwelling catheter inserted before breast feeding. The infant was allowed to suckle between 5 and 10 minutes on each breast and a blood sample taken at 5 and 20 minutes after completion of suckling. Prolactin (HP) levels are expressed in terms of ng/ml of the Pasteels human prolactin.

fed for at least four hours before each experiment to ensure maximum suckling. The infants were allowed to suckle between 5 and 10 minutes on each breast and a sample was taken five minutes after completion of suckling. A further sample was then taken 20 minutes after suckling (Fig. 1). Ten subjects showed increased levels of prolactin five minutes after suckling and of these four showed a further increase by 20 minutes. Four subjects showed a definite decline at 20 minutes after suckling. In only one subject did prolactin show a decline five minutes after suckling. From experiments in sheep and goats it would seem that a prolactin peak would be expected between 2 and 10 minutes after suckling.

More frequent sampling is required to establish peak levels but it is apparent that like endogenous ovine prolactin, the half-life of human prolactin is less than ten minutes. The maximum levels obtained were approximately 2000 ng/ml, which is likely to be a gross overestimate and to reflect the low percentage purity of the standard available.

Plasma growth hormone concentrations in this series of plasmas did not exceed 8 ng/ml, undetectable in the Pasteels prolactin—anti-prolactin system.

Plasma prolactin levels in patients treated with phenothiazines

Six patients for whom phenothiazine therapy was indicated were studied before and 60 minutes after receiving 5 mg of a fluphenazine preparation by mouth. The results (Fig. 2) suggest that in at least three

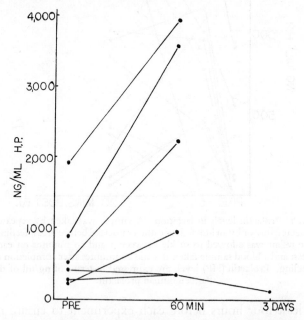

FIG. 2. Prolactin (HP) levels in patients treated with phenothiazines. Six patients for whom phenothiazine treatment was indicated were treated with 5 mg of a fluphenazine preparation by mouth. Samples were collected by venepuncture prior to the drug and 60 minutes after injection.

subjects the drug caused a marked increase in plasma prolactin. It is apparent that a more intensive study with multiple sampling after an intravenous dose is required.

Plasma prolactin levels in patients with galactorrhoea

Plasma samples were kindly supplied by Dr B. W. Webster, Toronto General Hospital, from glucose tolerance tests on a patient presenting with acromegaly and galactorrhoea, the latter of some four years' duration, the former, by history of acral change, sweating and decreased libido, of about two years' duration. Glucose tolerance tests were made both before and three weeks after surgical removal of the tumour and plasma growth hormone and prolactin were measured (Fig. 3). Growth hormone levels

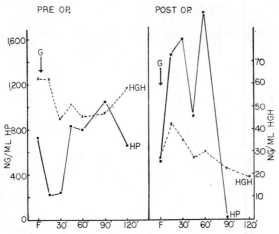

FIG. 3. Prolactin levels during glucose tolerance tests in a patient with acromegaly and galactorrhoea. Glucose tolerance tests were made on the patient before and after operation for removal of a pituitary tumour.
HGH, ×----×; human prolactin (HP), ●——●.

show a general decline rather than absence after hypophysectomy. The prolactin levels prior to the operation rose after glucose at 90 minutes and declined by 120 minutes. However, after hypophysectomy the prolactin response to glucose was much faster and of a greater magnitude, reaching post-suckling levels by 60 minutes. It is apparent that on two criteria much pituitary tissue remained after operation.

The same patient was given an insulin tolerance test before removal of the tumour (Fig. 4). Prolactin poured out into the blood by 30 minutes, with a peak value of 8000 ng/ml. The significance and reproducibility of the prolactin response to glucose and insulin must await further studies. Prolactin release is a sensitive index of stress in the sheep and goat but insulin abolished prolactin secretion in these ruminants. Our interpretation at the present time is that human prolactin also rises in response to stress and may also rise after glucose only when there is an element of stress in the test.

A second patient of Dr Webster's with a pituitary tumour and galactorrhoea was studied. The prolactin response to intravenous oxytocin is shown in Fig. 5. A modest peak level of 165 ng/ml was reached 45 minutes after the injection. Growth hormone levels throughout this experiment

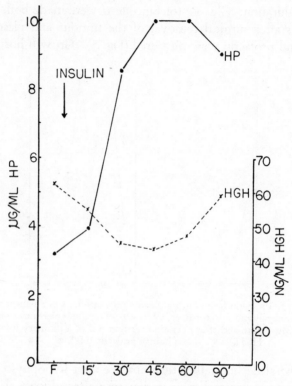

FIG. 4. Prolactin levels during an insulin tolerance test made before operation in a patient with acromegaly and galactorrhoea. HGH, expressed as ng/ml, x ---- x; human prolactin (HP), expressed as μg/ml, ●——●.

were all $1 \cdot 0$ or $< 1 \cdot 0$ ng/ml. By comparison with the levels after suckling and phenothiazine stimulation, this is a very modest rise of prolactin and could represent a mild stress effect. This is further suggested by the results after an intravenous injection of hypertonic saline solution (Fig. 5). This caused a massive release of prolactin at 50 minutes. The growth hormone levels on these samples showed a marginal increase at 10 minutes to $3 \cdot 0$ ng/ml, reaching a peak of $4 \cdot 0$ ng/ml at 20 minutes and then declining to $< 1 \cdot 0$ ng/ml at 40 minutes. Parallel studies in the goat and sheep after intravenous hypertonic saline show a marked release of prolactin. Non-specific stress or a specific release have not yet been distinguished.

Plasma prolactin levels in normal subjects on oral glucose

Four healthy middle-aged subjects were given an oral glucose tolerance test and the samples assayed for growth hormone and prolactin (Fig. 6). One subject showed evident signs of distress during the test and a massive release of prolactin at 60 minutes was observed, whereas the others showed either a fall or a very small rise. The large rise is of similar magnitude and time scale to that shown by the patient in Fig. 3 before removal of the

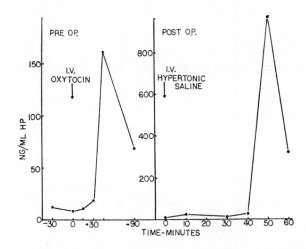

FIG. 5. Prolactin levels in a patient with galactorrhoea. *Left*: Prolactin response to intravenous oxytocin before removal of the tumour. Growth hormone levels throughout this experiment were all $1 \cdot 0$ or $< 1 \cdot 0$ ng/ml. *Right*: Prolactin response to intravenous hypertonic saline solution. Growth hormone levels in these samples showed a marginal increase at 10 minutes to $3 \cdot 0$ ng/ml, reaching a peak at 20 minutes of $4 \cdot 0$ ng/ml and then declining to $1 \cdot 0$ ng/ml at 40 minutes.

tumour. It seems evident that, as in the ruminant (Bryant, Linzell and Greenwood 1970), prolactin secretion is stimulated by stress in man. Like human growth hormone, ACTH, cortisol and the catecholamines, it would seem that prolactin plays some role in glucose homeostasis.

Attempted correlation of biological and radioimmunoassay

Dr A. G. Frantz of Columbia University kindly provided ten plasma samples bioassayed by the method of Frantz and Kleinberg (1970) against ovine prolactin (NIH-P-S-6). Samples were taken from patients with galactorrhoea or healthy postpartum women, sent coded and radio-immunoassayed using Pasteels human prolactin as standard. Results are presented as ranking of the ten plasmas by bioassay and the ranking obtained by radioimmunoassay (Table I). The latter gave consistently higher values and the scales have been adjusted to allow for the evident difference in standards. The standard deviation of the bioassay results allowed some samples to be ranked in more than one range. The radioimmunoassay

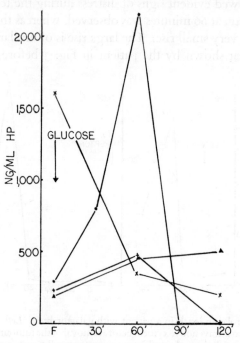

FIG. 6. Plasma prolactin levels in normal subjects on oral glucose. Four healthy middle-aged subjects were given an oral glucose tolerance test and the samples assayed for prolactin. Prolactin (HP) is expressed as ng/ml Pasteels human prolactin.

results are of unknown precision and a repeat assay confused rather than clarified the ranking. The results here are the best possible interpretation

TABLE I

ATTEMPTED CORRELATION OF BIOLOGICAL AND RADIOIMMUNOLOGICAL ASSAYS OF HUMAN PROLACTIN IN PLASMA

Bioassay	0–50*	50–100	100–150	150–200	200
Ranking of plasmas 1–10	1.2.3.4	5.6	6.7†.8.9	7†.9	9.10†
Radioimmunoassay	50–100**	100–150	150–250	200	300
Ranking of plasmas 1–10	1.2.3.4.10†	5	6.8.9	—	7†

* Bioassay (Frantz and Kleinberg 1970), expressed as ng/ml ovine prolactin.
** Radioimmunoassay (Bryant et al. 1971), expressed as ng/ml Pasteels human prolactin.
† Anomalies.

and show two obvious anomalies (plasmas 7 and 10). The results may be regarded as encouraging if we regard the radioimmunoassay as imprecise but showing specificity for the material measured in the biological assay. In neither assay can the results be ascribed to plasma growth hormone.

DISCUSSION

The results reported provide additional evidence that the radioimmunoassay developed for Pasteels human prolactin measures some material in human plasma which is without doubt not human growth hormone but which physiological and experimental manipulation of its concentrations show could fulfil the role of a human prolactin. Like better-characterized prolactins, the human hormone responds vigorously to the stimuli of stress, suckling and phenothiazine. Unlike the prolactins of sheep and goat, injection of insulin stimulates secretion rather than suppresses it and we find this not surprising in view of the importance of fatty acid rather than glucose in the ruminant. Its presence in plasma without a prior or concurrent secretion of immunoreactive growth hormone make it unlikely that we are measuring metabolically altered endogenous growth hormone. Undoubtedly the standard is not homogeneous and the antiserum currently available is both heterogeneous and of low affinity. Even in the present state the assay is probably as precise and as sensitive as the biological assay, less subject to interference by high growth hormone levels and easier to carry out as a routine. The results amply confirm the extensive studies of Professor J. L. Pasteels and his colleagues showing that the foetal pituitary in culture releases a prolactin distinct from a growth hormone. Our results simply show the presence of this material in those plasmas expected on other grounds to contain a human prolactin.

It is customary, and more evident than usual in this work, to conclude that more work is necessary to confirm and extend our findings. A bulk extraction is under way, as described in the companion paper (p. 207), with the production of higher affinity antisera and the amino acid sequence of the purified peptide as endpoints. Concurrent routine assays will be used to confirm and extend the present studies on the stimuli causing a prolactin release in the human.

SUMMARY

Further evidence for the specificity of a radioimmunoassay for prolactin in human plasma is presented. Measurements made after suckling, phenothiazine treatment and stress show that these stimuli, as in animals, cause a marked release of human prolactin. Simultaneous measurements of ten plasmas were made by the biological assay of Frantz and Kleinberg and by the present radioimmunoassay. The results suggest that the same material, human prolactin, is measured and that this material is unrelated to human growth hormone levels.

Acknowledgements

We are grateful to Dr H. Vu for providing us with human lactation plasmas and to Dr J. Pierson for the plasma of patients being treated with phenothiazines.

This work was supported by a contract grant (NIH-69-2190), research grant (NIAMD AM 13217) and a research grant from G. D. Searle and Co., Chicago.

REFERENCES

BRYANT, G. D., CONNAN, R. M. and GREENWOOD, F. C. (1968) *J. Endocrinol.* **41,** 613–614.
BRYANT, G. D. and GREENWOOD, F. C. (1971) In *Methods in Investigative and Diagnostic Endocrinology—Peptide Hormones*, ed. Yalow, R. S. and Berson, S. A. Amsterdam: North Holland. To be published.
BRYANT, G. D., LINZELL, J. L. and GREENWOOD, F. C. (1970) *Hormones* **1,** 26–35.
BRYANT, G. D., SILER, T. M., GREENWOOD, F. C., PASTEELS, J. L., ROBYN, C. and HUBINONT, P. O. (1971a) In *Radioimmunoassay Methods*, pp. 218–222, ed. Kirkham, K. E. and Hunter, W. M. Edinburgh and London: Churchill Livingstone.
BRYANT, G. D., SILER, T. M., GREENWOOD, F. C., PASTEELS, J. L., ROBYN, C. and HUBINONT, P. O. (1971b) *Hormones* **2,** 139–152.
FRANTZ, A. G. and KLEINBERG, D. L. (1970) *Science* **170,** 745–747.
GREENWOOD, F. C., SILER, T. M., MORGENSTERN, L. L. and BRYANT, G. D. (1971) In *Growth Hormone (Proceedings of the Second International Symposium)*, ed. Pecile, A. and Müller, E. E. Amsterdam: Excerpta Medica Foundation. In press.
SILER, T. M., MORGENSTERN, L. L. and GREENWOOD, F. C. (1972) This volume, pp. 207–217.
STEPHENSON, F. A. and GREENWOOD, F. C. (1967) In *Abstracts, International Symposium on Growth Hormone*, p. 29, ed. Pecile, A., Müller, E. E. and Greenwood, F.C. International Congress Series 142. Amsterdam: Excerpta Medica Foundation.
STEPHENSON, F. A. and GREENWOOD, F. C. (1969) In *Protein and Polypeptide Hormones*, p. 28, ed. Margoulies, M. Amsterdam: Excerpta Medica Foundation.

[For discussion, see pp. 217–222.]

THE RELEASE OF PROLACTIN AND OTHER PEPTIDE HORMONES FROM HUMAN ANTERIOR PITUITARY TISSUE CULTURES

THERESA M. SILER, L. L. MORGENSTERN* AND F. C. GREENWOOD

Department of Biochemistry and Biophysics, University of Hawaii and Department of Obstetrics and Gynecology,* Tripler General Hospital, Honolulu

OUR interest in measuring peptide hormones released into the culture media by foetal anterior pituitary tissue stemmed from the observation of Pasteels and his colleagues (Pasteels, Brauman and Brauman 1963; Pasteels 1969) that human growth hormone and human prolactin could be separately identified in such media. Concurrent radioimmunoassay studies on a Pasteels human prolactin preparation and an antiserum to this material have been reported (Bryant *et al.* 1971), culminating in the development of a radioimmunoassay and its application to human plasma (Bryant and Greenwood 1972). The human prolactin material initially available for radioiodination and for use as standard was in short supply and the antiserum was shown to limit the sensitivity and precision of the radioimmunoassay. Since radioimmunoassay requires a few nanogrammes of labelled antigen it was considered desirable to label culture media and choose a protein fraction on the basis of two criteria—the adsorption of labelled peptide hormones to paper and the ability of labelled prolactin to bind to the Pasteels anti-human prolactin antiserum.

A bulk preparation of human prolactin from foetal pituitaries in culture would allow more vigorous attempts to generate a higher affinity antiserum, and permit a more sensitive and precise radioimmunoassay than is at present possible. A greater supply of prolactin would also allow its characterization by bioassay, physicochemical and homogeneity studies including studies of the amino acid sequence—the ultimate test of the separate identity of a human prolactin. From the inter- and intramolecular homologies of human growth hormone, human placental lactogen and ovine prolactin (Niall *et al.* 1971) it seems likely that the human prolactin will resemble human growth hormone, being a single chain peptide with about 200 amino acids and 2–3 disulphide bridges.

Since the bulk preparation required many foetal pituitaries in culture for long periods of time with frequent changes of medium, it was decided to obtain information in addition to the release of human prolactin. A study of the other peptide hormones, FSH, LH, HGH, TSH and ACTH, was undertaken as a function of time in culture and as a function of the age of the foetus when the pituitary was removed and placed in culture. Since these experiments are continuing, this paper will be in the form of a progress report.

Abbreviations. Human prolactin (HP) refers here to material isolated from or measured in media from foetal anterior pituitary tissue or adult pituitary tumour tissue, capable of reacting with an antiserum (anti-PHP) to a human prolactin (PHP) isolated by Professor J. L. Pasteels.

MATERIALS AND METHODS

Tissue cultures

The results are based on 39 human foetal pituitaries obtained within one hour of death of the foetus. The anterior lobe was dissected and cut into explants of approximately 0·5 mm³ which were immediately placed in 3–5 culture dishes. The culture medium (1·0 ml per dish) was as described by Pasteels (1969): 80 per cent Medium 199, 10 per cent chick embryo extract, 10 per cent rabbit serum plus streptomycin, penicillin, mycostatin and oestrogen. The cultures were incubated in an atmosphere of 5 per cent CO_2 and 95 per cent O_2 at 37°C and the medium was renewed twice weekly. The medium was collected at each change and stored at −20°C until assayed for the pituitary hormones or used for the isolation and characterization of human prolactin.

Tissue from two human pituitary tumours from galactorrhoea patients was kindly supplied by Dr B. W. Webster, Toronto General Hospital. The tissue was cut into small pieces and sent at room temperature in 20 ml of culture medium. Upon arrival in Honolulu (approximately 12 hours later) the tissue pieces were placed in culture under the conditions described.

Radioimmunoassays

HGH and HP were measured according to the methods described by Bryant and co-workers (1971). Radioimmunoassays for TSH were done in Dr B. W. Webster's laboratory. FSH and LH levels were assayed by a modification of the double antibody method recommended by Dr W. D. Odell in the reagent kit obtained from the National Pituitary Agency. Some LH assays were measured by a solid-phase procedure using the antiserum and coating procedure provided by Dr R. A. Donald, Medical Unit, The Princess Margaret Hospital, Christchurch. ACTH was measured according to the method of Landon and Greenwood (1968) as modified by Dr Lesley Rees. The antibody used was raised to synthetic 1-24 ACTH and cross-reacts completely with α-MSH and at approximately 1 per cent with β-MSH. A more detailed account of the assays used with the tissue culture media will be published elsewhere. In each case suitable controls of complete tissue culture medium have been studied and the specificity of the assays established.

Isolation and radioiodination of human prolactin from tissue culture medium

The incubation medium from foetal pituitary 4 from days 36-39 was assayed for HGH and a 3-ml sample was fractionated on Sephadex G-75 (55×1 cm) in $0 \cdot 05$M-PO_4^{3-} buffer, pH $7 \cdot 6$. The protein eluted corresponding to a molecular weight of approximately 20 000 was concentrated and radioiodinated (Hunter and Greenwood 1962). After purification on Sephadex G-75 the two fractions with the least chromatoelectrophoretic damage were chromatographed on DEAE Sephadex, A-50 (10×1 cm) at pH $8 \cdot 4$, in $0 \cdot 05$M-BO_4^{4-} with a salt gradient of 0–$0 \cdot 30$M-NaCl. The eluates were analysed by chromatoelectrophoresis and reacted with antiserum to Pasteels human prolactin (anti-PHP). A fraction was used as the label with anti-PHP for a radioimmunoassay of standard PHP, of plasma from normal subjects before and after receiving fluophenazine, and of a lactation plasma sample.

After 37 days in culture a pituitary tumour (T_1) was incubated in protein-free Medium 199 for 10 hours. After incubation the medium was concentrated and purified on a Sephadex G-75 column (55×1 cm) in $0 \cdot 05$M-PO_4^{3-} buffer, pH $7 \cdot 6$ and eluates corresponding to a molecular weight greater than 15 000 were collected and concentrated. (*a*) The concentrate ($0 \cdot 1$ ml) was radioiodinated and purified on a Biogel P-30 column ($220 \times 0 \cdot 9$ cm) in $0 \cdot 05$M-BO_4^{4-} buffer, pH $8 \cdot 4$. The labelled proteins were analysed by chromatoelectrophoresis and a fraction (T_1-1) was reacted with anti-PHP and anti-HGH. (*b*) The concentrate ($1 \cdot 0$ ml) was further purified on a Sephadex G-75 column (53×1 cm) in $0 \cdot 05$M-PO_4^{3-} buffer, pH $7 \cdot 4$. The OD_{280} was measured and the protein concentration calculated against a standard preparation of ovine prolactin (NIH-PS-6). Fraction 12 (T_1-2) was stored in microcaps for routine radioiodination and the HP immunoactivity was measured.

RESULTS

Release of HGH and human prolactin in selected cultures

HGH was measured in 35 of the pituitaries in culture for up to 323 days in culture. The amount released in the medium in the first five days in culture may be greater than 2 µg/pituitary/day. In the first four weeks of culture the HGH concentrations gradually declined to a basal level of approximately 2 ng/pituitary/day. Small fluctuations appeared in the following few weeks but after ten weeks in culture HGH continued at the basal level. Small amounts of HGH were present in the pituitaries of foetuses of 5, 6, 8 and 10 weeks of gestation and its presence in some cultures could still be detected after three and a half weeks in culture.

Studies on the release of prolactin have been completed for 15 pituitary cultures. All the cultures had a high concentration in the first two weeks in culture. The amount of HP then declined but gradually increased in the following months. Fig. 1 shows the release of HGH and HP into the

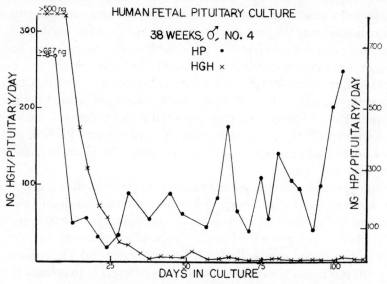

FIG. 1. Release of human prolactin (HP) and HGH into the culture medium by pituitary tissue taken from a human male foetus at 38 weeks of gestation. The amount of HGH and HP released with increasing time in culture is shown.

medium from a normal male foetal pituitary in culture. As with the HGH, a fluctuating pattern of release of prolactin is noted during this period. In cultures older than six months the release of HP is always more than 1 μg/pituitary/day. Pituitaries from foetuses of 5, 8 and 10 weeks of gestation contained HP when first cultured. An analysis of the total amount of HGH and HP released in relation to the age and sex of the pituitary has not been completed.

Release of FSH, LH, TSH and ACTH-MSH

FSH was measured in 36 foetal pituitary cultures and was always present in the first change of medium. It then decreased to amounts below 30 m i.u./pituitary/day in cultures from male foetuses. In the female FSH is released for 6–20 days. It may be noted that FSH has been detected in foetal pituitary cultures obtained at only eight and ten weeks of gestation. One pituitary had a secondary increase of FSH after six weeks in culture and then the concentration decreased again in the following three weeks (Fig. 2). A few other instances of FSH being detected after its initial release have also been noted.

LH was studied in 23 pituitary cultures. Its release did not always parallel the FSH release. One pituitary released LH over three months in culture (Fig. 2). Other pituitaries showed sporadic release of LH after its initial

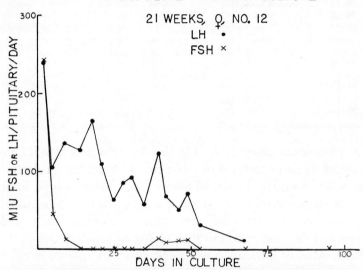

FIG. 2. Release of LH and FSH into the culture medium by pituitary tissue taken from a human female foetus at 21 weeks of gestation. The amount of LH and FSH released is shown with increasing time in culture.

release. All pituitaries studied except the cultures from a five and an eight and a half-week foetus initially had detectable LH, although one pituitary culture from a six-week foetus has released large quantities of LH throughout the 20 days of culture thus far studied. The sex of the early foetuses is now being determined by histological studies of the gonads.

TSH was measured in the first few weeks in culture from five pituitaries over a span of three months in culture. The results showed that TSH was released in the first 28 days in culture, except in a five and an eight-week foetal pituitary culture where it was less than 2 μu./pituitary/day. 14 700 μu. of TSH/pituitary/day were released from a hydrocephalic, 42-week gestation age. Table I gives the release of FSH, LH and TSH into the culture medium from pituitary 4. This pattern is representative of a male pituitary removed at term and then cultured.

TABLE I

RELEASE OF FSH, LH AND TSH IN THE CULTURE MEDIUM FROM A NORMAL MALE PITUITARY WITH INCREASING TIME IN CULTURE

Days in culture	FSH m i.u./pituitary/day		LH m i.u./pituitary/day		TSH μ u./pituitary/day	
0–3	38·4	(1)*	315	(1)*	—	(0)
3–7	<12·5	(1)	<31	(1)	—	(0)
7–10	<16·6	(1)	<41	(1)	—	(0)
10–24	<12·5	(4)	<3	(4)	—	(0)
24–28	<12·5	(1)	<9·4	(1)	18·2	(1)
28–35	<12·5	(2)	<31	(2)	—	(0)
35–38	<16·6	(1)	<41	(1)	6·1	(1)
38–52	<12·5	(4)	<31	(4)	—	(0)
52–57	<10	(1)	<50	(1)	<6	(1)
57–71	<12·5	(4)	<50	(4)	—	(0)
71–75	<12·5	(1)	<50	(1)	<7·5	(1)
75–88	<12·5	(4)	<50	(4)	—	(0)
88–92	<7·8	(1)	<50	(1)	<7·5	(1)
92–106	<7·8	(4)	<50	(4)	—	(0)
106–109	<7·8	(1)	<20	(1)	<9·2	(1)

* Number of estimations.

The radioimmunoassay of ACTH also detects α and β-MSH and the latter are likely to be secreted in tissue culture since, like prolactin, they appear to be normally under inhibition by a hypothalamic inhibiting factor. A scattergram from 34 pituitaries in culture was completed. Initial amounts averaged 5 μg/pituitary/day and these levels gradually decreased and then increased to values of 1–7 μg/pituitary/day by 20 days in culture. Marked fluctuation was evident after six months in culture.

Release of peptide hormones from a human pituitary tumour in culture

A pituitary tumour from a male with galactorrhoea was cultured and its peptide hormone release studied. Fig. 3 illustrates the release of HGH and HP with time in culture. FSH and LH were undetectable. Initially the amount of HP released was more than 10 μg/pituitary/day. In the following three weeks there was a sharp drop in HP release which fluctuated thereafter between 0·5 and 3 μg/tumour/day. HGH levels dropped below 10 ng/tumour/day by the third week of culture.

Isolation and radioiodination of human prolactin from tissue culture medium

The HGH concentration in the medium from the foetal pituitary culture used for the isolation and labelling of HP was 10 ng/ml; therefore, the total HGH content of the 3 ml used for purification was 30 ng. The radioiodinated protein following purification on Sephadex G-75 yielded four fractions (4, 5, 6 and 7) with a chromatoelectrophoretic 'damage' of respectively 48·8, 65·5, 69·0, and 65·0 per cent. After DEAE-Sephadex chromatography of fractions 4 and 5 there were two areas of labelled

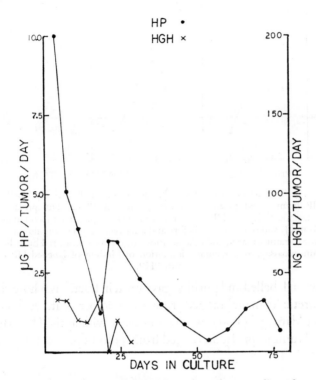

FIG. 3. Release of HP and HGH into the culture medium from a pituitary tumour (T_1) taken from a 24-year-old male with galactorrhoea. The amount of HP and HGH released is shown with increasing time in culture.

protein which bound to anti-PHP serum. The chromatoelectrophoretic 'damage' of these pools (4 and 6) was 35 and 13 per cent. The binding of pool 4 to anti-PHP serum was inhibited by plasma from two normal women, a normal man and a lactating woman (Fig. 4). Plasma taken from the normal women after receiving 1·2 mg fluphenazine intramuscularly caused a three to four-fold increase in the inhibition of binding of labelled tissue culture medium prolactin to anti-PHP serum. The concentration of HP in the plasma from a lactating woman was the highest of the samples measured.

The concentrate obtained from the pituitary tumour culture had a protein concentration estimated to be 8·2 mg/ml. A sample of this

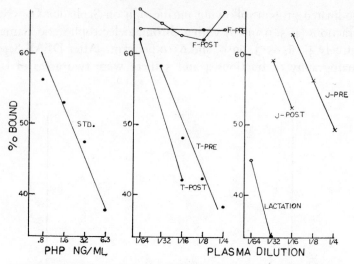

Fig. 4. The detection of plasma prolactin using radioactively labelled HP from foetal pituitary culture 4 and anti-PHP. The prolactin immunoactivity of PHP (std.) and of plasma samples from two females (T and J) and one male (F) before and after receiving 1·2 mg fluphenazine intramuscularly, and of a lactation sample, was measured by radioimmunoassay, as percentage inhibition of binding of labelled HP to anti-PHP.

material when labelled and purified gave protein fractions whose chromato-electrophoretic 'damage' ranged from 10 to 60 per cent. A fraction with 15 per cent 'damage' (T_1-1) was used to react with anti-PHP. Binding of the labelled tumour prolactin ranged from 33 to 15 per cent (Fig. 5).

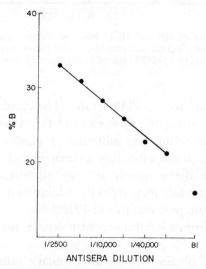

Fig. 5. Titration of protein fraction T_1-1 against anti-PHP. B, binding.

The concentrate was purified further on a Sephadex G-75 column and 12 µg of a fraction (T_1-2) was radioiodinated. A fraction with 10 per cent 'damage' was obtained after iodination. This fraction binds to anti-PHP serum and the binding was inhibited by plasma from a lactating woman. The potency of the fraction was compared in terms of weight with the Pasteels standard (PHP) (Fig. 6). The fraction was found to have 10 per cent of the immunoactivity of this standard.

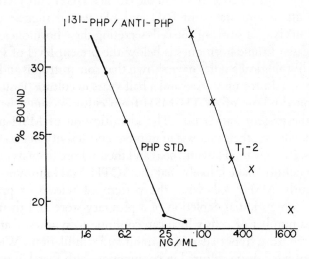

FIG. 6. Antigenic relationship of tumour prolactin and PHP. A purified protein fraction (T_1-2) from the culture medium of a pituitary tumour (T_1) was compared to the standard PHP for inhibition of binding of labelled PHP to anti-PHP.

DISCUSSION

The results presented confirm the observation by Pasteels and his colleagues of the continued release of a human prolactin and a diminished secretion of growth hormone when the human anterior pituitary from a foetus is held in long-term tissue culture. Our complementary studies show the existence of this prolactin material in plasma and permit its provisional identification as the human prolactin. The amino acid sequence of the prolactin molecule when purified and its homologies and heterologies with respect to human growth hormone would provide additional evidence.

The release of the peptide hormones measured here from the foetal pituitary cultures also supports the current concept that the synthesis and release of pituitary hormones are under the control of the hypothalamic releasing and inhibiting factors. The early decrease in FSH, LH, HGH and TSH release on culturing suggests that these hormones are dependent on extrapituitary factors for continued synthesis and release. The addition of hypothalamic extract and TRH (kindly supplied by Dr M. S. Anderson, Scientific Division, Abbott Laboratory) to the medium and the effect on hormone release is under study. The low amounts of HGH present throughout the culture period may represent a basal secretion of this hormone or a cross-reaction of HP in the labelled HGH/anti-HGH system. The intermittent appearance of FSH and LH would suggest either the fortuitous survival of some of the cells secreting these hormones in culture or that release of these hormones is below the present level of sensitivity. However, histological studies have shown that somatotrophs and gonadotrophs may persist for up to two and a half years in culture (Pasteels 1969). The continued release of ACTH/MSH from cultures cannot be dissected out with the present antiserum. The identification of MSH-producing basophils in culture, their increase in number and activity in excess oxygen (Hermanus, Pasteels and Herlant 1964) and their release from hypothalamic restraint in culture make it likely that the ACTH/MSH immunoactivity is predominantly MSH. Likewise, the pattern of release of prolactin is consistent with an initial depletion of a pituitary store and then the persistence of a number of cells which can continue to synthesize and release prolactin free from hypothalamic restraint and stimulation. We are now studying pituitary tissue cultured in conjunction with foetal hypothalamic extracts in order to obtain further evidence that the material is prolactin, by its response to hypothalamic prolactin-inhibiting factor.

The presence of these hormones as early as 5, 6, 8 and 10 weeks of gestation gives new data in the field of pituitary differentiation. The early presence of FSH and LH suggests that the foetal pituitary may play an early role in foetal sex differentiation.

The gentle isolation and radioiodination of material, not growth hormone, from foetal and tumour pituitary cultures, its reaction with Pasteels antiserum and the inhibition of this reaction by plasma containing immunoreactive Pasteels human prolactin, provide more evidence that the Pasteels material is not an isolation artifact. It also affords a useful source of a labelled peptide for routine radioimmunoassays. The antigenic identity of Pasteels human prolactin and the human tumour prolactin has been indicated. It is possible that the final fraction, certainly free of growth hormone, obtained from the tumour media after a pulse of protein-free medium, contains peptide of similar size to prolactin released by the tumour and is responsible for the low specific immunoactivity of this fraction.

SUMMARY

The release of prolactin, growth hormone, FSH, LH, TSH and ACTH-MSH by human foetal adenohypophysial tissue maintained in culture for up to eight months has been studied. Release of hormones from pituitary tissue was demonstrated as early as five weeks of gestation. Human prolactin obtained from the culture media of the foetal pituitaries and from media of adult pituitary tumour tissue in culture has been used in a radioimmunoassay for human prolactin.

Acknowledgements

We would like to thank Mrs C. G. Lino and Miss J. Lamosse for their skilled technical assistance.

This work was supported by a contract grant (NIH-69-2190), research grant (NIAMD AM 13217), National Institutes of Health training grant GM 1039 and Ford Foundation grant 66202.

REFERENCES

BRYANT, G. D. and GREENWOOD, F. C. (1972) This volume, pp. 197–206.
BRYANT, G. D., SILER, T. M., GREENWOOD, F. C., PASTEELS, J. L., ROBYN, C. and HUBINONT, P. O. (1971) *Hormones* 2, 139–152.
HERMANUS, J. P., PASTEELS, J. L. and HERLANT, M. (1964) *C. R. Hebd. Séance Acad. Sci., Sér. D* 259, 6530–6532.
HUNTER, W. M. and GREENWOOD, F. C. (1962) *Nature (Lond.)* 194, 495–496.
LANDON, J. and GREENWOOD, F. C. (1968) *Lancet* 1, 273–276.
NIALL, H. D., HOGAN, M. L., SAUER, R., ROSENBLUM, I. Y. and GREENWOOD, F. C. (1971) *Proc. Natl. Acad. Sci. (U.S.A.)* 68, 866–869.
PASTEELS, J. L. (1969) *Mém. Acad. R. Méd. Belg.* 7, 1–45.
PASTEELS, J. L., BRAUMAN, H. and BRAUMAN, J. (1963) *C. R. Hebd. Séance Acad. Sci., Sér. D.* 256, 2031–2033.

DISCUSSION

Turkington: Some of the blood prolactin levels in your subjects before phenothiazine treatment were fairly substantial, in the 100 ng range. Were you confident that they were not stressed? We have run into the problem of stress also; we have seen patients increase their serum prolactin concentrations up to several thousand nanogrammes.

Greenwood: I would agree that after a single venepuncture the chances of being stressed are very high.

Frantz: Do you measure human growth hormone during the glucose tolerance test, and does that show a similar rise in situations where there seems to be high stress?

Bryant: We have measured this and there was no rise in growth hormone above 8 ng/ml.

Greenwood: But I was the subject, and I can only get up to 10 ng/ml of HGH after two hours of exercise in the fasting state.

Friesen: What was the maximum binding of your ^{131}I-labelled 'prolactin' tracer with anti-ovine prolactin?

Bryant: It was ten per cent in either case. The maximum binding with the homologous system is something like 25 per cent.

Friesen: Were your data expressed in terms of the Pasteels standard? If so, what sort of correction must be made in terms of Dr Frantz's bioassay results?

Greenwood: These results were on the Pasteels standard. I would like to wait to answer the second question until we all assay say Dr Griff Ross's plasma stock and start expressing our results in terms of that. Roughly speaking we have to divide our values by ten, because we are well up in Dr Frantz's ranges.

Friesen: The amount of prolactin found in your plasma after stress would imply that your pituitary has an especially large amount of prolactin! To reach and maintain a prolactin concentration of 1000–2000 ng/ml for an hour your pituitary would have to secrete at least 10 mg of prolactin.

Turkington: I also generate that amount of biologically active material in plasma after stress.

Sherwood: One wonders about the function of so much prolactin in male subjects!

Friesen: I don't understand these fantastic amounts of prolactin detected by bioassay when surgical or post-mortem pituitary tissue contains so little prolactin. Even in monkeys or in humans after surgical hypophysectomy we never found more than 100 µg per pituitary.

Nicoll: Perhaps prolactin is highly susceptible to the action of pituitary proteases.

Greenwood: I think there is a problem there, but the even bigger problem is that one will now have to measure plasma cortisol every time one does a prolactin estimation. I am sure that human prolactin is going to be as sensitive as ACTH to stress.

Friesen: I don't quarrel with the response to stress; it's the magnitude of the response that puzzles me.

Wilhelmi: The apparent magnitude of the response is a function of the purity of the standard preparation. It now seems that we are very close to the point at which a good reference preparation containing human prolactin, presumably a postpartum serum, will be a highly desirable thing to share around.

Greenwood: Dr Griff Ross at the National Institutes of Health has a large amount of plasma from a patient with galactorrhoea. Dr Turkington has assayed this.

Turkington: We obtained a value of 660 ng of ovine equivalent prolactin and 3–4 ng of immunoreactive human growth hormone.

Beck: A supply of a standard preparation is a very critical thing at this time for those involved in measuring prolactin by either bioassay or immunoassay.*

Sherwood: Dr Greenwood, your organ cultures included serum to maintain viability. How were you able to purify hormone from the tissue culture medium in the presence of serum proteins?

Greenwood: To make human prolactin *in situ* for radioiodination the *foetal* pituitaries are cultured in a medium containing chick embryo extract and rabbit serum. However the pituitary tumour is incubated in protein-free Medium 199 for six to eight hours. We can produce enough to use for radioimmunoassay after radioiodination. I might add that this approach has become academic as we have increased our stock of purified, unlabelled prolactin from bulk isolations.

Spellacy: I am interested in the high prolactin levels that you found in the foetus and Dr Friesen's high levels in the neonate; could you speculate on prolactin's function in the foetus and neonate? Is this also a response to high circulating oestrogen levels within the foetal compartment, as has been speculated on for growth hormone, or is there a specific function, other than breast development and breast secretion, for this hormone?

Greenwood: We know it targets the breast in combination with other hormones when present, but the breast must be only one of its targets. As I said before, I don't like having a hormone secreted in stress which has no physiological effect. We have wondered whether this is an inappropriate reaction, and that there isn't any target tissue in the body at that time which is accepting prolactin, so that it is biologically inactive. I can't believe that, because growth hormone released in stress has been shown to affect carbohydrate tolerance and hence to be physiologically active (Conn *et al.* 1954; Landon and Greenwood 1969). I think it is a question of trying to identify target tissues for prolactin other than the mammary gland. In the foetus the levels simply reflect the pituitary cells that Dr Herlant and Dr Pasteels have identified; what the role is, I haven't any idea.

Pasteels: We have evidence of prolactin cells in the hypophysis of human neonates (by differential staining). At this period of life there is a transitory mammary stimulation. However, we do not know whether this prolactin secretion has any important physiological significance or not.

---

* For further discussion of standard preparations, see General Discussion, pp. 396–400.

Beck: In lower vertebrates there is evidence for a role in salt and water metabolism, as Dr Nicoll will be describing (pp. 304–306).

Meites: What about the breast growth and lactation that occur in newborn babies? Dr Lyons (1937) showed that newborns had levels of prolactin in their urine which he could detect by the local pigeon crop test. If oestrogens from the mother cross the placenta, they could stimulate prolactin secretion by the foetal pituitary, and this could be responsible for the breast development and secretion that occurs. Prolactin from the mother may also act on the foetal breast. The prolactin in the urine of the babies appears only for a few days after birth.

May I ask Dr Greenwood to clarify the question of oxytocin and prolactin release? At various times he has reported that oxytocin stimulates prolactin secretion in goats and sheep, but at a recent Workshop on Prolactin (at the National Institutes of Health, January 1971) he appeared to take it all back!

Greenwood: This is because it is not simply a yes-no situation. If you inject sheep or goats with oxytocin using the carotid artery you can sometimes get a pulse of prolactin in the jugular vein, and we have recently obtained the same effect in man. The *magnitudes* of the responses obtained are nothing in comparison with the suckling stimulus, so the answer is not that suckling is producing oxytocin which causes the release of prolactin. All our data can be interpreted in terms of suckling producing oxytocin and *also* producing prolactin. The direct effect of intravenous oxytocin on prolactin is a non-specific one.

Beck: While we are talking about interpretation of data, how convincing is the evidence that an infusion of hypertonic saline raised the levels of prolactin?

Nicoll: A word of caution should be given about effects with hypertonic saline. Some of my neurophysiological colleagues emphasize that relatively small infusions of hypertonic saline can cause massive discharge of neurons in the brain. This could presumably cause non-specific discharge of pituitary hormones.

Friesen: Has Dr Forsyth or Dr Bryant looked at serum prolactin levels in the newborn or the foetus in either the goat or the sheep, and were the levels elevated? And what about prolactin in the rat foetus, Dr Meites?

Bryant: When we first developed the ovine assay we did random samples in sheep; among these were a few over parturition. We also took blood samples from the lambs and they were elevated; however they were much lower than in the mother, but she had extremely high levels at parturition.

Nicoll: Do you think prolactin could be a foetal growth hormone?

Bryant: It is the stress of being born, perhaps!

Sherwood: Growth hormone secretion is elevated during the neonatal period.

Forsyth: Dr H. L. Buttle has some results on prolactin concentrations in pregnant sheep and their foetuses from 80 days to just before term (D. A. Nixon and H. L. Buttle, personal communication). Maternal prolactin levels were similar to those in the pregnant goat, remaining quite low until just before parturition. In the foetus significant prolactin concentrations were found only within 12 days of delivery and were always lower than those in the mother.

Meites: We examined prepubertal prolactin levels in male and female rats in the pituitary and in the blood and they are very low before the onset of puberty (Voogt, Chen and Meites 1970). They rise markedly when oestrogen starts to be secreted. We have not looked at the foetus.

Frantz: Dr Forsyth and Dr Bryant have found that prolactin concentrations do not rise during pregnancy in either sheep or goats; they increase only at the time of parturition. It would be interesting to look at this in cows, because it is common in dairying to maintain cows in a state of lactation while they are pregnant, and there's no reason to believe that lactation decreases the ability of cows to become pregnant. Presumably a cow can be kept in continual lactation right through pregnancy, parturition and the next pregnancy. I would be interested to know if the concentrations of prolactin in a lactating pregnant cow are different from those in a non-lactating pregnant cow.

Greenwood: Dr W. D. Odell and his colleagues have studied bovine prolactin levels by radioimmunoassay (personal communication). They may have the data that Dr Frantz seeks. However, the difficulties in getting non-stressed levels in the cow are the same as in the sheep and goat.

Forsyth: Dr Buttle and I have measured prolactin by both radioimmunoassay and bioassay in only two multiparous pregnant goats, one of which had stopped lactating and the other was lactating right up to term in the subsequent pregnancy. The animal that was lactating had higher plasma prolactin levels, measured by radioimmunoassay, than either the non-lactating multiparous goat or any of the primiparous goats we have studied (see p. 159).

Meites: We measured prolactin in the pituitaries of pregnant lactating rabbits (Meites and Turner 1948). If you reduce the young of postpartum rabbits to about two per mother, they continue to lactate and will become pregnant if you breed them on the day of parturition. The prolactin levels in the pituitary are just as high as in postpartum animals that are lactating and not pregnant. These are pituitary values, but prolactin pituitary values often correspond to serum values under many conditions.

Frantz: In fact the correspondence is much better for prolactin than for growth hormone; is that correct?

Friesen: In our experience in man, we find no good correlation between pituitary content and serum concentrations of prolactin for the pituitaries that were studied after surgical hypophysectomy.

Meites: In the rat under many conditions—pregnancy, lactation, ovariectomy, oestrogen administration—the two go together very well. Not under every condition, of course, particularly under acute stimulation (suckling).

Frantz: I think it's possible, although this cannot be tested, that there might be a correlation between acute depletion of pituitary prolactin in man and the rise in plasma levels; the kind of thing that has not always been shown for growth hormone in rats and other animals.

REFERENCES

CONN, J. W., FAJANS, S. S., LOUIS, L. H., SELTZER, H. S. and KAINE, H. D. (1954) *Recent Prog. Horm. Res.* **10,** 471–488.
LANDON, J. and GREENWOOD, F. C. (1969) In *Progress in Endocrinology*, p. 595, ed. Gual, C. Amsterdam: Excerpta Medica Foundation.
LYONS, W. R. (1937) *Proc. Soc. Exp. Biol. & Med.* **37,** 207–209.
MEITES, J. and TURNER, C. W. (1948) *Res. Bull. Mo. Agric. Exp. Stn.* nos. 415, 416.
VOOGT, J. L., CHEN, C. L. and MEITES, J. (1970) *Am. J. Physiol.* **218,** 396–399.

IMMUNOASSAY OF HUMAN PLACENTAL LACTOGEN: PHYSIOLOGICAL STUDIES IN NORMAL AND ABNORMAL PREGNANCY

William N. Spellacy

Department of Obstetrics and Gynecology, University of Miami Medical School, Florida

In 1962 Josimovich and MacLaren noted that human retroplacental blood contained a protein which was immunologically similar to human growth hormone and had the biological activity of both growth hormone and prolactin. This protein hormone, human placental lactogen (HPL), has since been isolated and purified. Subsequent investigations have included studies of the chemistry and synthesis of HPL, its biological activity and function, and variations in the concentrations of circulating hormone in normal and abnormal pregnancies (Spellacy 1969). It is this latter problem which has been a prime concern in our laboratory and the purpose of this paper is to describe a method of assay for HPL and the results of measurements of HPL in normal and abnormal pregnancies.

IMMUNOASSAY

A purified HPL protein was obtained and mixed with complete Freund's adjuvant and rabbits were injected subcutaneously at weekly intervals. After four injections the animals were bled from the ear artery and the antiserum was frozen at $-20°C$. The gamma globulin fraction of the serum was later separated using 18 per cent sodium sulphate and then resuspended in $0.1M$-sodium carbonate. The HPL antibody was then chemically coupled to cyanogen-activated G-25 Sephadex as described by Wide, Axén and Porath (1967) and the Sephadex–HPL antibody complex (S-HPL) was washed. A more highly purified HPL preparation* was labelled with ^{131}I using the method of Greenwood, Hunter and Glover (1963). The labelled protein was placed over a G-100 Sephadex column (60 × 1 cm) using $0.07M$-barbitone buffer at pH 8.6. Three protein peaks were

* The HPL used as a standard in these assays was Purified Placental Protein Lot 438-7C-125 which was kindly supplied by Dr Paul H. Bell of the Lederle Research Laboratories of Pearl River, New York.

separated by this procedure, as shown in Fig. 1. A test of the immunological reactivity of each peak was made, measuring the ability of HPL antibody to bind with the labelled protein. As can be seen from Fig. 1, the third peak consistently gave the highest immunological binding (64·3 per cent). This third fraction was used as the tracer ($[^{131}I]$HPL) in the

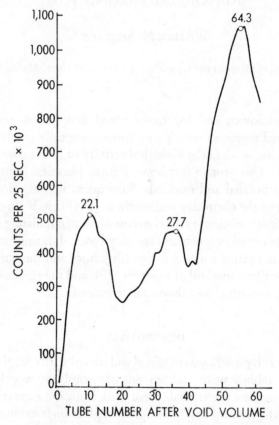

FIG. 1. The purification of $[^{131}I]$HPL on a second G-100 Sephadex column gives three distinct peaks. The immunological reactivity of each peak is given as the percentage binding to antibody and the third peak contains the most reactive protein (64·3 per cent).

assay. The S-HPL was then diluted to a concentration which would bind 50 per cent of the $[^{131}I]$HPL during four hours of incubation with continuous shaking. The S-HPL was usually diluted to 1:400 or 1:500. The assay involved the addition of standards or unknown serum, labelled $[^{131}I]$HPL and S-HPL. In each assay of fifty duplicate serum samples, there was a standard curve varying from 0·5 to 50 μg of HPL per ml and high and low control sera for quality control. The mixture was placed on a shaker and incubated for four hours and then centrifuged for five minutes at 2000 rev./min. The supernatant was discarded and the Sephadex precipitate was counted in a well-type gamma counter. A typical standard curve is shown in Fig. 2. An increase in the serum HPL would

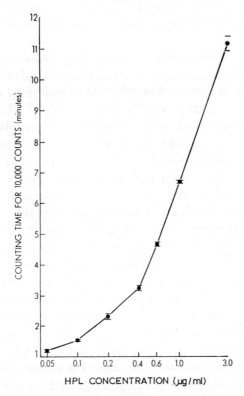

FIG. 2. Typical standard curve for HPL using the Sephadex solid-phase radioimmunoassay.

compete for the antibody binding sites with [^{131}I]HPL and ultimately displace the [^{131}I]HPL from the S-HPL; thus the time to reach a certain number of counts would be longer.

The solid-phase radioimmunoassay provides a rapid assay system where one technician can obtain the results on 100 serum samples in one day. The sensitivity of the assay expressed as the smallest amount of hormone which can be distinguished from no hormone is 0·05 µg/ml. The coefficient of variation is 7·2 per cent and the accuracy of the method as tested by experiments on the recovery of HPL added to male serum is 93–107 per cent between 1 and 50 µg/ml.

CONTROLLING FACTORS FOR HPL

The first studies were designed to try to discover what factors controlled the levels of HPL in maternal serum so that physiological studies could be undertaken in both normal and abnormal pregnancies. It was found that the concentration of HPL was unaffected by the time of day, sex of the foetus, presence of congenital anomalies in the foetus or maternal haemoglobin concentration (Spellacy, Carlson and Birk 1966; Spellacy et al. 1971a).

Since it has been shown that serum prostaglandin levels are elevated during pregnancy (Karim 1968), and also that prostaglandin can alter the secretion of human growth hormone (MacLeod and Lehmeyer 1970) the effects of these substances on the serum levels of HPL were studied. Term pregnant women were placed on bed-rest and an indwelling catheter was placed in the antecubital vein of each arm. One of the catheters was connected to an infusion pump for delivery of the test substance. The other catheter was kept open with heparinized saline and was used for the withdrawal of blood. Since the prostaglandin solutions would cause uterine contractions which might alter the level of HPL, a control group was included who received oxytocin and they, too, had uterine contractions during the infusions. During the first sixty minutes, the women received only a saline infusion and after this baseline was established, infusion of the test substance was begun and the dosage was increased at specified time intervals. Eight serial blood samples were drawn during the five-hour infusion. A total of forty-two women were studied. There were sixteen who received prostaglandin F_{2a} (2·5 to 20 μg/minute), six who received prostaglandin E_2 (0·3 to 2·4 μg/minute)* and twenty who received oxytocin (0·5 to 4 mU/minute). The results of these studies are shown in Table I and Fig. 3. It can be seen that there were no significant changes in the maternal levels of HPL with any of the three drug infusions.

The next area of control to be investigated was the effects of eating on HPL. Since HPL has some biological activity related to growth hormone and may, therefore, be involved in metabolism, and since the major metabolic fuel for the foetus is the glucose crossing the placental barrier by facilitated diffusion, the effects of varying concentrations of glucose on the maternal level of HPL were studied. Blood samples were drawn in a fasting resting state and then the maternal blood glucose concentration was acutely altered by either the intravenous injection of 25 g of glucose as a 50 per cent solution ($N=27$), or the intravenous injection of insulin

---

* The prostaglandin solutions were generously supplied by the Upjohn Pharmaceutical Company of Kalamazoo, Michigan.

TABLE I

STATISTICAL STUDIES OF SERUM LEVELS OF HUMAN PLACENTAL LACTOGEN IN RESPONSE TO INTRAVENOUS INFUSIONS OF PROSTAGLANDIN $F_{2\alpha}$ OR E_2 OR OXYTOCIN

Serum HPL
μg/ml

Drug infused	Time in minutes:	Control			Drug				
		0	30	60	90	120	180	240	300
Prostaglandin $F_{2\alpha}$ ($N=16$)	Mean	8·4	7·3	7·4	8·1	8·3	7·9	7·7	8·2
	S.D.	1·7	1·8	1·6	1·9	2·0	1·5	1·9	1·9
	t	—	—	—	1·798	1·947	1·422	0·607	1·111
	P	—	—	—	N.S.	N.S.	N.S.	N.S.	N.S.
Prostaglandin E_2 ($N=6$)	Mean	8·4	7·7	8·4	8·2	7·7	7·6	8·1	8·5
	S.D.	2·8	2·1	2·1	1·4	2·6	2·2	2·3	2·3
	t	—	—	—	0·293	1·951	1·785	0·769	0·352
	P	—	—	—	N.S.	N.S.	N.S.	N.S.	N.S.
Oxytocin ($N=20$)	Mean	8·4	8·4	8·3	8·8	8·3	8·1	8·3	8·7
	S.D.	2·2	2·7	2·2	2·4	2·4	2·6	1·9	2·2
	t	—	—	—	1·529	0·401	0·218	0·171	1·259
	P	—	—	—	N.S.	N.S.	N.S.	N.S.	N.S.

N.S., not significant.

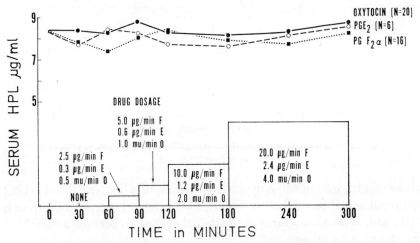

FIG. 3. The mean HPL levels during the intravenous infusion of prostaglandins $F_{2\alpha}$ or E_2 or oxytocin in 42 term pregnant women.

(0·1–0·3 units/kg) to create a period of hypoglycaemia ($N=23$). Serial blood samples were drawn during these glucose changes and it was found that the serum concentration of HPL changed significantly in a pattern which was opposite to that of the glucose change. Thus, when the blood glucose rose from 81·1±1·9 to 243·7±4·0 mg per cent the serum HPL fell from 7·4±0·4 to 6·9±0·4 μg/ml ($t=3·214$; $P<0·01$). Conversely, when the blood glucose fell from $75 \cdot 6 \pm 1 \cdot 9$ to $41 \cdot 8 \pm 2 \cdot 2$ mg per cent the serum HPL rose from $6 \cdot 8 \pm 0 \cdot 5$ to $7 \cdot 7 \pm 0 \cdot 5$ µg/ml ($t = 2 \cdot 876$; $P < 0 \cdot 01$) (Spellacy et al. 1971a). These findings have been supported by similar work with hyperglycaemia by Burt, Leake and Rhyne (1970) and with the hypoglycaemia of starvation by Tyson, Austin and Farinholt (1971). It would appear then that the level of metabolites reaching the placental-foetal unit regulates in part the concentration of HPL found in the maternal serum. The usefulness of this might be that if placental-foetal hypo-

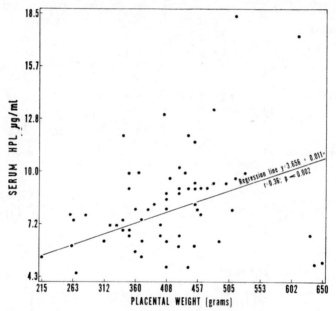

FIG. 4. Correlation of placental weight and serum HPL levels 7–28 days before delivery ($N = 69$).

glycaemia triggers in some manner the release of a lipolytic factor (HPL) so that maternal fat catabolism occurs, then maternal glucose utilization is decreased. This allows for more maternal glucose to become available to the placental–foetal unit.

The relationship of the placental–foetal mass to the circulating level of HPL was also investigated. Serum was serially drawn from normally pregnant women. At the time of delivery the placenta was preserved and the umbilical cord, membranes and blood clots were removed and it was weighed. The neonate was also weighed and correlations were then made between the serum HPL level 7–28 days before delivery and these two weights. In Fig. 4 the placental weight–HPL results and the regression line for 69 subjects are shown. It can be seen that there is a significant relationship between the variables (correlation coefficient $r=0.36$; $P<0.002$). Since the infant weight is usually directly related to the placental weight, it is not surprising to see from the data in Fig. 5 that there is also a significant

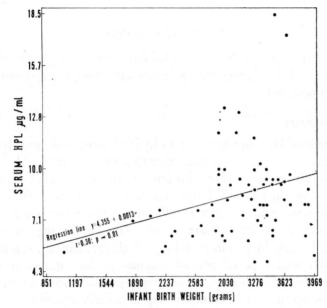

FIG. 5. Correlation of infant birth weight and serum HPL levels 7–28 days before delivery ($N=69$).

relationship between infant weight and serum HPL ($r=0.30$; $P<0.01$). The second important factor regulating HPL levels then appears to be the functioning placental mass.

NORMAL PREGNANCY

The levels of HPL were determined in normal pregnant women, and these can be seen in Fig. 6. There were 377 samples from 204 women. The hormone is detectable early in gestation (five weeks). The mean curve rises progressively until about 34–36 weeks of gestation and then it slowly decreases. Also plotted on the graph are the results from 13 twin gestations, and it can be seen that these are generally elevated, which is not surprising since the placental mass was also greater. There is an asymmetrical distribution of values around the mean curve in the last trimester so it is meaningless to describe the standard deviations. It should be noted, however, that there are very few low values ($< 4 \mu g/ml$) in late pregnancy (after 30 weeks). With the assay and standards used, low values were considered abnormal and potentially indicative of placental failure of hormone production and perhaps indicative of foetal danger. Thus, this low area has been termed the abnormal foetal danger (FD) zone and it will be discussed later in relation to the late-pregnancy complications.

ABNORMAL PREGNANCY

The abnormal pregnancies are divided into three major groups: placental tumours; first half of pregnancy problems (abortion); and second half of pregnancy problems.

Placental tumours

The levels of HPL in women with hydatidiform mole and choriocarcinoma tumours are very low in relation to the mass of trophoblastic tissue present and also to the levels of chorionic gonadotropin (HCG) which are reached. Thus, the mean HPL concentration in 11 women with hydatidiform mole was 0.066 ± 0.0088 µg/ml and for eight women with choriocarcinoma it was 0.077 ± 0.0033 µg/ml. This finding suggests that the tumour cell is primitive in its protein synthetic mechanisms, as is the early villus trophoblast type of normal pregnancy, since it produces HCG rather than HPL. The discrepancy between these two hormones in a patient with a large uterus may be clinically helpful in differentiating between the conditions of multiple gestation and placental trophoblastic tumour.

Abortion

Serial blood samples have been drawn routinely from women in weeks 5 to 20 of gestation. The normal pattern of HPL is shown in Fig. 6. The woman who has vaginal bleeding in early pregnancy may demonstrate one of two patterns, as seen in Fig. 7. The normal HPL levels (L.W.) indicate that the placenta is alive and functioning and the pregnancy may or may not abort. The low levels (M.C.) indicate that the placenta is no longer functioning properly and that an abortion will probably occur soon. Thus, the measurement of serum HPL levels in the early pregnant woman who has bleeding may be a useful prognostic aid to the physician.

Late pregnancy complications

Extensive studies have been done on women with pregnancy complications which have proved to be of high risk to the foetus. Three of these are blood factor (Rh) sensitization, diabetes mellitus and hypertensive toxaemia. The levels of HPL in the Rh-sensitized or diabetic pregnant woman were similar to those of normal pregnancy, although on occasion if the placenta was excessively large the values were high. No relationship was found between the occurrence of foetal or neonatal death and the HPL values. When the hypertensive group was studied, it became clear that abnormally low values were often encountered. Indeed, in an original group of 656 samples from 239 women 12 per cent of the values were in the

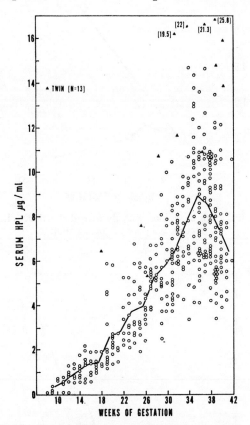

FIG. 6. Levels of serum HPL in normally pregnant women with the mean curve for the data also drawn. In addition, the values for 13 twin pregnancies are also shown.

FD zone (22·5 per cent of the women). Generally, the higher the maternal blood pressure, the lower were the HPL values (Spellacy et al. 1971b). In addition, most of the foetal deaths occurred in those women who had low (FD zone) HPL values (Spellacy, Teoh and Buhi 1970). Three examples from this group of subjects with FD zone values are shown in Fig. 8.

Case 1: M.C. was a 41-year-old gravida 12, para 9–2–0–7* Negro female whose expected date of confinement was 27/5/69. She had had hypertension for several years. During the current pregnancy her blood pressures

* Full-term infants–premature infants–abortions–surviving children.

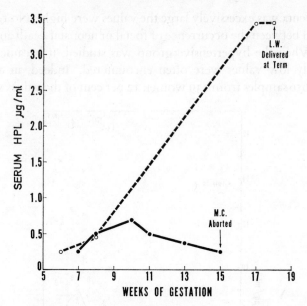

FIG. 7. HPL levels in women with threatened abortion. The low flat curve is a poor prognostic sign.

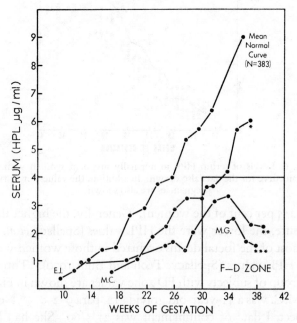

FIG. 8. Serial HPL values in three women with toxaemia. The management and outcome of the cases were: *, spontaneous labour, infant survived; **, induced labour, infant survived; ***, foetal death *in utero*.

ranged from 135/90 to 180/120 mmHg and she was treated with sodium restriction, diuretics, barbiturates and antihypertensives. Her laboratory studies included a negative urine albumin, and a blood type O Rh positive, negative serology, serum creatinine 0·7 mg per cent and a fasting blood glucose of 75 mg per cent. The serial values of HPL are shown in Fig. 8. Although these values were in the FD zone for a brief period, they soon rose above it to a normal zone. She spontaneously ruptured membranes on 1/5/69 and after three hours of labour, delivered a 2353 g normal male infant over a midline episiotomy using pudendal block anaesthesia. The one and five-minute APGAR scores were 5 and 8 respectively. The placenta weighed 652 g. The patient had an uneventful postpartum course and she was discharged home with her infant on 5/5/69. Her blood pressure six months later was 160/105 mmHg.

Case 2: E.J. was a 39-year-old Negro female gravida 5, para 2-2-0-4 whose expected date of confinement was 30/8/68. The patient had had hypertension in her last pregnancy in 1965. Her blood pressures during the current pregnancy ranged between 140/90 and 170/104 mmHg, and she was treated with sodium restriction, diuretics, barbiturates and antihypertensives. Her laboratory studies included a urine albumin of one plus, and a blood type B Rh positive, negative serology, serum creatinine 0·9 per cent and a fasting blood glucose of 70 mg per cent. Distal femoral epiphyses were not seen on a foetogram and the pelvis was of normal size by X-ray pelvimetry measurements. The serial serum HPL values are shown in Fig. 8. Her values persisted in the FD zone. On 28/8/68 labour was induced with an intravenous infusion of oxytocin and after 2 hours 25 minutes she spontaneously delivered a 2934 g normal male infant while using nitrous oxide and oxygen anaesthesia. The one and five-minute APGAR scores were 9 and 10 respectively. The placenta weighed 482 g. On the sixth postpartum day an abdominal tubal ligation was performed. The patient had an uneventful postoperative course and was discharged with her infant on 7/9/68.

Case 3: M.G. was a 39-year-old Negro female gravida 17, para 12-0-4-12 whose expected date of confinement was 27/1/68. She had essential hypertension diagnosed four years before her pregnancy. During the current pregnancy her blood pressures ranged from 130/90 to 190/135 mmHg and she was treated with low sodium diet, diuretics and barbiturates. Her urine was free of albumin and her blood studies showed a blood type O Rh positive, negative serology, serum creatinine 0·8 mg per cent, blood urea nitrogen of 12 mg per cent and a fasting blood glucose of 80 mg per cent. Serial amnioscopy studies were normal. Normal foetal heart tones were heard on 20/10/68 but not afterwards. The serial HPL values while the foetus was alive are shown in Fig. 8 and it can be seen that they remained in the FD zone. After the foetal death the HPL values fell even lower but they are not shown on the graph. On 27/10/68 the patient spontaneously began labour and after 2 hours 58 minutes she spontaneously delivered a macerated 1930 g male infant and a 250 g placenta containing many small infarcts. On the third postpartum day an abdominal tubal ligation was performed. The patient had an uneventful postoperative course and she was discharged home on 5/11/68. Her blood pressure one year later was 148/100 mmHg.

These data seem to be well correlated with the morphometric studies of the placenta. Aherne and Dunnill (1966) have shown that the placental volume is significantly reduced with maternal hypertension. Since placental mass and HPL levels are correlated, the findings of low HPL levels in some women with complicating hypertension is not surprising. The clinical usefulness of this information seems to be in the fact that serial measurements of HPL appear to be a method to detect the woman who has a deficient functioning placental mass. It would appear from the foetal death data that it is in this group of women that most of the foetal deaths occur. Thus, instead of giving intensive care to all pregnant women with the complication of hypertension and thus diluting our resources, we could select a much smaller group ($\sim$ 22·5 per cent) on the basis of HPL values in the FD zone and these women could then receive other biochemical and biophysical studies of the foetal-placental status. Whether such screening and management can reduce perinatal mortality by detecting cases for early delivery is still unknown but it is currently under study.

SUMMARY

In order to study blood human placental lactogen (HPL) concentrations, an immunoassay system had to be developed. A purified sample of HPL was labelled with ^{131}I, rabbit antibodies to HPL were chemically coupled to Sephadex, and a solid-phase assay was developed. This assay has proved to be rapid, sensitive, precise and reproducible.

Studies were made of several possible controlling factors for HPL levels in maternal blood. The correlation coefficient between the serum HPL concentration and the placental and foetal mass was significant. The concentrations of HPL were inversely related to the circulating maternal blood glucose levels. The levels were unaffected by the time of day, foetal sex or the presence of congenital anomalies. The intravenous infusion of prostaglandin E_2 or $F_{2\alpha}$ did not alter HPL levels.

The studies of abnormal pregnancies were in three areas: placental tumours, problems during weeks 1-20 (abortion), and problems from week 20 until the end of pregnancy. Whereas the chorionic gonadotropin levels are high with trophoblast tumours, the HPL levels were very low. In normal pregnancy there is a progressive rise in the mean HPL level until approximately 34-36 weeks and then there is a gradual decline. During the first half of pregnancy, despite vaginal bleeding, a normal rising concentration of HPL indicated a favourable outcome. In late pregnancy the disease complications of Rh factor blood sensitization, anaemia or diabetes mellitus did not alter the HPL pattern. Maternal hypertension did significantly lower the level of HPL in direct proportion to the amount of blood pressure elevation. The prognosis for the foetus appears to be predictable from the HPL pattern. A low HPL value indicates a failing placenta and may result in foetal distress or death.

Acknowledgements

The author would like to express his appreciation to E. S. Teoh, W. C. Buhi, S. A. Birk, S. A. McCreary, B. A. Singer and K. K. Holsinger for their help in carrying out these studies. The studies were supported in part from funds from The Children's Bureau, Department of Health, Education and Welfare, Washington, D.C., and from the Florida Regional Medical Program, Tampa, Florida.

REFERENCES

AHERNE, W. and DUNNILL, M. S. (1966) *Br. Med. Bull.* **22**, 5-8.
BURT, R. L., LEAKE, N. H. and RHYNE, A. L. (1970) *Obstet. & Gynecol. (N.Y.)* **36**, 233-237.
GREENWOOD, F. C., HUNTER, W. M. and GLOVER, J. S. (1963) *Biochem. J.* **89**, 114-123.
JOSIMOVICH, J. B. and MACLAREN, J. A. (1962) *Endocrinology* **71**, 209-220.
KARIM, S. M. M. (1968) *Br. Med. J.* **4**, 618-621.
MACLEOD, R. M. and LEHMEYER, J. E. (1970) *Clin. Res.* **18**, 366.
SPELLACY, W. N. (1969) *Southern Med. J.* **62**, 1054-1057.
SPELLACY, W. N., BUHI, W. C., SCHRAM, J. D., BIRK, S. A. and MCCREARY, S. A. (1971a) *Obstet. & Gynecol. (N.Y.)* **37**, 567-573.
SPELLACY, W. N., CARLSON, K. L. and BIRK, S. A. (1966) *Am. J. Obstet. & Gynecol.* **96**, 1164-1173.
SPELLACY, W. N., TEOH, E. S. and BUHI, W. C. (1970) *Obstet. & Gynecol. (N.Y.)* **35**, 685-689.
SPELLACY, W. N., TEOH, E. S., BUHI, W. C., BIRK, S. A. and MCCREARY, S. A. (1971b) *Am. J. Obstet. & Gynecol.* **109**, 588-598.
TYSON, J. E., AUSTIN, K. L. and FARINHOLT, J. W. (1971) *Am. J. Obstet. & Gynecol.* **109**, 1080-1082.
WIDE, L., AXÉN, R. and PORATH, J. (1967) *Immunochemistry* **4**, 381-386.

DISCUSSION

Friesen: I would like to congratulate Dr Spellacy on this fascinating work, and especially on having the courage to undertake a controlled study. This sort of study is very badly needed, as many obstetricians are already using HPL as an index of placental function and being guided in their practice of obstetrics by serum HPL levels.

Frantz: How does HPL compare with other indices of placental function?

Spellacy: We have measured oestriol in many of these women. From an obstetrical viewpoint oestriol is helpful in people with hypertensive toxaemia. It is not a good index of rhesus incompatibility or with certain diabetic mothers. Although oestriol is presumably an index of many links on a long chain, including the foetus and placenta and maternal kidneys, it doesn't seem to work that way. When we compare HPL to oestriol, it agrees very favourably in the hypertensive group. The advantage of HPL is that you don't have to collect a 24-hour sample of urine and you can run 100 samples in four hours.

Turkington: Have you found cross-reactivity of your anti-HPL to ovine prolactin?

Spellacy: We haven't looked at that.

Greenwood: What standard were you using?

Spellacy: We used Purified Placental Protein Lot 438-7C-125 from Lederle. It's a 90–95 per cent pure preparation. We have compared this to the new NIH standard preparation, and immunologically it is almost identical.

Cotes: The interim standard 70/144 (Bangham and Cotes 1971) being distributed from the Division of Biological Standards, National Institute for Medical Research, London, is derived from Lederle's Purified Placental Protein (Batch E-12 lot 1359) estimated by Dr P. H. Bell to be approximately 75 per cent pure. Bearing in mind the possibility that there may be dimers as well as contaminants in the various preparations of HPL available at present, it would be helpful to have some discussion of what sort of preparation would be suitable for provision as a common standard. This would be of value in the comparison of data obtained in different centres which use the HPL assay as a routine guide to the management of pregnancy.

Wilhelmi: The Lederle material sent to Dr Bangham at the National Institute for Medical Research is a less pure Lederle preparation, different from the one NIH is distributing. The latter is 95 per cent pure, which would make it almost indistinguishable from the 90–95 per cent pure material, as Dr Spellacy says.

Sherwood: We have not found any difference in the immunological activity of dimer and monomer.

Greenwood: Do they label as well?

Sherwood: I have not studied them systematically for this property.

Greenwood: From Dr Spellacy's iodination curves, it looks as though you are putting the monomer in the actual assay.

Li: For the standard, I would rather use the monomer; there could be differences that we haven't detected yet. From Dr Spellacy's curve he has polymer, dimer and monomer. The monomer will have some growth-promoting activity as well as lactogenic activity.

Greenwood: In fact Dr Spellacy answered my question in his presentation. If I read his Sephadex G-100 column correctly, he finds that the polymer and the dimer bind less well to his antibody than the monomer.

Nicoll: Is there any evidence that the ovarian steroids modify HPL secretion *in vivo* or *in vitro*? There is good evidence that they can change adenohypophysial prolactin secretion.

Spellacy: We have given acute injections of oestrogens and progesterone to pregnant women and have not seen any change in HPL, within a 24-hour period.

Greenwood: Dr Lauritzen has reported that injecting dehydroepiandrosterone into pregnant women produced a fall in HCG (Lauritzen 1971).

MacLeod: The fall in HPL near term is very interesting. I have never understood the relatively late increase in plasma prolactin as pregnancy approaches term. Some of us are intrigued by the idea of the auto-feedback mechanism, whereby plasma prolactin inhibits its own secretion: I just wonder if HPL inhibits the pituitary gland from synthesizing and releasing prolactin?

Spellacy: The only data on pituitary prolactin that I am aware of are those Dr Friesen presented here (pp. 96–100). He showed a progressive rise of prolactin in pregnancy which is similar to the curve for HPL except during the last four or five weeks. Since both prolactin and HPL are rising there does not appear to be any feedback interplay.

Frantz: Didn't Dr Friesen find that in monkeys there's no rise in prolactin in pregnancy, whereas monkey placental lactogen follows the same curve as the human?

Friesen: Yes.

Swanson Beck: Have you studied circulating levels of HPL in mothers of anencephalic foetuses? Are they normal?

Spellacy: The group of women with infants having congenital abnormalities included eight anencephalics, and the HPL levels were normal in those mothers.

Turkington: I was also interested in the mean rise and fall of HPL that you found. Have you followed single individuals as well, to substantiate this?

Spellacy: Yes, and generally they follow the same pattern as the group.

Turkington: We have also measured a few patients throughout pregnancy in terms of biological activity of HPL and a significant proportion fell below 4 μg/ml during the last trimester, which would be quite a bit below the immunoreactivity values. In order to compare them one would have to assay the same sample by both techniques. The difference could be methodological, but it suggests the question of whether some of the material which is immunoreactive may be biologically inactive.

Spellacy: I think the mean curve is interesting too, because it relates to what we have known for a long time, that the placenta reaches its peak function at a time generally before delivery and then it begins to age. If you follow the curve out too far, into the post-maturity period beyond 43 weeks' gestation, there is commonly such a loss of reserve that foetal mortality begins to rise significantly. Secondly, I think it is very hard to compare different levels from different laboratories until we have a common standard. We have not done bioassay work but it is interesting that this may give different results.

Greenwood: Professor J. Landon at St Bartholomew's Hospital has automated an HPL assay and is doing a similar prospective study. We just use a four-hour charcoal assay; we don't have the numbers to play with. But if you are following a pregnant woman, what time-scale do you recommend: sampling at weekly intervals for three weeks, or sampling say Monday, Tuesday and Wednesday and assay on Wednesday?

Spellacy: I work in a hospital with more than 400 deliveries a month, and most of the women with clinically recognizable high-risk problems come to a central clinic. Before they are seen by a physician they go through the laboratory to have a blood sample drawn. If they have an even-numbered hospital chart, the senior resident will receive the HPL values from that visit. If they have an odd-numbered chart they serve as a control group and he never finds out about the HPL results. The women are usually seen at one to two-week intervals but if their HPL value becomes low, they are seen more frequently. We also have a test of foetal lung maturity, measuring certain amniotic fluid phospholipids. If the HPL test indicates a failing placenta and if the amniotic fluid indicates that the respiratory tract of the foetus is mature so that it can carry on respiration in a nursery, we deliver the infant. Whether this approach will be successful in lowering perinatal mortality remains to be proved.

Beck: You showed minor changes with respect to glucose and insulin hypoglycaemia. Have you looked at any of the other circulating metabolites that might influence this, such as free fatty acids, β-hydroxybutyrate, acetoacetate, or amino acids?

Spellacy: Only a few. If we infuse arginine the HPL level does not change.

Meites: Dr Spellacy, what do you consider to be the biological role of HPL in pregnancy?

Spellacy: There may be many functions for this protein. It may prepare the breasts for nursing. It may serve as a regulator of maternal lipid metabolism and thus allow the maternal glucose to be spared for foetal use. It may aid in salt and water regulation. A final answer must await further experimentation.

Friesen: How do you interpret the fluctuations in HPL that you occasionally see in the same patient? How do you relate that to functional placental mass? In some instances, there was more than a 50 per cent change in a one-week period; surely placental mass doesn't change by 50 per cent in one week?

Spellacy: I don't believe that this represents interassay variation, since it occurs even if we take all the samples from a given patient and study them in the same assay. It may indicate that there are other regulatory mechanisms controlling the secretion of HPL besides the placental mass. This is why I was curious about your *in vitro* incubation of placenta where, when time was extended, the HPL levels rose. Could it be that the tissue was using up all the available glucose and then it made more HPL?

Friesen: We have not investigated this.

Frantz: Is there any evidence that HPL levels change in sleep?

Spellacy: We studied the effect of position and sleep, because when one goes into bed in the lateral position the uterine blood flow increases and during sleep there are bursts of growth hormone release. These factors do not seem to affect HPL.

Sherwood: The main factor in patients with toxaemia seems to be reduced blood flow to the placenta.

Beck: Would you accept that interpretation, as an obstetrician?

Spellacy: I would interpret the low HPL values as indicating a decreased functional placental mass which may or may not be related to uterine blood flow. For example, some placental area may infarct and not secrete HPL and yet the uterine blood flow remains unchanged.

Greenwood: Have you ever separated a pregnancy plasma on Sephadex G-200? Does the endogenous HPL obey the 21 000 molecular weight?

Friesen: We have done that experiment and it comes out at the same elution volume as our standard does, with an elution volume/void volume ratio of two, which is the same as Dr Li finds with his monomer. I can't quite understand this difference between Dr Li's results and our own.

REFERENCES

BANGHAM, D. R. and COTES, P. M. (1971) In *Radioimmunoassay Methods*, pp. 345–368, ed. Kirkham, K. E. and Hunter, W. M. Edinburgh and London: Churchill Livingstone.

LAURITZEN, C. (1971) In *Proceedings of the Third International Congress on Hormonal Steroids* (Hamburg 1970). Amsterdam: Excerpta Medica. In press.

MORPHOLOGY OF PROLACTIN SECRETION

J. L. Pasteels

Laboratoires d'histologie et de microscopie électronique, Faculté de Médecine et de Pharmacie, Université Libre de Bruxelles

It is generally agreed, on the basis of a considerable body of evidence, that the anterior hypophysis functions as a multiple endocrine gland. Each of its hormones is secreted by specific cells, dispersed among those which secrete the other hormones. This concept is supported by the differential staining of various categories of cells and by electron microscopy. Circumstantial evidence for their specific function was first derived from the careful study of numerous physiological or experimental conditions (reviewed by Herlant 1964). Later on, the use of immunofluorescence techniques demonstrated clearly the presence of one single hormone in each functional category of cells. It was shown that the 'granules' characterizing the hypophysial cells are in fact secretory material, representing the hormone itself (reviewed by Herlant and Pasteels 1967). Still more recently, the functional specificity of cells belonging to the pars distalis has been assessed by radioautographic studies (Goluboff *et al.* 1970).

My aim in this paper is to correlate the results of these various morphological techniques in the specific case of prolactin secretion, with the intention of evaluating their usefulness in endocrinological work.

DIFFERENTIAL STAINING

Techniques proposed

The older distinction between 'acidophilic', 'basophilic' and 'chromophobic' cells is obsolete. In fact, the so-called 'basophilic' cells are not truly basophilic, but contain glycoprotein material. 'Acidophils' are cells secreting simple proteins: growth hormone, prolactin and ACTH. When present in sufficient amounts, their secretory material can be stained by acid dyes such as eosin, erythrosin or orange G. But very active cells are degranulated: their secretion leaves the cell without significant storage. Moreover, in such cells the ribosomes are numerous enough to account for a true cytoplasmic basophilia. As for the 'chromophobes', their description resulted mainly from a lack of proper identification of the secretory material in partly degranulated cells. Only undifferentiated cells might be considered as true chromophobes. But when very sensitive techniques are applied to the hypophysis, such as electron microscopy or immunofluorescence, it becomes exceedingly difficult to find hypophysial glandular cells completely devoid of secretory granules (even in the foetus). Thus, the so-called 'chromophobes' were in fact a confusing mixture of very active, degranulated cells and of resting cells. For all these reasons it is apparent that the proper method of identifying a functional category of hypophysial cells should be based only upon the properties of their secretory material (i.e. the hormone that they contain) rather than on other cytological features (such as cell shape, size or cytoplasmic organelles). The latter vary depending upon the functional conditions of the cell. Some earlier attempts (Cleveland and Wolfe 1932; Dawson 1946) provided valuable results in this regard but they are now superseded by more specific and reliable techniques. To our knowledge, at least three different staining procedures provide a clear identification of prolactin cells:

Herlant's tetrachrome staining (Herlant 1960; Herlant and Pasteels 1967).

Methasol blue–PAS–orange G (Herlant and Grignon 1961; Dubois and Herlant 1968).

Brookes' stain (Brookes 1968; Goluboff and Ezrin 1969).

In our laboratory we regularly use Herlant's tetrachrome and methasol blue–PAS–orange G.

Herlant's tetrachrome staining

The differential staining of serous cells (those which secrete simple proteins; Herlant 1964) is based upon competition between erythrosin and orange G in acid solutions, with subsequent differentiation in aqueous and ethanol solutions of phosphomolybdic acid. The glycoprotein-secreting cells are stained with aniline blue. Prolactin cells are stained bright red by erythrosin whereas somatotrophs appear orange-yellow. The secretory material of corticotrophs is weakly erythrosinophilic, but cannot be confused with prolactin granules, because ACTH granules are smaller and scarcer (this is the reason why corticotrophs have long been considered to be 'chromophobes' or 'basophils'). Herlant's tetrachrome staining provides a valid, although empirical identification of prolactin granules. The differential staining mechanism is probably based upon discrete differences in the isoelectric points of growth hormone, ACTH and prolactin. Successful results have been obtained in the hypophysis of every vertebrate studied so far (for details of the technique see Herlant and Pasteels 1967).

Methasol blue–PAS–orange G

The PAS-positive material of glycoprotein-secreting cells is stained purple by this procedure. Owing to their specific affinity for methasol blue the somatotrophs appear bluish-green. It has been pointed out by Racadot (1963b) that prolactin granules are slightly PAS positive (this reaction can result from the staining of their lipoprotein membrane as well as from properties of prolactin itself). This is why they appear reddish orange with this technique. They are stained intensely by orange G and slightly by PAS. They contrast satisfactorily with the corticotrophs, which are nearly chromophobic with sparse and thin orange-yellow secretory granules. A detailed description of this technique can be found in Dubois and Herlant (1968).

Discussion of results

Empiricism

An absolute standardization of the differential staining techniques can never be achieved, for obvious reasons. First, different batches of staining material, even with the same trademark, are far from identical. Second, the solutions themselves change with time. As can be inferred from its dark colour, a 'good' solution of phosphomolybdic acid does not contain only water and phosphomolybdic acid.

Failure to obtain satisfactory results comes mainly from the use of inappropriate histological fixation, such as Bouin's fluid or even formalin alone. One should use Bouin–Hollande sublimate or, for the hypophysis of man and large mammals, Helly's mixture (see Herlant and Pasteels 1967). Even in such optimal conditions a worker experimenting with new reagents has to proceed by trial and error before obtaining good results. However, once the appropriate conditions are found, the results are quite reproducible and reliable. One should then continue to use these familiar solutions even if they become dirty-looking; they will provide constant results for weeks or months. Moreover, it should be pointed out that when 'good' differential staining is not achieved, it is obvious to the experienced cytologist because of the poor tinctorial quality of the preparation. Accordingly, failures should not give misleading results and categories of cells should not be confused with one another.

Reliability and specificity for prolactin

Once a suitable procedure is found its results are completely reliable. This is clearly shown by the concordant results obtained by different workers in different laboratories. In the rat, we have been able to assess the specificity of Herlant's tetrachrome stain in various physiological and experimental conditions: pregnancy; lactation; administration of reserpine, oestradiol, progesterone and androgen; pituitary grafts and tissue culture. Not only was specific identification of prolactin cells possible in every case, but the accumulation of erythrosinophilic granules within prolactin cells provided a fair estimate of the hormone content of the hypophysis (Pasteels 1963a, b). We made use of Grosvenor and Turner's controlled lactation experiments. By bioassays Grosvenor and Turner (1958) demonstrated that when a female rat was separated from her litter for ten hours on the sixth day *post partum*, pituitary prolactin content reached a maximal value. When the pups were returned to their mother, half an hour of suckling was enough to deplete the anterior hypophysis of most of its prolactin. We did the same experiment, in order to study its morphological equivalent. In the same conditions we observed a significant depletion of the erythrosinophilic granules from the prolactin cells (Fig. 1). When the same experiment was done on the 21st day of the lactation period, no significant depletion of bioassayable prolactin occurred. Once again, there is good agreement between assays and morphology, for we observed no reduction of the secretory material in the prolactin cells.

In the human we obtained evidence for the secretion of a separate prolactin, distinct from human growth hormone, from the study of tissue cultures of the anterior hypophysis (Pasteels 1962, 1963a; Pasteels, Brauman and Brauman 1963). Prolactin cells could be identified in the hypophyses of pregnant women (Pasteels 1963a, b; Herlant and Pasteels 1967) and in prolactin-secreting tumours (Herlant *et al.* 1965).

Earlier, a satisfactory identification of prolactin cells was derived from accurate descriptions of the cyclic changes in the hypophysis of the bat (Herlant 1953, 1956), the cat (Herlant and Racadot 1957) and the mole (Herlant 1959). Results were first obtained using the Cleveland-Wolfe staining method and were confirmed later with Herlant's tetrachrome.

Similar work provided suggestive evidence of prolactin cells in the rat (Herlant and Pasteels 1959; Pasteels and Herlant 1962). With the techniques mentioned here, prolactin cells were unequivocally demonstrated by various authors, working in different laboratories, in the hypophyses of numerous mammals: the badger (Herlant and Canivenc 1960), *Galemys pyrenaicus* (Peyre and Herlant 1961), pig (Bugnon 1963; Bugnon and Racadot 1963), sheep (Racadot 1963a), rabbit, guinea pig (Racadot 1963b), hamster (Girod, Dubois and Curé 1964), monkey (*Macacus irus*) (Girod 1964), hedgehog (Girod, Dubois and Curé 1965), dog (Carlon 1967) and cattle (Dubois and Herlant 1968). In fact, no mammal was found which did not exhibit conspicuous erythrosinophilic or orange-staining (by the

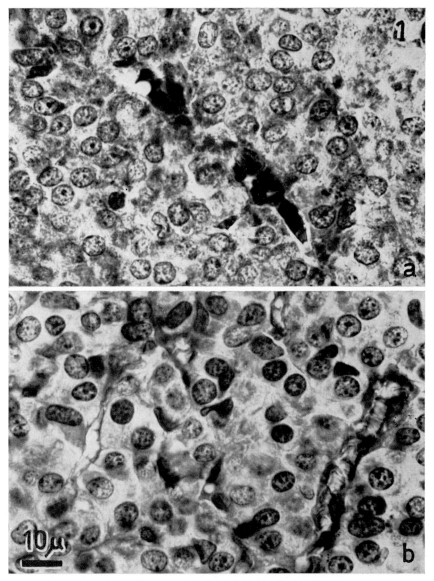

FIG. 1. Details of the anterior lobe of the hypophysis of lactating rats, killed on the sixth day *post partum*. Herlant's tetrachrome stain. (*a*) The hypertrophied prolactin cells are filled with erythrosinophilic granules when the female is separated from the pups for 10 hours. (*b*) The lactating rat had been previously separated from the pups and was then allowed to nurse them for 30 minutes before being killed. The prolactin cells are depleted of their granular material. This is in good agreement with the prolactin release which occurs in such circumstances.

(*facing page 244*)

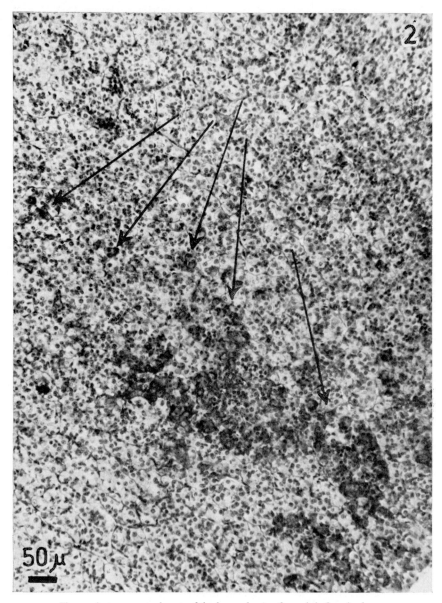

Fig. 2. Antero-ventral part of the hypophysis of an adult female rhesus monkey. Herlant's tetrachrome stain. The prolactin cells are recognizable on this photograph by the dark staining of their cytoplasm (arrows). Their distribution among the other pituitary cells is not uniform. Large clusters of prolactin cells may be observed in addition to an irregular scattering of small clusters and individual cells.

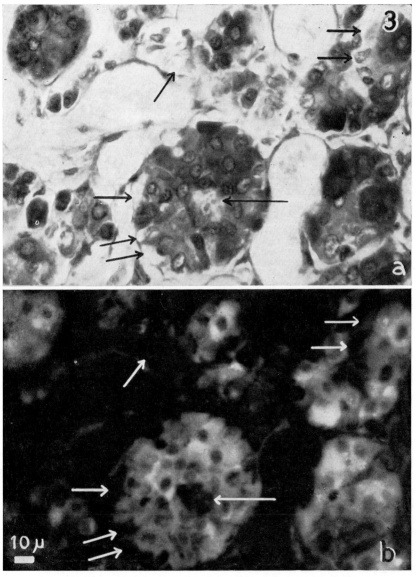

FIG. 3. Detail of the anterior hypophysis of a pregnant woman (seventh month). (*a*) Tetrachrome stain. The numerous prolactin cells are recognizable by the dark staining of their cytoplasm, in contrast to the pale somatotrophs (arrows). (*b*) Before tetrachrome staining the same section was processed for immunofluorescence, using antibodies to NIH ovine prolactin. Compare to (*a*). The erythrosinophilic cells are fluorescent whereas somatotrophs (arrows) are not.

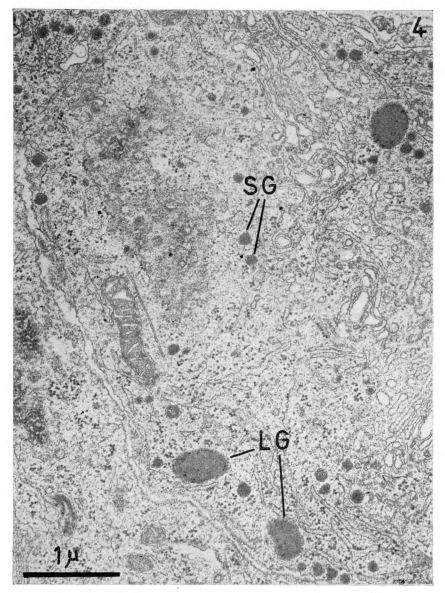

FIG. 4. Part of the cytoplasm of a prolactin cell. Short-term (18 hour) organ culture of rat hypophysis. Large granules (LG), typical of prolactin cells, are identical to those observed in the hypophysis *in situ*, but numerous small granules (SG) appear in response to the organ culture conditions. We believe they are a specific feature of highly active cells, where the release of the hormone (as a consequence of suppression of hypothalamic prolactin-inhibiting factor) becomes too rapid to allow the storage of large secretory granules.

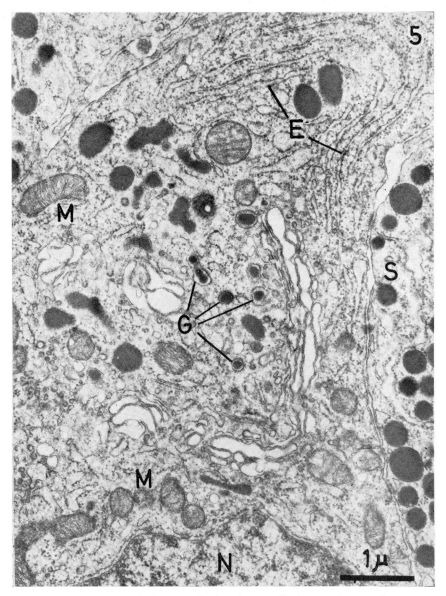

FIG. 5. Detail of a hypertrophied prolactin cell taken on the seventh day of mechanically induced pseudopregnancy. E, ergastoplasm; M, mitochondrion; N, nucleus; G, segregation of secretory granules within the Golgi area. Compare the prolactin granules (irregular shape) to those of the neighbouring somatotroph (S).

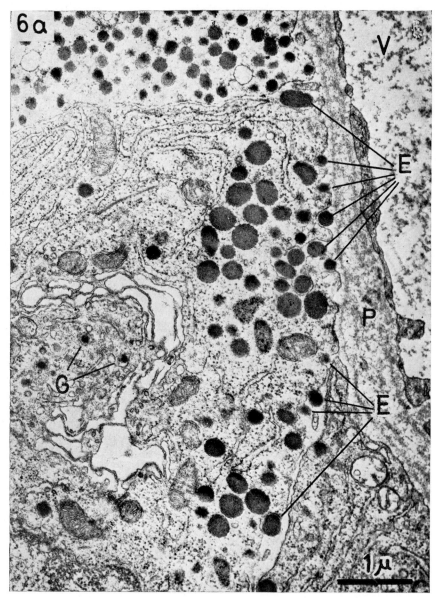

FIG. 6(a). Massive exocytosis of prolactin granules in an oestrus rat killed by ether anaesthesia. E, released granules; P, perisinusoidal space; V, blood vessel; G, new granules undergo segregation within the Golgi vacuoles. The neighbouring cell, above, is a gonadotroph.

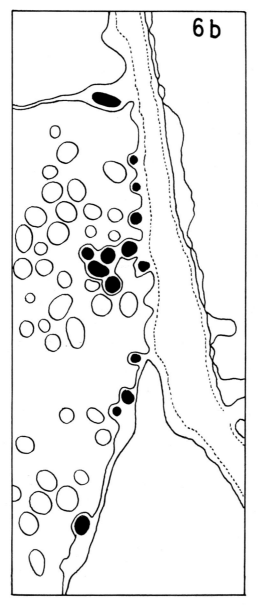

Fig. 6 (b). Schematic drawing of the apical membrane of the cell shown in (a). The released granules are stained black.

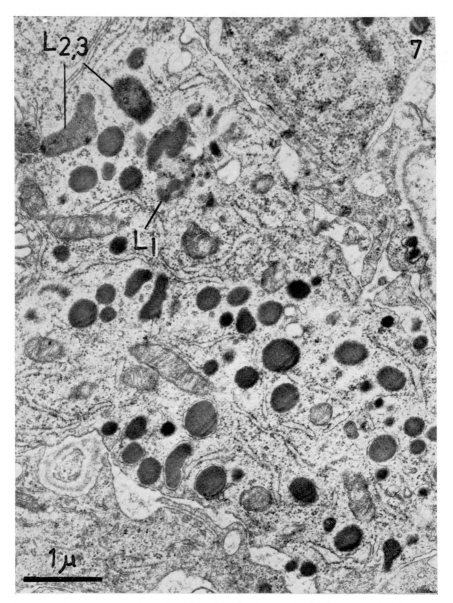

FIG. 7. Destruction of unreleased prolactin granules by lysosomes in the hypophysis of a pseudopregnant rat treated previously with ergocornine. In lysosome L1, the partly destroyed secretory granules can still be recognized. Lysosomes L2 and L3 appear as 'dense bodies'.

methasol blue technique) prolactin cells in its adenohypophysis during pregnancy or lactation.

With the same histological techniques, prolactin cells have been described in the hypophyses of birds: duck (Tixier-Vidal, Herlant and Benoit 1962; Gourdji 1970), pigeon and quail (Gourdji 1970), and possibly in the turtle (Herlant and Grignon 1961), in amphibians (Pasteels and Herlant 1960; van Oordt 1963) and in teleost fish (Olivereau and Ball 1964, 1966; Mattheij and Sprangers 1969).

Possibility of making cell counts

A first approach could be to count all the cells within a well-defined area of a histological section. In our opinion, this cannot provide satisfactory results. The distribution of prolactin cells in the anterior hypophysis is not uniform. Large clusters of prolactin cells occur in specific regions of the hypophysis (in primates they are localized in the antero-ventral part of the pars distalis; in other mammals their localization differs) and irregular scatterings of other prolactin cells occur among the remaining part of the anterior lobe (Fig. 2). To obtain a statistically valid estimate of the proportion of prolactin cells within a single section, one would therefore have to count every cell in this section.

Fortunately, an ingenious technique was devised by Chalkley to prevent bias from counting too many cells in a restricted area (Chalkley 1943-1944). Chalkley made use of a microscopic eye-piece which is marked with a few dots. Fields in the section are chosen randomly and only the cells under the dots are counted. In this laboratory, Desclin (1961) improved Chalkley's technique by the use of a special microscope (Reichert 'Lanameter'). The picture is projected at high magnification on the opaque glass screen of this instrument. A transparent sheet with 25 points distributed at random is placed over the projected image and the coincidence of the points with tissue components is noted. Desclin developed this method for the histometric study of the submandibular glands in mice but we found it very convenient for counting anterior pituitary cells. What it actually measures is the percentage of the total area of a section occupied by the various categories of hypophysial cells. The results thus depend both on the number of differentiated prolactin cells and on their functional hypertrophy, and should therefore correlate with the hormone content of the hypophysis. In collaboration with C. S. Nicoll, P. Varavhudi, F. Ectors, P. Gausset and A. Danguy we are making regular use of this technique in the study of prolactin secretion in rhesus monkey pituitaries. Our first results are that identifiable erythrosinophilic prolactin cells represent 0 per cent (less than 1/2000) of anterior pituitary cells in infant monkeys. In adult males and non-lactating females, prolactin cells range from 2·5 to 11·1 per cent (in the mediosagittal sections: the richest in prolactin cells). In lactating females, there are 18·0 to 32·0 per cent of prolactin cells. In contrast to this variation, the percentage of somatotrophs remains constant in the various categories of animals. This is further evidence for the existence of separate prolactin cells in primates.

Conclusion: *specific advantages of differential staining of prolactin cells in endocrinological studies*

Even with the help of cell counts the differential staining of prolactin cells cannot be expected to supplant the results of modern assays of the hormone in the hypophysis, but it provides supplementary data. The hormone content of a cell depends on its synthesis, storage and release. Morphological techniques remain valuable because they can distinguish between atrophic resting cells and degranulated hyperactive cells (both with moderate hormone content). In the latter case, the enlargement of the whole cytoplasm, of the ergastoplasm and Golgi apparatus, is easily seen even with ordinary microscopy. With Herlant's tetrachrome stain, erythrosinophilic secretory material, ergastoplasm (stained bluish by alizarin and aniline blue) and the Golgi area (unstained, contrasting with the remaining cytoplasm) can be observed in hypertrophied prolactin cells, along with neighbouring cells responsible for other endocrine functions. In this respect, differential staining provides more information than immunofluorescence. As will be described, electron microscopy gives an even more accurate picture of prolactin secretion, but this technique also has its drawbacks, deriving principally from the sampling of the small parts of the anterior lobe chosen for electron microscopy. Differential staining procedures remain invaluable for the general screening of the anterior hypophysis prior to electron microscopy, and also for analysing the glands of a larger number of experimental animals.

IMMUNOFLUORESCENCE

Immunofluorescence techniques provide conclusive evidence of the presence of endogenous prolactin within anterior pituitary cells, provided that highly specific immune sera can be obtained. Fortunately some parts of the prolactin molecules from different species appear to be similar, for antibodies to ovine prolactin cross-react with endogenous prolactin of numerous mammals: ox (Rümke and Ladiges 1965; Nayak, McGarry and Beck 1968; Dubois 1971), pig (Nayak, McGarry and Beck 1968), cat (Emmart, Spicer and Bates 1963), rabbit (Shiino and Rennels 1966), rat

(Emmart et al. 1963; Rümke and Ladiges 1965; Shiino and Rennels 1966), mouse (Rümke and Ladiges 1965) and monkey (Herbert and Hayashida 1970). Antibodies to ovine prolactin have even been used successfully to localize prolactin in the hypophysis of a fish (Emmart 1969). By the use of immune sera to growth hormone it was confirmed that prolactin and somatotropin are secreted by different acidophilic cells.

From our experience we can conclude that immunofluorescence is an extremely sensitive means of localizing even small amounts of endogenous hormone. Positive results are obtained for several hormones in the hypophysis of the foetus or of the newborn (Herlant and Pasteels 1967), when differential staining showed no or little evidence of secretory granules. But it cannot provide useful information on the amount of hormone present within pituitary cells, especially when Coons' indirect procedure (Coons, Leduc and Conolly 1955) is applied. The intensity of the fluorescence reaction depends on optimal concentrations of both antigen and antibodies. With a weak or diluted immune serum, reacting on cells containing little secretory material, a bright fluorescence can be observed.

Another limitation of immunofluorescence is that it gives no information on the secretion of other hormones by neighbouring pituitary cells. However, the simultaneous use of two different immune sera has been proposed. The antibodies to one hormone are labelled with rhodamine (red fluorescence), while the antibodies to another hormone are processed by the indirect fluorescent antibody technique, with fluorescein isothiocyanate labelling (green fluorescence). By this procedure Shiino and Rennels (1966) were able to demonstrate beautifully the localization of prolactin and growth hormone in separate cells of the same section of the rabbit anterior pituitary. Nevertheless one could hardly hope to do simultaneous differential fluorescent staining of the cells responsible for the secretion of each of the anterior pituitary hormones. For the best results immunofluorescence should be combined with subsequent differential staining. We have previously reported (Robyn et al. 1964; Herlant and Pasteels 1967) that immunofluorescence can give excellent results on hypophyses fixed with Bouin–Hollande sublimate or Helly's mixture, provided that the tissue is thoroughly rinsed in water before dehydration and paraffin embedding. The sections are photographed for immunofluorescence. They are then differentially stained, and a comparison of the results of both techniques can thus be made on the same cells.

Immunofluorescence is also an especially interesting tool for differentiating hormones whose existence as separate entities remains uncertain. This is the case for prolactin and growth hormone in the human. We have shown previously that antibodies to human growth hormone did not localize on erythrosinophilic cells of the hypophysis of pregnant women (Herlant and Pasteels 1967), nor on those of organ cultures of foetal pituitaries (Pasteels 1969). This was confirmed by Herbert and Hayashida (1970), using Brookes' differential staining, for prolactin cells in the hypophysis of pregnant monkeys. Moreover, Herbert and Hayashida demonstrated that specific antibodies to ovine prolactin did not localize on somatotrophs, but provided a positive fluorescence on 'carminophils' (the equivalent of our erythrosinophils).

We are now using this technique regularly in further studies to differentiate prolactin and growth hormone in primate pituitaries. In collaborative experiments with C. S. Nicoll, P. Varavhudi, F. Ectors, P. Gausset and A. Danguy, we are making a systematic study of the hypophysis of infant, adult non-lactating and lactating rhesus monkeys, and of human hypophyses in equivalent conditions, with the help of immunofluorescence and differential staining. So far three rabbits have been immunized with human growth hormone (Raben), and six rabbits with ovine prolactin. No cross-reaction could be demonstrated between human growth hormone and ovine prolactin by the Ouchterlony agar diffusion technique. In both human and monkey, somatotrophs fluoresce with antibodies to human growth hormone (by the Coons indirect technique). They do not react to antibodies to prolactin. In both human and monkey, erythrosinophilic prolactin cells, hypertrophied during the postpartum period, fluoresce with antibodies to ovine prolactin but not to human growth hormone (Fig. 3). Prolactin and growth hormone are thus different molecules secreted by different cells in primates as well as in other mammals.

ELECTRON MICROSCOPY

The pioneer work of Rinehart and Farquhar (1953; Farquhar and Rinehart, 1954a, b) has been followed by numerous studies of the anterior hypophysis by electron microscopy. Owing to the recent progress in this technique it has become a very efficient tool for endocrinological work, especially for the study of prolactin secretion.

Material and methods

We prefer the glutaraldehyde–osmium double fixation procedure, as described by Grégoire (1963), with Epon embedding. Sections are cut with diamond knives (Dupont de Nemours, angle 41–43°) on an LKB ultramicrotome. The sections are double stained with uranyl acetate and lead citrate (Reynolds 1963). Siemens Elmiskop I microscopes are used, preferably at rather low magnification (direct: 5000 to 10 000, with a subsequent three-fold magnification of the photographs). This magnification enables the various categories of pituitary cells to be distinguished most clearly and is sufficient for assessing their functional condition.

Results and discussion

Identification of prolactin cells

So far no satisfactory differential 'staining' of prolactin cells has been achieved with the electron microscope, as it has been with light microscopy. It might be possible to develop an equivalent of immunofluorescence, labelling antibodies with ferritin (De Petris and Karlsbad 1965) or peroxidase (Avrameas and Bouteille 1968), but the use of such techniques is greatly hindered by the fact that in properly preserved hypophyses the secretory granules remain enclosed by a membrane impenetrable to foreign proteins.

Fortunately, prolactin cells can be recognized by their very distinctive morphological features. First, the lamellar, often concentric pattern of the rough endoplasmic reticulum is typical of cells secreting growth hormone or prolactin. Second, in numerous species, it has been pointed out that prolactin granules can be distinguished from somatotropin granules by their larger size. This was shown in the rat (Hedinger and Farquhar 1957; Hymer, McShan and Christiansen 1961; Rennels 1962; Pasteels 1963a, b; Petrovic 1963; Pantic and Genbacev 1969; Zambrano and Deis 1970), in man (Pasteels 1963a,b; Foncin 1966; Peake et al. 1969; Salazar et al. 1969), hamster (Petrovic 1963; Girod, Curé and Dubois 1964), hedgehog (Girod, Dubois and Curé 1965) and guinea pig (Petrovic 1963). By contrast in the mole (Herlant and Klastersky 1961; Herlant 1963, 1964) and mouse (Barnes 1962) prolactin granules appear to be smaller than growth hormone granules. It should be stressed that such secretory granules reach their maximal size only when the storage of the hormone is sufficient. We found that in tissue or organ culture of rat hypophyses, when prolactin cells are in a highly active condition, their prolactin granules are small (Fig. 4). An even more reliable feature for recognizing prolactin cells is the extreme polymorphism of their secretory granules. Within the same cell they vary considerably in size and in shape. Unlike what is observed in other pituitary cells, prolactin granules often become elongated, oval, triangular or of even more irregular shape (Fig. 5). This feature was observed in all the species mentioned above with the exception of the hedgehog. It was also confirmed in birds (Tixier-Vidal and Picart 1967; Gourdji 1970).

Although the distinctive features of prolactin cells vary according to species and to their functional condition, they are characteristic enough to be recognized in all the animals studied.

Functional study

No technique can provide more accurate data on the function of prolactin cells than electron microscopy. Within the same section the synthesis of the hormone, its segregation into secretory granules and its release, storage or destruction can be ascertained simultaneously.

Synthesis. A conspicuous ergastoplasm is evidence of synthesis of this protein hormone. The rough cisternae of the endoplasmic reticulum are then filled with amorphous material, as described in other protein-secreting cells, such as in the pancreas (Tashiro et al. 1966). Evidence for this function of the ergastoplasm in prolactin secretion was obtained in radioautographic experiments by Tixier-Vidal and Picart (1967).

Segregation. In most glandular cells secreting protein the amorphous material synthesized in the ergastoplasm is transferred to the cisterns of the Golgi system. There it may be chemically transformed (especially in the case of glycoprotein secretion). More generally, it is concentrated into secretory granules, coated with a membrane coming from the Golgi vesicle itself (Farquhar and Wellings 1957; Caro 1961; Zeigel and Dalton 1962). Conclusive evidence for this mechanism was provided by radioautographic studies by Caro (1961) on the pancreas and by Tixier-Vidal and Picart (1967) on prolactin secretion by the hypophysis in organ culture.

In highly active prolactin cells, a hypertrophied Golgi system with concentration of granules within its cisternae (Fig. 5) may thus be considered a reliable criterion for enhanced glandular secretion. We observed it in pseudopregnant and lactating rats, and also in cultures of rat and human hypophyses.

Release of hormone. The release of prolactin granules can also be clearly demonstrated by electron microscopy. It was described by Sano (1962) in lactating mice and by us (Pasteels 1963a) during controlled lactation in the rat, following Grosvenor and Turner's (1958) experimental schedule. This enabled us to conclude that prolactin was released in large amounts by the hypophysis when the animals were killed for electron microscopy.

The secretory granules are driven to the cell membrane, probably by some active mechanism involving cytoplasmic microtubules (see Rasmussen 1970). The lipoprotein membrane of the granule (supplied by the Golgi system) coalesces with the cell membrane. A hole appears in the centre of the area of coalescence so that the hormone can be discharged outside the cell, into the perisinusoidal space, without breaking the integrity of the plasma membrane. Once outside the cell, prolactin, now devoid of a membrane coating, quickly dissolves and penetrates the blood vessels by pinocytosis. The study of this exocytotic process seems ideally suited for experiments on prolactin release. However, one should be careful in work on living animals; it was recently shown by Neill (1970) using radioimmunoassay that the stress of ether anaesthesia is sufficient to induce substantial discharge of prolactin. We can confirm this from electron microscopy (Fig. 6).

Storage. When the release of prolactin is inhibited (or not stimulated) the secretory granules accumulate in the cytoplasm of the prolactin cells. A cell heavily loaded with secretory material is a resting cell. This is clearly demonstrated in lactating rats separated for some hours from their litters (Pasteels, 1963a, b). It seems that in such conditions the synthesis of hormone becomes inhibited by some 'ultra-short' feedback mechanism, for heavily loaded prolactin cells show little evidence of ergastoplasm and Golgi function.

Destruction of excess secretory material. A clear description of lysosomal function in the regulation of the secretory process in prolactin cells was provided by Smith and Farquhar (1966). In postpartum rats separated from their litters for several days, they found that the excess secretory granules, as well as mitochondria and ribosomes, were enclosed and digested within lysosomal vesicles. Such lytic bodies are easily recognizable in the electron microscope. They appear first as vacuoles containing partly destroyed secretory granules and later as 'dense bodies' (Fig. 7). They are responsible for the return of prolactin cells to the resting condition when the release of hormone is inhibited for a period. We obtained evidence of this mechanism in pseudopregnant rats and in organ cultures of rat hypophyses treated with ergocornine.

Use of electron microscopy in the study of prolactin secretion: inhibition of secretion by ergocornine

Ergocornine methane sulphonate is a potent inhibitor of prolactin secretion in the whole animal. It might act in one of two ways, either by a direct effect on prolactin cells (Zeilmaker and Carlsen 1962) or by an action on the hypothalamus (Lindner and Shelesnyak 1967). To explore these alternatives we made studies of the hypophysis in organ culture, where it is removed from hypothalamic control. We found that ergocornine methane sulphonate (10 μg/ml culture medium) or its analogue 2-bromo-α-ergocryptine (1-5 μg/ml) significantly inhibited the release of prolactin by the cultures. Prolactin was assayed on the pigeon crop gland (Pasteels and Ectors 1970; Pasteels *et al.* 1971). The cultures were examined morphologically in order to ascertain that the inhibition of prolactin secretion was not due to destruction of the cultures. In control cultures there was evidence of very active synthesis of prolactin, of its segregation into secretory granules, and of immediate release of the hormone, without significant storage. In cultures treated for four or 18 hours with ergocornine no evidence of prolactin secretion was found. Storage of the hormone within the cells was pronounced. This was confirmed by their higher prolactin content, as measured by bioassay. Destruction of the granules within lysosomal vesicles was observed. However, the ergastoplasm was still in a highly active condition, and new granules were found in the Golgi cisternae. From this electron microscopic study we can conclude that the immediate effect of ergocornine on prolactin cells is to inhibit the hormone's release, without directly interfering with its synthesis.

CONCLUSIONS

The morphological study of prolactin secretion is not only interesting in itself, but it is also a very effective tool for physiological experiments. Each of the three main techniques—differential staining, immunofluorescence and electron microscopy—has its own advantages and shortcomings, which have been discussed here.

Electron microscopy should be combined with light microscopy (differential staining) to control the sampling of the material studied. It will then provide valuable information on synthesis, concentration in granules, storage or intracellular destruction of prolactin. The immunofluorescence procedures are especially helpful in validating the results of differential staining and in specific immunological work, such as the study of primate prolactin and growth hormone.

SUMMARY

Prolactin cells can be recognized as a separate category of acidophilic cells by several differential staining procedures. The results of two recent techniques, Herlant's tetrachrome and methasol blue–PAS–orange G, are shown. They provide a clear identification of prolactin cells in the hypophysis of every mammal studied so far, including man. After such differential staining, quantitative results may be expected from cell counts, or better, from the measurement of the total area of prolactin cells within a section of the pituitary gland, and an improvement of Chalkley's method is described. Evidence that acidophilic granules of prolactin cells are the hormone itself comes from the study of various physiological conditions and also from immunofluorescence studies. New results are presented of the study of the hypophysis of monkeys (made with C. S. Nicoll) and of pregnant women, using the immunofluorescence of prolactin cells and somatotrophs.

Prolactin cells can be recognized in the electron microscope independently of the staining procedures mentioned, owing to the irregular shape of their secretory granules and, when storage of the hormone is significant, owing to the large size of the granules. Electron microscopy is the only technique simultaneously providing evidence of synthesis of the hormone, its segregation in secretory granules, and their storage and release or destruction by lysosomes.

Acknowledgements

We are pleased to acknowledge the financial support of the Belgian Fonds de la Recherche Scientifique and Fonds de la Recherche Scientifique Médicale, grants of purified ovine prolactin from the National Institutes of Health, Bethesda and the gift of human growth hormone by Dr Raben. The author wishes to thank Dr C. S. Nicoll warmly for improvements in the writing of the manuscript.

REFERENCES

AVRAMEAS, S. and BOUTEILLE, M. (1968) *Exp. Cell Res.* **53**, 166–176.
BARNES, BARBARA G. (1962) *Endocrinology* **71**, 618–628.
BROOKES, L. D. (1968) *Stain Technol.* **43**, 41–42.
BUGNON, C. (1963) *Arch. Anat. (Strasbourg)* **56**, 395–407.
BUGNON, C. and RACADOT, J. (1963) *C. R. Séance Soc. Biol.* **157**, 1934–1937.
CARLON, N. (1967) *Z. Zellforsch. & Mikrosk. Anat.* **78**, 76–91.
CARO, L. G. (1961) *J. Biophys. & Biochem. Cytol.* **10**, 37–49.
CHALKLEY, H. W. (1943–1944) *J. Natl. Cancer Inst.* **4**, 47–53.
CLEVELAND, R. and WOLFE, J. M. (1932) *Anat. Rec.* **51**, 409–413.
COONS, A. H., LEDUC, E. H. and CONOLLY, J. M. (1955) *J. Exp. Med.* **102**, 49–60.
DAWSON, A. B. (1946) *Am. J. Anat.* **78**, 347–402.
DE PETRIS, S. and KARLSBAD, G. (1965) *J. Cell Biol.* **26**, 759–778.
DESCLIN, J. C. (1961) *Anat. Rec.* **141**, 305–314.
DUBOIS, M. P. (1971) *C. R. Hebd. Séance Acad. Sci., Sér. D* **272**, 433–435.
DUBOIS, M. P. and HERLANT, M. (1968) *Ann. Biol. Anim. Biochem. Biophys.* **8**, 5–26.
EMMART, E. W. (1969) *Gen. & Comp. Endocrinol.* **12**, 519–525.
EMMART, E. W., SPICER, S. S. and BATES, R. W. (1963) *J. Histochem. & Cytochem.* **11**, 365–373.
FARQUHAR, M. G. and RINEHART, J. F. (1954a) *Endocrinology* **54**, 516–541.
FARQUHAR, M. G. and RINEHART, J. F. (1954b) *Endocrinology* **55**, 857–876.
FARQUHAR, M. G. and WELLINGS, S. R. (1957) *J. Biophys. & Biochem. Cytol.* **3**, 319–322.
FONCIN, J. F. (1966) *Path. & Biol. (Paris)* **14**, 893–902.
GIROD, C. (1964) *C. R. Hebd. Séance Acad. Sci., Sér. D* **258**, 5079–5081.
GIROD, C., CURÉ, M. and DUBOIS, P. (1964) *C. R. Hebd. Séance Acad. Sci., Sér. D.* **258**, 6244–6247.
GIROD, C., DUBOIS, P. and CURÉ, M. (1964) *C. R. Hebd. Séance Acad. Sci., Sér. D* **258**, 6536–6538.
GIROD, C., DUBOIS, P. and CURÉ, M. (1965) *C. R. Hebd. Séance Acad. Sci., Sér. D* **261**, 5660–5663.

Goluboff, L. G., and Ezrin, C. (1969) *J. Clin. Endocrinol. & Metab.* **29**, 1533–1538.
Goluboff, L. G., Mac Rae, E., Ezrin, C. and Sellers, E. A. (1970) *Endocrinology* **87**, 1113–1118.
Gourdji, D. (1970) Ph.D. Thesis, Paris; Laboratoire de Biologie Moléculaire, Collège de France.
Grégoire, A. (1963) *J. Microscopie (Paris)* **2**, 613–620.
Grosvenor, C. E. and Turner, C. W. (1958) *Endocrinology* **63**, 535–539.
Hedinger, C. E. and Farquhar, M. G. (1957) *Schweiz. Z. Allg. Pathol. & Bakteriol.* **20**, 766–768.
Herbert, D. C. and Hayashida, T. (1970) *Science* **169**, 378–379.
Herlant, M. (1953) *Ann. Soc. R. Zool. Belg.* **84**, 87–116.
Herlant, M. (1956) *Arch. Biol. (Brussels)* **67**, 89–180.
Herlant, M. (1959) *C. R. Hebd. Séance Acad. Sci., Sér. D* **248**, 1033–1036.
Herlant, M. (1960) *Bull. Micr. Appl.* **10**, 37–44.
Herlant, M. (1963) In *Coll. Int. 128 La cytologie de l'adénohypophyse*, pp. 74–94. Paris: CNRS.
Herlant, M. (1964) *Int. Rev. Cytol.* **17**, 299–382.
Herlant, M. and Canivenc, R. (1960) *C. R. Hebd. Séance Acad. Sci., Sér. D* **250**, 606–608.
Herlant, M. and Grignon, G. (1961) *Arch. Biol. (Brussels)* **72**, 97–151.
Herlant, M. and Klastersky, J. (1961) *C. R. Hebd. Séance Acad. Sci., Sér. D* **253**, 2415–2418.
Herlant, M., Laine, E., Fossati, P. and Linquette, M. (1965) *Ann. Endocrinol. (Paris)* **26**, 65–71.
Herlant, M. and Pasteels, J. L. (1959) *C. R. Hebd. Séance Acad. Sci., Sér. D* **249**, 2625–2626.
Herlant, M. and Pasteels, J. L. (1967) In *Methods and Achievements in Experimental Pathology*, vol. 3, pp. 250–305, ed. Bajusz, E. and Jasmin, G. Basel: Karger.
Herlant, M. and Racadot, J. (1957) *Arch. Biol. (Brussels)* **68**, 217–248.
Hymer, W. C., McShan, W. H. and Christiansen, R. G. (1961) *Endocrinology* **69**, 81–90.
Lindner, H. R. and Shelesnyak, M. C. (1967) *Acta Endocrinol. (Copenh.)* **56**, 27–34.
Mattheij, J. A. M. and Sprangers, J. A. P. (1969) *Z. Zellforsch. & Mikrosk. Anat.* **99**, 411–419.
Nayak, R., McGarry, E. E. and Beck, J. C. (1968) *Endocrinology* **83**, 731–736.
Neill, J. D. (1970) *Endocrinology* **87**, 1192–1197.
Olivereau, M. and Ball, J. (1964) *Gen. & Comp. Endocrinol.* **4**, 523–532.
Olivereau, M. and Ball, J. (1966) *Proc. R. Soc. B* **164**, 106–129.
Pantic, V. and Genbacev, O. (1969) *Z. Zellforsch. & Mikrosk. Anat.* **95**, 280–289.
Pasteels, J. L. (1962) *C. R. Hebd. Séance Acad. Sci., Sér. D* **254**, 4083–4085.
Pasteels, J. L. (1963a) *Arch. Biol. (Brussels)* **74**, 439–553.
Pasteels, J. L. (1963b) In *Coll. Int. 128 La cytologie de l'adénohypophyse*, pp. 137–148. Paris: CNRS.
Pasteels, J. L. (1969) *Mém. Acad. R. Méd. Belg.* **7**, 1–45.
Pasteels, J. L., Brauman, H. and Brauman, J. (1963) *C. R. Hebd. Séance Acad. Sci., Sér. D* **256**, 2031–2033.
Pasteels, J. L., Danguy, A., Frérotte, M. and Ectors, F. (1971) *Ann. Endocrinol. (Paris)* **32**, 188–192.
Pasteels, J. L. and Ectors, F. (1970) *Arch. Int. Pharmacodyn. & Ther.* **186**, 195–196.
Pasteels, J. L. and Herlant, M. (1960) *Anatom. Anzeiger.* 109 Ergänz. Verhandl. 1st Europ. Anat. Kongr., 764–767.
Pasteels, J. L. and Herlant, M. (1962) *Z. Zellforsch. & Mikrosk. Anat.* **56**, 20–39.
Peake, G. T., McKeel, D. W., Jarett, L. and Daughaday, W. H. (1969) *J. Clin. Endocrinol. & Metab.* **29**, 1383–1393.
Petrovic, A. (1963) In *Coll. Int. 128 La cytologie de l'adénohypophyse*, pp. 121–136. Paris: CNRS.
Peyre, A. and Herlant, M. (1961) *C. R. Hebd. Séance Acad. Sci., Sér. D* **252**, 463–465.

RACADOT, J. (1963a) C. R. Séance Soc. Biol. **157**, 729-732.
RACADOT, J. (1963b) In Coll. Int. 128 La cytologie de l'adénohypophyse, pp. 33-49. Paris: CNRS.
RASMUSSEN, H. (1970) Science **170**, 404-412.
RENNELS, E. G. (1962) Endocrinology **71**, 713-722.
REYNOLDS, E. S. (1963) J. Cell Biol. **17**, 208-213.
RINEHART, J. F. and FARQUHAR, M. G. (1953) J. Histochem. & Cytochem. **1**, 93-113.
ROBYN, C., BOSSAERT, Y., HUBINONT, P. O., PASTEELS, J. L. and HERLANT, M. (1964) C. R. Hebd. Séance Acad. Sci., Sér. D **259**, 1226-1228.
RÜMKE, P. and LADIGES, N. C. J. J. (1965) Z. Zellforsch. & Mikrosk. Anat. **67**, 575-583.
SALAZAR, H., McAULAY, M. A., CHARLES, D. and PARDO, M. (1969) Arch. Path. (Chic.) **87**, 201-211.
SANO, M. (1962) J. Cell Biol. **15**, 85-97.
SHIINO, M. and RENNELS, E. G. (1966) Texas Rep. Biol. & Med. **24**, 659-673.
SMITH, R. E. and FARQUHAR, M. G. (1966) J. Cell Biol. **31**, 319-347.
TASHIRO, Y., NAGATA, T., MORIMOTO, T. and MATSUURA, S. (1966) Proc. VI Int. Congress for Electron Microscopy, vol. 2, pp. 609-610, ed. Uyeda, R. Tokyo: Maruzen.
TIXIER-VIDAL, A., HERLANT, M. and BENOIT, J. (1962) Arch. Biol. (Brussels) **73**, 317-368.
TIXIER-VIDAL, A. and PICART, R. (1967) J. Cell Biol. **35**, 501-519.
VAN OORDT, P. G. W. J. (1963) In Coll. Int. 128 La cytologie de l'adénohypophyse, pp. 301-313. Paris: CNRS.
ZAMBRANO, D. and DEIS, R. P. (1970) J. Endocrinol. **47**, 101-110.
ZEIGEL, R. F. and DALTON, A. J. (1962) J. Cell Biol. **15**, 45-54.
ZEILMAKER, G. H. and CARLSEN, R. A. (1962) Acta Endocrinol. (Copenh.) **41**, 321-325.

[For discussion, see pp. 277-286.]

SECRETION OF PROLACTIN AND GROWTH HORMONE BY ADENOHYPOPHYSES OF RHESUS MONKEYS *IN VITRO*

Charles S. Nicoll

Department of Physiology-Anatomy, University of California, Berkeley

The question of the existence of prolactin as a separate hormonal entity of the pars distalis of primates has been reviewed by a number of authors in recent years (Forsyth 1969; Lyons 1969; Pasteels 1969; Nicoll et al. 1970; Sherwood 1971; and see other papers in this volume). Accordingly, all the pertinent literature will not be reviewed again here.

All attempts to isolate human prolactin free of growth hormone activity have been unsuccessful, and all preparations of primate growth hormone have prolactin activity, including synthetic human growth hormone (HGH) (Li and Yamashiro 1970). Accordingly, it has been suggested that prolactin does not exist as such in man and HGH alone regulates the physiological processes which are controlled by both hormones in other species (Bewley and Li 1970). This state of affairs is somewhat disquieting because prolactin has been isolated from the adenohypophyses of several mammalian species and shown to be a hormonal principle distinct from growth hormone. Furthermore, recent work indicates that prolactin and growth hormone are chemically distinct molecular species in the adenohypophyses of non-mammalian vertebrates, including chondrichthyeans, teleosts, amphibians, reptiles and birds (Nicoll and Nichols 1971; Nicoll and Licht 1971; Nicoll, Licht and Clarke 1971). Accordingly, if humans and other primates did not have prolactin as part of their 'pituitary equipment' they would indeed be exceptional among the various vertebrate groups.

Despite the unsuccessful attempts by biochemists to isolate a homogeneous protein from primate pituitary glands which can be unequivocally labelled prolactin, several fractions have been prepared which have high prolactin activity relative to growth hormone (Lyons 1969; Tashjian, Levine and Wilhelmi 1965; Freeman and Wilhelmi 1969). It remains uncertain, however, whether this represents partial separation of prolactin from growth hormone or whether the fractions with different ratios of somatotropic to prolactin activities merely represent altered forms of a single hormonal species. Guyda and Friesen (1971) claim to have separated monkey prolactin from monkey growth hormone by affinity chromatography.

In vitro studies have provided physiological evidence that prolactin, as measured by bioassay, and growth hormone, as measured by radioimmunoassay, can be secreted independently by the pituitary glands of humans (Pasteels, Brauman and Brauman 1963) and monkeys (Nicoll *et al.* 1970; Channing *et al.* 1970). These physiological results are supported by clinical evidence from various sources. For example, acromegalic subjects with high circulating growth hormone titres do not have high serum prolactin levels, as determined by bioassay (Roth, Gorden and Bates 1968). In addition, blood growth hormone levels are not elevated in lactating women (Board 1968; Benjamin, Casper and Kalodny 1969; Spellacy, Buhi and Birk 1970). The fact that atelotic dwarf women can successfully lactate after normal pregnancy without detectable plasma growth hormone provides particularly cogent evidence that prolactin and growth hormone are separate entities in humans (Rimoin *et al.* 1968).

Various objections can be raised to these clinical and physiological findings. For example, it could be asserted that a basic growth hormone–prolactin molecule of primate adenohypophyses can be secreted in different forms. In some cases the molecule could have relatively more lactogenic than growth activity, and *vice versa*. Accordingly, definitive proof of the existence of a primate prolactin requires its isolation and chemical characterization and the demonstration that it is distinct, physically, chemically and biologically, from primate growth hormone. This paper includes data which show that in the rhesus monkey, considerable progress has been made towards this end.

METHODS

In order to substantiate and extend our previous work on the secretion of prolactin and growth hormone *in vitro* by adenohypophyses of rhesus monkeys (Nicoll *et al.* 1970) we have cultured the glands of adult male and female animals, and of lactating females and their infants. An organ culture procedure was used to study the secretory activity of these glands and to obtain a large quantity of medium for separation of their secretory products. Samples of the medium were processed by polyacrylamide disc gel electrophoresis (Nicoll *et al.* 1969; 7 per cent gel, pH 9·5, tris–glycine system) throughout the duration of the cultures to monitor the progress of the experiments and to obtain information on the proteins secreted by the explants. In addition, fresh tissue collected from the animals immediately

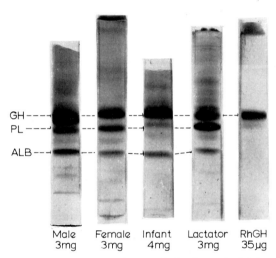

FIG. 1. Disc electrophoretic columns of adenohypophysial tissue from rhesus monkeys compared with highly purified RhGH (Peckham 1967). The growth hormone (GH), presumptive prolactin (PL) and albumin (ALB) bands are indicated.

after death was processed by disc gel electrophoresis to obtain information on the protein patterns of the rhesus monkey adenohypophyses in different physiological conditions. The concentrations of prolactin and growth hormone in the tissue and medium samples were estimated by densitometry (Nicoll *et al.* 1969), using a standard curve prepared with ovine prolactin (NIH P-S-8) to convert the optical density readings to μg equivalents. It should be emphasized, however, that the homogeneity of the prolactin and growth hormone bands of the disc electrophoretic columns has not yet been verified, owing to inadequate availability of material. Hence, the estimated levels of both hormones in the tissue and medium samples must be viewed as tentative.

RESULTS

The protein patterns obtained with adenohypophyses of the four types of monkeys are shown in Fig. 1, together with a gel column containing homogeneous rhesus monkey growth hormone (RhGH) prepared by Dr W. D. Peckham (1967). It can be seen that all four glands contain a protein with the same mobility as the RhGH. The fastest migrating prominent protein in all four columns of adenohypophysial tissue corresponds to serum albumin, as determined by comparing the electrophoretic patterns of the pituitaries with monkey serum. The glands of the adult male and of the lactating and non-lactating females contain a prominent protein which migrates just in front of the growth hormone. This protein has been tentatively identified as rhesus monkey prolactin on the basis of several lines of evidence. It appears from Fig. 1 that its concentration in the adenohypophysis of the adult male* and female is about the same, whereas it is virtually absent from the glands of the infants. However, it is particularly prominent in the adenohypophysis of the lactating female. The preparation of RhGH shows a faint band running just ahead of the major component and its mobility corresponds to that of the presumptive rhesus prolactin in the tissue preparations. This may be deamidated growth hormone (Lewis, Cheever and Hopkins 1970), because it is not apparent when the protein is electrophoresed in an acidic (pH $3 \cdot 8$) acrylamide system (Peckham 1967).

Additional data which support the presumption that this protein is rhesus prolactin are shown in Fig. 2 where the concentrations of growth hormone and prolactin in the different glands are presented. The infant glands contained substantially more growth hormone than the glands of

* This particular male monkey adenohypophysis had an atypically high concentration of prolactin.

any of the adult monkeys and all of the latter had about the same concentration of growth hormone. Furthermore, the infant adenohypophyses contained about as much prolactin as the glands of the adult males and both had less than the glands of the adult females. The adenohypophyses of the

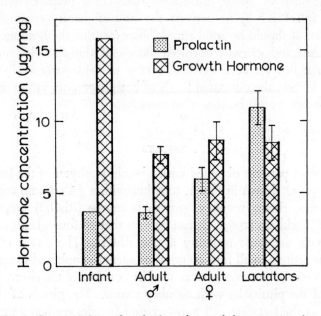

FIG. 2. Concentrations of prolactin and growth hormone in adenohypophyses of rhesus monkeys as estimated densitometrically from disc electrophoretic columns. Data on the infants were obtained from two pools of adenohypophysial fragments of three and nine glands. Five adult male and five adult female glands and adenohypophyses from seven lactating females were used for the other estimates. The data are presented as means ± S.E.M.

lactating females contained about twice as much presumptive prolactin as the pituitaries of the adult females.

These densitometric estimates of the concentrations of prolactin in the adenohypophyses of rhesus monkeys are consistent with differences reported to occur in the adenohypophyses of rats of different age, sex and reproductive condition (Minaguchi, Clemens and Meites 1968; Voogt, Chen and Meites 1970; Meites and Turner 1950; C. T. McKennee and C. S. Nicoll, unpublished observations). The densitometric estimate of the concentration of growth hormone in the glands of the adult monkeys of both sexes of about 8 µg/mg is in good agreement with the results of W. D. Peckham (personal communication) who found about 9·0 to 9·5 µg/mg in pooled adult glands, measured by immunoassay.

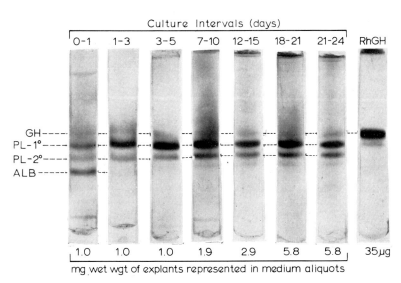

FIG. 3. Disc electrophoretic columns of medium samples collected at different intervals during 24 days of culture of an adenohypophysis from a lactating rhesus monkey compared with highly purified RhGH (Peckham 1967). The growth hormone (GH) and primary and secondary prolactin (PL-1° and PL-2°, respectively) and the albumin (ALB) bands are indicated. Note that the aliquots of medium were not uniform.

(*facing page 260*)

The presumptive rhesus prolactin band has been isolated from medium derived from organ cultures of adenohypophyses of lactating female rhesus monkeys by sucrose density gradient electrophoresis followed by Sephadex G-100 chromatography with subsequent isolation by polyacrylamide gel electrophoresis at pH 3·8. The protein isolated by these procedures had more than 8000 i.u. of prolactin (crop sac assay) per mg of radioimmunoassayable growth hormone (Peckham and Nicoll 1971). Accordingly, there is good reason to believe that this protein, which migrates just in front of RhGH in the pH 9·5 disc electrophoretic system, is authentic rhesus prolactin. This conclusion is substantiated by the physiological data described below. Nagy, Kurcz and Baranyai (1968) have also identified the same protein band as rhesus prolactin.

Secretion of prolactin and growth hormone by adenohypophyses of rhesus monkeys in organ culture

The adenohypophyses of five adult male and five adult female rhesus monkeys were incubated in organ culture for 12 days and the medium was changed at different intervals, ranging from one to three days. The glands of nine lactating females and their infants were similarly processed but the duration of culture was extended to 24 days. Aliquots of the medium from each culture, taken at each collection interval, were processed by a disc electrophoretic procedure to estimate the concentration of prolactin and growth hormone contained therein. The wet weight of the explants was determined at the end of incubation. Medium 199, containing insulin (1 μg/ml), penicillin (25 U/ml) and streptomycin (25 μg/ml), was used and the cultures were incubated at 35°C in an atmosphere of 5 per cent CO_2 in O_2.

The protein patterns obtained with medium samples from a culture of a gland from a lactating female are shown in Fig. 3, together with a column of purified RhGH (Peckham 1967).

Growth hormone was present as a faint band in all the medium samples taken throughout the 24 days of incubation. The albumin band, which is conspicuous in the first medium sample (days 0–1 of incubation), disappeared by the third incubation interval (days 3–5). Prolactin was present as a prominent band in all the medium samples collected throughout the course of the experiment. It should be noted that the columns shown in Fig. 3 represent aliquots of medium of different volume at the different incubation intervals. Accordingly, the columns are not directly comparable from one time to another. An additional protein band, running between prolactin and albumin, is evident in all the medium samples. This was seen in the samples taken from all the pituitary cultures and it is present, but less conspicuous, in the electrophoretic columns of the adenohypophysial tissue shown in Fig. 1. This protein may be deamidated prolactin (Lewis, Cheever and Hopkins 1970) because its occurrence and concentration is closely related to the level of the primary prolactin band in the gel columns, and not to the amount of growth hormone. Accordingly, this protein is considered to be a secondary prolactin band and the amount of it in each medium sample was estimated densitometrically and added to the estimated amount of the primary prolactin band.

The amounts of prolactin and growth hormone secreted by the adenohypophyses of the adult male and female monkeys are shown in Fig. 4.

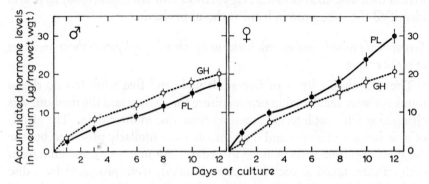

FIG. 4. Accumulated amounts of prolactin (PL) and growth hormone (GH) secreted by adenohypophyses of adult male and female rhesus monkeys during 12 days in organ culture. Densitometric estimates from disc electrophoretic columns. Vertical bars represent S.E.M.

The glands of both sexes secreted growth hormone at a fairly steady rate, and in comparable amounts throughout the 12 days of incubation. The male glands secreted slightly less prolactin than growth hormone, on the average, during all incubation intervals and the profile of prolactin secretion was not appreciably different from that of growth hormone. The female pituitaries secreted slightly more prolactin than growth hormone until about the eighth day of incubation. Thereafter, they secreted substantially more prolactin than growth hormone.

Presenting the data on prolactin and growth hormone secretion as average accumulated levels at the different incubation times obscures differences in the relative amounts of each hormone secreted by individual glands in the different incubation intervals. In a number of instances, substantially more prolactin than growth hormone was secreted by individual male pituitaries at different culture intervals, and the converse of this was also seen. The same phenomenon was observed with the female pituitaries during the first week of incubation.

The results obtained with the glands of the lactating females and their infants are shown in Fig. 5. After the third day the glands of the mothers

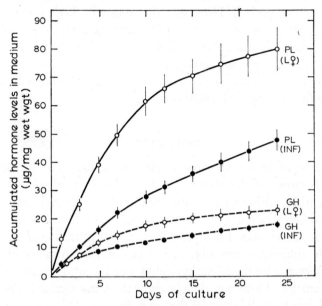

FIG. 5. Accumulated amounts of prolactin (PL) and growth hormone (GH) secreted by adenohypophyses of 9 lactating female (L♀) and 9 infant (INF) rhesus monkeys during 24 days in organ culture. Densitometric estimates from disc electrophoretic columns. Vertical bars represent S.E.M.

secreted significantly more growth hormone per mg wet weight of tissue than did those of their infants. However, by the 24th day of culture, this amounted to an increment in growth hormone secretion of only about 30 per cent above that of the glands of the infants. By the 12th day of culture, the glands of the lactators secreted about the same amount of growth hormone as did the adenohypophyses of either the adult males or non-lactating females over the same culture period (Fig. 3).

The adenohypophyses of the infant monkeys secreted surprisingly large amounts of prolactin *in vitro*. By the 12th day of incubation, they had secreted as much prolactin as did the glands of the adult non-lactating females and 65 per cent more than did the adult male adenohypophyses during the same period. Although the rate of prolactin secretion by the infant glands decreased after about the tenth day of culture, it continued at a fairly constant rate up to the 24th day.

The most striking feature of the data presented in Fig. 5 is the secretory performance of the adenohypophyses of the lactating females. By the

12th day of culture, these glands secreted more than twice as much prolactin as did the glands of either the infants or the adult non-lactating females and more than three times as much as did the male adenohypophyses over the same culture period. However, the unusually high capacity of the glands of lactating females to secrete prolactin was sustained for only about the first ten days of culture. Thereafter, the secretion rate was comparable with that of the infant glands during the same period, and with the secretion profiles obtained with the pituitaries of the adult male and female monkeys during the first 12 days of incubation (Fig. 3).

DISCUSSION

The results of these experiments clearly indicate that prolactin and growth hormone of the rhesus monkey adenohypophysis are discrete chemical entities which are separable by disc electrophoresis and that they are secreted independently of one another *in vitro*. Recent work indicates that they are also secreted independently *in vivo* (Nicoll et al. 1971). No appreciable difference was obtained between the amounts of growth hormone secreted by the adenohypophyses of the adult monkeys and the lactating females. However, the glands of the latter secreted substantially more prolactin than did those of the other groups. This is consistent with results previously obtained with adenohypophyses from lactating and non-lactating rats (Meites, Kahn and Nicoll 1961).

The concentrations of prolactin and growth hormone in the adenohypophyses of these monkeys are comparable to those we have previously found in the glands of immature, adult and lactating female rats (unpublished). In addition, the levels of prolactin in the glands of the adult males and females are comparable to those of a number of other mammalian species (Nicoll and Licht 1971). However, the concentration of growth hormone in the adult monkey glands, including the lactators, is less than half of that found in the glands of several common laboratory animals (Nicoll and Licht 1971). The significance of this is unknown.

An estimate of the relative degree of autonomy of prolactin and growth hormone secretion by the adenohypophyses of these different monkey groups can be obtained by expressing the amount of hormone secreted *in vitro* during a given time span as a function of the amount present in the tissue at the beginning of the culture. Fig. 6 shows the amounts of prolactin and growth hormone secreted by the glands of the monkeys used in these experiments during the first three days of culture, expressed as a percentage of the initial gland content. This time period was selected to allow the results with the monkeys to be compared with other mammalian species, also shown in Fig. 6. When expressed in this manner, the data are essentially an expression of the autonomous turnover rate of the hormones.

It is evident from Fig. 6 that the infant monkey glands have a relatively high autonomous turnover rate for prolactin and a low one for growth hormone. This is due to the combined effect of their relatively high growth

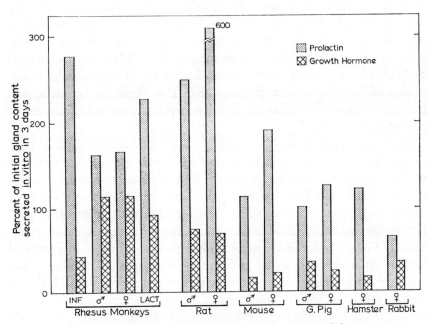

FIG. 6. Autonomous turnover rate of prolactin and growth hormone by adenohypophyses of rhesus monkeys and other mammalian species in organ culture. INF, infant; LACT, lactating.

hormone concentration and prolactin secretion rate, and to their low prolactin concentration and growth hormone secretion rate. The pituitaries of the lactating female monkeys also showed a very high autonomous turnover rate for prolactin *in vitro*. The glands of the adult male and female monkeys and of the lactating females had an unusually high rate of growth hormone turnover in comparison with the other mammalian species. Only the male and female rat pituitaries showed a growth hormone turnover rate independent of hypothalamic influence which approached that of the adult monkeys. It appears that although there is nothing unique about the capacity of the rhesus monkey adenohypophysis to secrete prolactin autonomously *in vitro*, the gland of this primate does have an unusually high capacity to secrete growth hormone independently of hypothalamic influence.

The electrophoretic mobility characteristics of the prolactins and somatotropins of a number of mammalian and non-mammalian vertebrate species in the alkaline (pH 9·5) disc electrophoresis system have now been determined (Nicoll and Nichols 1971; Nicoll and Licht 1971; Nicoll, Licht and Clarke 1971). In all species studied, prolactin was found to be a rapidly migrating protein (R_F value of 0·5 or greater) whereas growth hormone had low electrophoretic mobility (R_F less than 0·5). The R_F value of the RhGH in this system is 0·5 and that of rhesus prolactin is 0·57. In all the other vertebrate species examined, the two hormones were much more clearly separated. This suggests that RhGH and rhesus prolactin may be structurally more similar than are the prolactins and somatotropins of the other vertebrates examined. This would be consistent with the high degree of overlapping activity in primate growth hormone or prolactin preparations.

Three recent reports suggest that human prolactin may be unusual in its electrophoretic characteristics on polyacrylamide gel. Peake and co-workers (1969) have studied a protein from a prolactin-rich human pituitary tumour and Gala (1971) reported on the proteins secreted by a human pituitary in organ culture. They found that the presumptive prolactin migrated *behind* HGH in the alkaline disc gel electrophoretic system. Friesen, Guyda and Hardy (1970) studied protein synthesis in incubated human pituitaries and found that the incorporation of labelled amino acids was much higher in a protein which migrated *behind* HGH, than it was in the HGH itself. These reports suggest that human prolactin is indeed exceptional electrophoretically, inasmuch as we have found no other species in which prolactin migrated behind growth hormone (Nicoll and Licht 1971).

The experiments reported here show that the adenohypophyses of lactating monkeys, when placed in organ culture, secrete large quantities of prolactin into the medium for an extended period. This appears to be a very useful approach for harvesting reasonable quantities of primate prolactin.

SUMMARY

The adenohypophyses of adult male and female rhesus monkeys, and of lactating mothers and their infants, were incubated in organ culture for 12 or 24 days. The proteins secreted into the medium were separated by polyacrylamide disc gel electrophoresis and the amounts contained therein were quantified by densitometry. The rhesus growth hormone in the medium was identified by comparing the columns with ones containing purified rhesus growth hormone (from Dr W. D. Peckham). Rhesus prolactin was identified by several criteria, including isolating the suspected protein and demonstrating that it had high prolactin activity and virtually no growth hormone activity.

The adenohypophyses of all four types of monkeys secreted comparable amounts of growth hormone *in vitro*. However, the glands of the lactating females secreted much more prolactin than did those of the adult male or females or the infants. Aspects of these findings are discussed and it is suggested that culture of adenohypophyses from lactating females may be a useful way of isolating primate prolactin.

Acknowledgements

I am indebted to Miss Mary Tai, Miss Nan Nutt and Mr C. W. Nichols, Jr for technical assistance. This work was supported by a grant from the Population Council.

REFERENCES

BENJAMIN, F., CASPER, D. J. and KALODNY, H. M. (1969) *Obstet. & Gynecol. (N.Y.)* **34**, 34–39.
BEWLEY, T. and LI, C. H. (1970) *Science* **168**, 1361–1362.
BOARD, J. A. (1968) *Am. J. Obstet. & Gynecol.* **100**, 1106–1109.
CHANNING, C. P., TAYLOR, M., KNOBIL, E. and NICOLL, C. S. (1970) *Proc. Soc. Exp. Biol. & Med.* **135**, 540–542.
FORSYTH, I. A. (1969) In *Lactogenesis*, pp. 195–206, ed. Reynolds, M. and Folley, S. J. Philadelphia: University of Pennsylvania Press.
FREEMAN, R. and WILHELMI, A. E. (1969) *Program of the Endocrine Society, 51st Meeting*, p. 44 (abst. 27).
FRIESEN, H., GUYDA, H. and HARDY, J. (1970) *J. Clin. Endocrinol. & Metab.* **31**, 611–624.
GALA, R. R. (1971) *J. Endocrinol.* **50**, 637–642.
GUYDA, H. J. and FRIESEN, H. G. (1971) *Biochem. & Biophys. Res. Commun.* **42**, 1068–1075.
LI, C. H. and YAMASHIRO, D. (1970) *J. Am. Chem. Soc.* **92**, 7608–7609.
LEWIS, U. J., CHEEVER, E. V. and HOPKINS, W. C. (1970) *Biochim. & Biophys. Acta* **214**, 598–508.
LYONS, W. R. (1969) In *Lactogenesis*, pp. 223–227, ed. Reynolds, M. and Folley, S. J. Philadelphia: University of Pennsylvania Press.
MEITES, J., KAHN, R. H. and NICOLL, C. S. (1961) *Proc. Soc. Exp. Biol. & Med.* **108**, 440–443.
MEITES, J. and TURNER, C. W. (1950) In *Hormone Assay*, pp. 237–260, ed. Emmens, C. W. New York: Academic Press.
MINAGUCHI, H., CLEMENS, J. A. and MEITES, J. (1968) *Endocrinology* **82**, 555–558.
NAGY, I., KURCZ, M. and BARANYAI, P. (1968) *Acta Biochim. & Biophys. Acad. Sci. Hung.* **4**, 27–36.
NICOLL, C. S., BLAIR, S., NICHOLS, C. W., JR and TAYLOR, M. (1971) *Program of the Endocrine Society, 53rd Meeting*, p.24 (abst. 144).
NICOLL, C. S. and LICHT, P. (1971) *Gen. & Comp. Endocrinol.* (in press).
NICOLL, C. S., LICHT, P. and CLARKE, W. C. (1971) *Abstracts, Second International Symposium on Growth Hormone*, p. 25 (abst. 45). International Congress Series 236. Amsterdam: Excerpta Medica.
NICOLL, C. S. and NICHOLS, C. W., JR (1971) *Gen. & Comp. Endocrinol.* **17**, 300–310.
NICOLL, C. S., PARSONS, J. A., FIORINDO, R. P. and NICHOLS, C. W., JR (1969) *J. Endocrinol.* **45**, 183–196.
NICOLL, C. S., PARSONS, J. A., FIORINDO, R. P., NICHOLS, C. W., JR and SAKUMA, M. (1970) *J. Clin. Endocrinol. & Metab.* **30**, 512–519.

PASTEELS, J. L. (1969) In *Lactogenesis*, pp. 207–216, ed. Reynolds, M. and Folley, S. J. Philadelphia: University of Pennsylvania Press.
PASTEELS, J. L., BRAUMAN, H. and BRAUMAN, J. (1963) *C. R. Hebd. Séance Acad. Sci., Sér. D* **256**, 2031–2033.
PEAKE, G. T., MCKEEL, D. W., JARETT, L. and DAUGHADAY, W. H. (1969) *J. Clin. Endocrinol. & Metab.* **29**, 1383–1393.
PECKHAM, W. D. (1967) *J. Biol. Chem.* **242**, 190–196.
PECKHAM, W. D. and NICOLL, C. S. (1971) *Program of the Endocrine Society, 53rd Meeting*, p. 123 (abst. 161).
RIMOIN, D. L., HOLZMAN, G. B., MERIMEE, T. J., RABINOWITZ, D. and MCKUSICK, V. A. (1968) *J. Clin. Endocrinol. & Metab.* **28**, 1183–1188.
ROTH, J., GORDEN, P. and BATES, R. W. (1968) In *Growth Hormone*, pp. 124–128, ed. Pecile, A. and Müller, E. E. Amsterdam: Excerpta Medica Foundation.
SHERWOOD, L. M. (1971) *New Engl. J. Med.* **284**, 774–777.
SPELLACY, W. M., BUHI, W. D. and BIRK, S. A. (1970) *Am. J. Obstet. & Gynec.* **107**, 244–249.
TASHJIAN, A. M., JR, LEVINE, L. and WILHELMI, A. E. (1965) *Endocrinology* **77**, 1023–1036.
VOOGT, J. H., CHEN, C. H. and MEITES, J. (1970) *Am. J. Physiol.* **218**, 396–399.

[For discussion, see pp. 277–286.]

TISSUE CULTURE OF HUMAN HYPOPHYSES: EVIDENCE OF A SPECIFIC PROLACTIN IN MAN

J. L. Pasteels

Laboratoires d'histologie et de microscopie électronique, Faculté de Médecine et de Pharmacie,
Université Libre de Bruxelles

The discovery of the inherent prolactin-like activities of human growth hormone (HGH) ten years ago led to serious controversies about the existence of human prolactin as a separate molecule (Chadwick, Folley and Gemzell 1961; Ferguson and Wallace 1961; Forsyth 1964, 1969, 1970; Forsyth, Folley and Chadwick 1965; Wilhelmi 1964; Lyons 1969; Pasteels 1969a). At the same time, it was found that rat pituitaries in tissue or organ culture secrete large amounts of prolactin (Meites, Kahn and Nicoll 1961; Pasteels 1961a). Evidence for inhibitory hypothalamic control of prolactin secretion was obtained by culturing pituitary and hypothalamic tissue together (Pasteels 1961b; Danon, Dikstein and Sulman 1963) or by adding hypothalamic extracts to the culture medium of pituitary tissue (Pasteels 1962a, 1963; Talwalker, Ratner and Meites 1963).

We felt that similar culture experiments on human hypophyses would be valuable. One would expect that when the human anterior lobe is removed from the influence of the hypothalamus, its prolactin secretion will increase, while the release of HGH will decrease, as in the rat. This approach should be helpful in making a distinction between HGH and prolactin.

TISSUE CULTURE

In a first set of experiments pituitary tissue was cultured by the hanging drop technique in a medium containing 25 per cent horse serum, 5 per cent chick embryonic extract and 70 per cent Tyrode solution (Pasteels 1963). From the hypophyses of three adult men (aged 56, 59 and 76 years), 128 cultures were prepared. From the whole anterior lobe of five foetuses, 173 cultures were obtained. The foetuses were from 3 to 7 months of gestation. In the first study, we found that adult or foetal hypophyses secreted increasing amounts of a pigeon crop gland-stimulating material (Pasteels 1962b). In a subsequent experiment, a comparison was made between the prolactin activity measured by the crop sac assay and growth hormone levels, estimated by complement fixation. A dramatic decline in HGH secretion was observed, in striking contrast to the increase in pigeon crop gland-stimulating activity (Pasteels, Brauman and Brauman 1963; Brauman, Brauman and Pasteels 1964). Moreover, adding hypothalamic extracts significantly inhibited the prolactin activity, while increasing the release of HGH twofold (Pasteels 1963).

Morphological controls showed that only the erythrosinophilic cells multiply in these conditions of tissue culture. The somatotrophs had disappeared by the end of the second week of culture (without hypothalamic extract). Accordingly a gradual selection of erythrosinophils occurred, leading to an apparently pure culture of prolactin cells. Unfortunately, such cultures could not be maintained in good condition for more than six weeks.

LONG-TERM ORGAN CULTURES

Organ cultures of the hypophyses of human neonates or foetuses were then made, in order to obtain more prolactin. Procedures were used which allow the pituitaries to remain in a healthy state for a longer period without much proliferation (this could be responsible for mutations, giving abnormal hypophysial cells). So far, explants of the whole hypophyses of 46 humans have been cultivated in such conditions for periods ranging from a few weeks to 30 months. The details of the technique have been described previously (Pasteels 1969b).

Morphology

Pituitary organ cultures of humans are very similar to grafts of the rat hypophysis made under the kidney capsule (Courrier et al. 1961). As pointed out by Petrovic (1963), the anterior lobe maintains its heterogeneous structure as well as its organotypic character. We used differential staining and immunofluorescence on our material. Somatotrophs were identified by an immune serum to HGH (Raben) and atrophic gonadotrophs by antibodies to human chorionic gonadotropin. Both these cell types were still present in hypophyses cultivated for more than two years (Pasteels 1969b). More recently, we have been able to confirm this by electron microscopy (unpublished results). However, such cells are poorly developed. The cultures are composed primarily of highly active prolactin cells, as shown by differential staining and electron microscopy (Fig. 1; cf. my paper on the morphology of prolactin secretion, this volume, pp. 241–255).

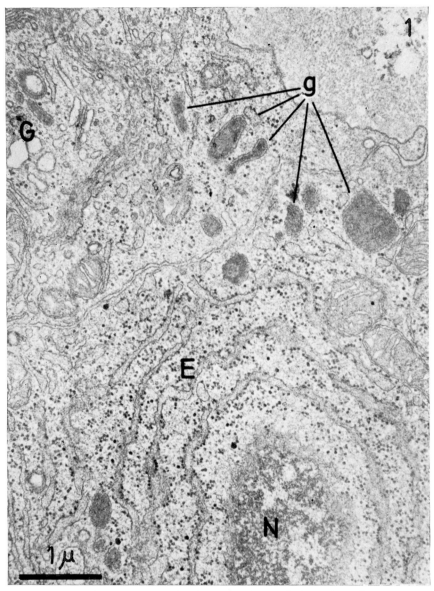

FIG. 1. Electron micrograph of human prolactin cells from a hypophysis of a 7-month-old foetus, maintained in organ culture for four weeks. N, nucleus; G, Golgi apparatus. As in other mammals, the human prolactin cells are characterized by their lamellar ergastoplasm (E) and the irregular shape of their secretory granules (g).

(*facing page 270*)

Evidence of prolactin secretion

(1) Pigeon crop gland assay

The presence of prolactin in organ culture media was first demonstrated by our modification (Pasteels 1963) of Lyons and Page's pigeon crop gland assay. However, the organ cultures also secrete some HGH, as shown by immunoassays (Pasteels, Brauman and Brauman 1965) and by the tibia test on hypophysectomized rats (Pasteels 1969b). It was therefore necessary to distinguish whether the pigeon crop gland stimulation resulted from HGH or from prolactin. Two experiments were done for this purpose.

Neutralization of growth hormone with an anti-HGH serum. Samples of organ culture medium containing the hypophysial secretion were divided into two equal parts. One half was incubated with antibodies to HGH (for details of the procedure see Pasteels 1967). The second half was used as a control. No significant difference in pigeon crop gland-stimulating activity was observed when the two fractions were injected on the two sides of the crop in the same pigeon (Table I).

TABLE I

ASSAY OF PROLACTIN ACTIVITY IN ORGAN CULTURE MEDIA OF HUMAN HYPOPHYSES

Diameter of stimulated area of pigeon crop sac gland
mm

Experiment	Control medium	Medium incubated with antibodies to HGH	P
1	$31 \cdot 83 \pm 1 \cdot 08$ ($n=6$)	$31 \cdot 50 \pm 0 \cdot 99$ ($n=6$)	N.S.
2	$26 \cdot 33 \pm 0 \cdot 80$ ($n=6$)	$27 \cdot 0 \pm 1 \cdot 24$ ($n=6$)	N.S.

Destruction of growth hormone by heating in neutral solution. Growth hormone is destroyed by heating in neutral solution, while prolactin is not (Geschwind and Li 1954). We made use of this finding. Samples of organ culture medium were divided into two equal parts. One half was used as a control, without heating. The second one was boiled for 20 minutes at pH $7 \cdot 2$. This neutralized most of the HGH activity (measured by increased body weight of hypophysectomized immature rats and by the tibia test for growth hormone) but did not impair the pigeon crop gland-stimulating activity (Table II).

(2) Mammogenic stimulation in hypophysectomized rats

The same 30 immature hypophysectomized rats that had been used for the assay of growth hormone in the culture medium provided evidence of mammary gland-stimulating activity. Mammary growth was especially conspicuous in the animals injected with untreated medium. Even after destroying HGH by heating, we observed an obvious budding of mammary alveoli. It is well known that growth hormone and prolactin have synergistic actions on the mammary gland (Forsyth 1964). From these results

TABLE II

COMPARISON OF THE HGH AND PROLACTIN ACTIVITY OF ORGAN CULTURE MEDIA WITH AND WITHOUT PREVIOUS HEATING IN NEUTRAL SOLUTION

Increase of body weight in immature hypophysectomized rats (g):

Untreated	Heated medium	Control medium
-0.82	-0.77	$+2.71$
± 0.36	± 0.65	± 0.80
$(n=10)$	$(n=10)$	$(n=10)$

N.S. (Untreated vs Heated); $P<0.001$ (vs Control)

Tibia test for growth hormone (width of the cartilage in μm):

Untreated	Heated medium	Control medium
131·3	170·2	218·1
± 6.7	± 5.2	± 4.6
$(n=10)$	$(n=10)$	$(n=10)$

$P<0.001$ (Untreated vs Heated); $P<0.001$ (vs Control)

Diameter of the pigeon crop gland stimulated area (mm):

Heated medium	Control medium
57·73	57·45
± 0.94	± 0.91
$(n=11)$	$(n=11)$

N.S.

it may be concluded that organ cultures of human hypophyses secrete some HGH, and a distinct human prolactin.

First attempts to purify human prolactin from the cultures

Large amounts of organ culture media containing the secretion of the hypophyses were first processed by our modification of Simkin and Goodart's (1960) acid-acetone precipitation for prolactin. We have shown that the recovery of prolactin activity is excellent with this technique. We compared the pigeon crop gland-stimulating activity of the original pool of unextracted medium and the acid-acetone precipitate diluted to its initial volume. No significant difference was observed (Pasteels 1967).

The soluble material was then dialysed for 36 hours against distilled water and lyophilized. Further purification was attempted by chromatography on a Sephadex G-200 column, on DEAE cellulose, or by preparative electrophoresis (Robyn and Pasteels, unpublished results). The purified fractions were assayed on the pigeon crop gland, with reference to NIH-OP-S-6 ovine prolactin.

The original acid-acetone precipitate, after dialysis and lyophilization, had a prolactin activity of 0·30 i.u./mg (95 per cent confidence limits: 0·18–0·46). This material was then purified by zone electrophoresis on

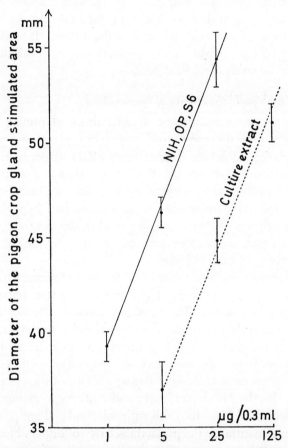

FIG. 2. Assay of the prolactin activity of an extract prepared from organ culture medium of human hypophysis compared with NIH-OP-S-6 ovine prolactin. The specific activity of the semi-purified human prolactin preparation was 2·95 i.u./mg (range 1·92–4·55).

horizontal 'Pevicon' blocks (Pevicon, Phosphatbolaget, Uppsala, Sweden) with veronal buffer pH 8·6, ionic strength 0·05. The electrophoresis was continued for 20 hours at 350–400 V and about 15 mA in a cold room. The Pevicon block was then cut into segments 1 cm wide. Each segment was eluted with 5 ml distilled water and the protein concentration of each fraction was estimated by spectrophotometry, reading at 780 nm.

Seventy per cent of the prolactin activity was recovered in a 'prealbumin' fraction, with a specific activity of 2·95 i.u./mg of prolactin (range 1·92–4·55) (Fig. 2). With this material, a clean response of the pigeon crop gland was observed, with typical 'ridging' of the epithelium and without inflammatory reactions.

The fact that a *crude* prealbumin electrophoretic fraction has a specific prolactin activity equivalent to that of purified HGH is further evidence for the existence of a specific prolactin in the human. It seems likely that the presence of some highly active prolactin in the prealbumin fraction accounts for the results described here.

Comparison with prolactin activity of human blood

Our experiments on tissue or organ culture of human hypophyses provide evidence for the secretion of a pigeon crop gland- and rat mammary gland-stimulating substance distinct from HGH. However, it could be objected that this 'prolactin secretion' is an artifact of culture, without physiological equivalent in normal humans. Such an objection would be far-fetched, for it is known that other mammalian hypophyses secrete prolactin in culture. Nevertheless, we tried to demonstrate an immunological correspondence between prolactin secreted by the cultures and the prolactin activity of human blood.

For this purpose we immunized rabbits with medium from the organ culture of human hypophyses. The immune sera thus obtained are not immunologically specific. They contain antibodies to the horse serum and the chicken embryo extract present, in addition to the hormone in the culture medium. However, the primary human material injected into the rabbits is prolactin. Such immune sera proved to be effective in neutralizing the pigeon crop gland-stimulating activity of the culture medium, even when absorbed by horse serum and embryonic extract. We called them 'antiprolactin' I and II. Both antiprolactin I and antiprolactin II sera were shown to neutralize the prolactin activity of human blood, collected either *post partum* or in women suffering from amenorrhoea–galactorrhoea. As a control, we used potent immune sera to HGH (Raben). They did not neutralize the prolactin activity of culture media or of human blood. The details of these experiments have been reported previously (Pasteels 1967, 1969a).

Strong criticisms have been made of the use of the pigeon crop gland local assay (as in our experiments) to measure prolactin activity in human blood (Bahn and Bates 1956). We admit that even after acetone extraction following Simkin and Goodart's procedure (Simkin and Goodart 1960; Pasteels 1967) the prolactin response of the crop could be partly confused with a non-specific inflammatory reaction. Such a technique is definitely not proposed for clinical use. In our neutralization experiments, however, the inhibition of the crop gland stimulation by antibodies to prolactin demonstrates that the crop reaction was specific for prolactin.

More recently (Bryant *et al.* 1971) our human prolactin preparation and its antiserum have been used to develop a radioimmunoassay for prolactin in human plasma. Human plasma samples were found to contain a material immunologically similar to the prolactin fraction isolated from the organ cultures of foetal pituitaries. No cross-reaction was found with HGH or placental lactogen. The plasma prolactin content was shown to be increased in a lactating woman and after the injection of a phenothiazine.

DISCUSSION

Our initial findings on the secretion of a specific prolactin, distinct from growth hormone, by cultures of human hypophyses, have been confirmed by Nicoll and co-workers (1970) in the monkey, and possibly by Friesen, Guyda and Hardy (1970) in short-term incubation of human pituitaries. They also agree with recent clinical findings showing that in normal or pathological cases HGH is not secreted in sufficient amounts to account for mammary gland stimulation (Rimoin *et al.* 1968; Board 1968; Spellacy, Carlson and Schade 1968; Benjamin, Casper and Kolodny 1969; Peake *et al.* 1969; Frantz and Kleinberg 1970). We believe that the existence of a separate primate prolactin has now been well demonstrated.

However, prolactin still eludes chemical extraction from human hypophyses. We have already proposed a morphological explanation for this. In hypophyses from non-pregnant and non-lactating women or from men, prolactin cells are exceptionally few and small by comparison with the hypophyses of other mammals. This conclusion was based on the study of individual hypophyses by differential staining (Pasteels 1963, 1969*a*). Recent studies with immunofluorescence, using antibodies to NIH ovine prolactin on human hypophyses, have confirmed this conclusion.

It is obvious that extraction of human prolactin should be started from a material containing significant amounts of the hormone, such as the hypophyses of pregnant or postpartum women, or from cultures of human pituitaries. Such material is not readily available.

CONCLUSION

Experiments with tissue or organ culture of human hypophyses provide convincing evidence for the existence of a human prolactin, distinct from human growth hormone.

SUMMARY

The fact that when mammalian hypophyses are removed from hypothalamic control, they secrete increasing amounts of prolactin while their other endocrine functions are reduced, provides an exceptional opportunity to identify human prolactin as a separate entity, distinct from human growth hormone.

In a first set of experiments we have shown that human hypophyses in tissue culture secreted a pigeon crop gland-stimulating hormone, whereas the release of growth hormone (assayed immunologically) decreased with time in culture. When rabbits were immunized against this prolactin secreted *in vitro* the immune sera neutralized the prolactin activity of human blood, collected *post partum* or from cases of amenorrhoea-galactorrhoea.

In our latest experiments we have succeeded in obtaining organ cultures of human hypophyses which survived for two to three years. Their prolactin activity has been assayed on the pigeon crop gland and also on the mammary glands of hypophysectomized rats. Comparisons have been made with the bioassay of growth hormone in the same animals. These results are discussed with reference to confirmatory evidence of other authors and also to clinical and histological findings suggesting that human prolactin exists as a separate hormone.

Acknowledgements

We are pleased to acknowledge the financial support of the Belgian Fonds de la Recherche Scientifique and Fonds de la Recherche Scientifique Médicale, grants of purified ovine prolactin from the U.S. National Institutes of Health, and the gift of human growth hormone by Dr Raben. The author wishes to thank Dr C. S. Nicoll for improvements in the writing of the manuscript.

REFERENCES

BAHN, R. and BATES, R. (1956) *J. Clin. Endocrinol. & Metab.* **16**, 1337–1346.
BENJAMIN, F., CASPER, D. J. and KOLODNY, H. H. (1969) *Obstet. & Gynecol. (N.Y.)* **34**, 34–39.
BOARD, J. A. (1968) *Am. J. Obstet. & Gynecol.* **100**, 1106–1109.
BRAUMAN, J., BRAUMAN, H. and PASTEELS, J. L. (1964) *Nature (Lond.)* **202**, 1116–1118.
BRYANT, G. D., SILER, T. M., GREENWOOD, F. C., PASTEELS, J. L., ROBYN, C. and HUBINONT, P. O. (1971) *Hormones* **2**, 139–152.
CHADWICK, A., FOLLEY, S. J. and GEMZELL, C. A. (1961) *Lancet* **2**, 241–243.
COURRIER, R., COLONGE, A., HERLANT, M. and PASTEELS, J. L. (1961) *C. R. Hebd. Séance Acad. Sci., Sér. D* **252**, 645–647.
DANON, A., DIKSTEIN, S. and SULMAN, F. G. (1963) *Proc. Soc. Exp. Biol. & Med.* **114**, 366–368.
FERGUSON, K. A. and WALLACE, A. L. (1961) *Nature (Lond.)* **190**, 632–633.
FORSYTH, I. A. (1964) *J. Endocrinol.* **31**, xxx–xxxi.
FORSYTH, I. A. (1969) In *Lactogenesis: the Initiation of Milk Secretion at Parturition*, pp. 195–205, ed. Reynolds, M. and Folley, S. J. Philadelphia: University of Pennsylvania Press.

FORSYTH, I. A. (1970) *J. Endocrinol.* **46**, iv–v.
FORSYTH, I. A., FOLLEY, S. J. and CHADWICK, A. (1965) *J. Endocrinol.* **31**, 115–126.
FRANTZ, A. G. and KLEINBERG, D. L. (1970) *Science* **170**, 745–747.
FRIESEN, H., GUYDA, H. and HARDY, J. (1970) *J. Clin. Endocrinol. & Metab.* **31**, 611–624.
GESCHWIND, I. I. and LI, C. H. (1954) In *The Hypophyseal Growth Hormone, Nature and Actions*, pp. 28–53, ed. Smith, R. W., Gaebler, O. H. and Long, C. N. H. New York: McGraw-Hill.
LYONS, W. R. (1969) In *Lactogenesis: the Initiation of Milk Secretion at Parturition*, pp. 223–228, ed. Reynolds, M. and Folley, S. J. Philadelphia: University of Pennsylvania Press.
MEITES, J., KAHN, R. H. and NICOLL, C. S. (1961) *Proc. Soc. Exp. Biol. & Med.* **108**, 440–443.
NICOLL, C. S., PARSONS, J. A., FIORINDO, R. P., NICHOLS, C. W., JR and SAKUMA, M. (1970) *J. Clin. Endocrinol. & Metab.* **30**, 512–519.
PASTEELS, J. L. (1961a) *C. R. Hebd. Séance Acad. Sci., Sér. D* **253**, 2140–2142.
PASTEELS, J. L. (1961b) *C. R. Hebd. Séance Acad. Sci., Sér. D.* **253**, 3074–3075.
PASTEELS, J. L. (1962a) *C. R. Hebd. Séance Acad. Sci., Sér. D* **254**, 2664–2666.
PASTEELS, J. L. (1962b) *C. R. Hebd. Séance Acad. Sci., Sér. D* **254**, 4083–4085.
PASTEELS, J. L. (1963) *Arch. Biol. (Brussels)* **74**, 439–553.
PASTEELS, J. L. (1967) *Ann. Endocrinol. (Paris)* **28**, 117–126.
PASTEELS, J. L. (1969a) In *Lactogenesis: the Initiation of Milk Secretion at Parturition*, pp. 207–216, ed. Reynolds, M. and Folley, S. J. Philadelphia: University of Pennsylvania Press.
PASTEELS, J. L. (1969b) *Mém. Acad. R. Méd. Belg.* **7**, 1–45.
PASTEELS, J. L., BRAUMAN, H. and BRAUMAN, J. (1963) *C. R. Hebd. Séance Acad. Sci., Sér. D* **256**, 2031–2033.
PASTEELS, J. L., BRAUMAN, H. and BRAUMAN, J. (1965) *C. R. Hebd. Séance Acad. Sci., Sér. D* **261**, 1746–1748.
PEAKE, G. T., MCKEEL, D. W., JARETT, L. and DAUGHADAY, W. H. (1969) *J. Clin. Endocrinol. & Metab.* **29**, 1383–1393.
PETROVIC, A. (1963) *Coll. Int. 128 La cytologie de l'adénohypophyse*, pp. 121–136. Paris: CNRS.
RIMOIN, D. L., HOLZMAN, G. B., MERIMEE, T. J., RABINOWITZ, D., BARNES, A. C., TYSON, J. E. A. and MCKUSICK, V. A. (1968) *J. Clin. Endocrinol. & Metab.* **28**, 1183–1188.
SIMKIN, B. and GOODART, D. (1960) *J. Clin. Endocrinol. & Metab.* **20**, 1095–1106.
SPELLACY, W. N., CARLSON, K. L. and SCHADE, S. L. (1968) *Am. J. Obstet. & Gynecol.* **100**, 84–89.
TALWALKER, P. K., RATNER, A. and MEITES, J. (1963) *Am. J. Physiol.* **205**, 213–218.
WILHELMI, A. E. (1964) *Metabolism* **13**, 1165.

DISCUSSION

Swanson Beck: Dr Pasteels, you say that your immunofluorescence staining system is giving specific localization of prolactin in the erythrosinophilic cells. What evidence do you have of the serological specificity of this reaction?

Pasteels: We prepared antibodies to ovine prolactin (as was done by Herbert and Hayashida 1970) and we hoped to find some cross-reaction between the immune sera thus obtained and Raben's human growth hormone, to be able to neutralize the antibodies to growth hormone by absorption with HGH, and thus to obtain only antibodies to prolactin. Unfortunately, from the six rabbits that we immunized, none showed cross-reaction with human growth hormone, and we found no cross-reaction between several rabbit immune sera to human growth hormone and the ovine prolactin used. At the present time the specificity of the antibodies has been assessed only on Ouchterlony plates and with immunoelectrophoresis. We know that such techniques are not very sensitive and we hope to achieve a still better demonstration of the specificity of the immune sera by radioimmunoassay; this is being done now. But we don't think it will change the results; clearly, the antibodies to human growth hormone are reacting with one kind of 'acidophil' and antibodies to ovine prolactin are reacting with different cells that undergo hypertrophy during pregnancy and lactation.

Swanson Beck: But it could be that your immune serum also contains an antibody to some other unidentified constituent of ovine prolactin. How then can you be certain that you are staining prolactin in the erythrosinophilic cells?

Pasteels: This constituent would have to be specific for pregnancy and lactation cells. I believe that this is rather far-fetched.

Frantz: We tested our anti-ovine prolactin against human growth hormone not only in bioassay, where we demonstrated some degree of neutralization, but also by double diffusion in agar gel, and were unable (as were Herbert and Hayashida 1970) to show any precipitin line. We did find, however, that when we labelled human growth hormone and submitted it to the antiserum to ovine prolactin in high concentrations, the antiserum did bind the label; about 30-40 per cent binding.

Greenwood: Dr Pasteels, have you used your antiserum to human prolactin in fluorescence studies in the human pituitary?

Pasteels: I have tried but it is too weak. For fluorescence you need high concentrations of antibodies, much more than for radioimmunoassay.

Greenwood: On theoretical grounds I would have argued the converse, that for a radioimmunoassay one would need a better antiserum, because in immunofluorescence there is essentially a solid phase with a tissue-fixed antigen and the reaction is 'amplified' by the fluorescent second antibody.

Swanson Beck: This is so. On the other hand, a fluorochrome dye is not nearly as efficient a label as a radioactive isotope: therefore a relatively large quantity of antibody must be attached before it can be detected with the immunofluorescence technique.

Greenwood: What about site amplification?

Swanson Beck: This might give four-fold amplification with the indirect technique. A further problem is that the staining reaction is generally done on fixed tissue and my impression is that there is considerable loss of antigen in histologically fixed tissue.

Pasteels: We have tried both procedures on the same pituitaries. In fixed tissue you have to remove the formalin before paraffin embedding. This is the critical point. If you do that by thorough rinsing after fixation you get practically the same result as on frozen material. We used the two halves of the same pituitary and found no difference in immunofluorescence between the fixed and unfixed halves.

Bryant: Dr Pasteels, do you see many lysosomes in your pituitary cultures? I wondered whether we are getting maximum secretion in culture or whether prolactin is being destroyed by lysosomes?

Pasteels: There can be lysosomes in any culture, because once the cells are being starved they start to digest their own components by means of lysosomes, to remain alive. This is a general biological mechanism. We did find lysosomes in long-term cultures, but we don't think this has any special significance for the problem of prolactin, because it could be the result of too much delay in changing the culture medium and so on. On the other hand, we saw lysosomes in short-term incubation experiments, as a response to ergocornine treatment: in those conditions we think they might be significant in relation to prolactin.

Sherwood: From your electron micrographs it looks as if you think the secretory process is similar to the one Lacy (1970) believes is present in the β-cell (that is, emmiocytosis, with the release of granules).

Pasteels: Exocytosis has already been frequently described for prolactin release (see Sano 1962; Pasteels 1963). I didn't stress this, therefore. I showed examples of this mechanism to illustrate that electron microscopy can be an efficient tool in endocrinology.

Sherwood: Dr MacLeod's observations on calcium and the idea that calcium is involved in microtubular function and granule ejection would fit here as well.

Pasteels: Yes. There are quite a lot of experiments on tissues other than the hypophysis on the role of microtubules, such as the Lacy experiments. I have no evidence yet for such a mechanism for secretion of prolactin. We see some microtubules in electron micrographs but probably this is a very general cytological feature after glutaraldehyde fixation.

Sherwood: Have you added colchicine to any tissues? Colchicine is known to inhibit microtubular function.

Pasteels: No.

Nicoll: We added colchicine and vinblastine to incubated rat pituitary glands and observed *stimulation* of early release of prolactin; subsequently no effect was noted. This doesn't support the suggestion that microtubules participate in the exocytic process.

Turkington: A feature of prolactin secretion which has impressed all of us who are measuring its concentration in plasma is how rapidly it can be secreted, and at very high levels for long periods of time, in women who are not pregnant or lactating and in males. Presumably in these people only a small percentage of the pituitary cells are prolactin cells, and I wonder if one can account for this huge outpouring of prolactin from such a small number of cells, or whether one must assume that some of the other cells also participate in prolactin synthesis.

Pasteels: This has never been satisfactorily demonstrated. Using immunofluorescence or electron microscopy after applying labelled antibody, as in the recent experiments of Nakane (1971), it's impossible to show that one pituitary cell is secreting a second hormone. In previous work (Pasteels 1963) we obtained tissue cultures that apparently contained only prolactin cells, and then after giving hypothalamic extracts we obtained evidence of PAS-positive cells that were probably secreting gonadotropins or TSH. We described this as a change of prolactin cells to another endocrine function, but we were unable to confirm this in other experimental conditions, so we cannot even ascertain whether a specific pituitary cell can change its function. However, one should not be surprised that those few prolactin cells, containing not very much secretory material, can release large quantities of prolactin. I think it's a question of turnover rates and that when you see a lot of prolactin cells filled with secretory material they are very often in a static condition, as probably are the somatotrophs. So it may not be surprising that a large amount of prolactin is released in comparison with growth hormone, at a time when the content of the somatotrophs is high.

Friesen: Dr Pasteels, what is the number of prolactin cells compared to the number of FSH, LH or TSH cells, in the non-pregnant female or male pituitary gland?

Pasteels: We cannot make absolute cell counts because in order to do so one must be able to identify every cell, and with differential staining one sees a lot of so-called 'undifferentiated cells' that apparently don't contain secretory material. If you then submit the same section to immunofluorescence you can recognize more differentiated cells. Even in foetal pituitaries (we did this on human foetuses by electron microscopy) practically all the cells contain some secretory material: so there are no real 'chromophobes'. What we have done (as I described in my paper, p. 245) is to estimate the area of a pituitary section occupied by prolactin cells, because the Chalkley procedure takes into account the number of cells and also their degree of hypertrophy. Using this method we found in adult male and non-lactating female monkeys a count ranging from 2·5 to 11·0 per cent. In lactating females it ranged from 18 to 32 per cent.

Friesen: How did that compare with the percentage of thyrotrophs or gonadotrophs?

Pasteels: We didn't count those cells on the same sections, so I cannot make an exact comparison.

Herlant: I think there are many prolactin cells which are impossible to recognize in the non-pregnant pituitary, because their granulations are too sparse.

Greenwood: Our foetal pituitaries seem to be producing about 1 μg each of prolactin a day, whereas Dr Nicoll's monkey pituitaries seem to be producing about 20 μg/mg; his best one produced 80 μg/mg cumulatively after 25 days. We don't weigh our glands in order to keep things sterile, so it's not easy to equate the yields.

Frantz: Dr Nicoll was using a completely synthetic medium for incubation, with only insulin added. Dr Pasteels, did you add foetal calf serum or some other protein to your media?

Pasteels: For short incubations (less than two days) we use synthetic Medium 199, but for long experiments (more than two days in culture) we need to have 10 per cent of serum present. We use rabbit serum since rabbits are being used for the production of the immune sera: it does not lead to the production of too many antibodies. We also need to include chick embryo extract, for survival of the cultures, in these very long-term cultures where we want to obtain the largest amounts of prolactin. The limiting factor being the supply of human hypophyses, we try to keep these as long as possible in a healthy condition.

Turkington: I was very interested in Dr Nicoll's electrophoretic analyses of growth hormone and prolactin. It looked to me that your rhesus monkey growth hormone standard always showed a band moving ahead of the main band. I wonder whether that material is also part of the prolactin you are measuring?

Nicoll: That concerned me also, but Dr Peckham pointed out that this band is not evident when the material is run on an acid acrylamide system. Accordingly, he believes that it is deamidated growth hormone. There is probably a minor amount of this deamidated growth hormone in the primary prolactin band of our culture medium samples.

Turkington: There is the question of whether you are not getting a growth hormone species in that band that has say a hundred times the specific activity of the other growth hormone species. That is a very theoretical objection but I was wondering if you had analysed the prolactin band further by radioimmunoreactivity to an anti-growth hormone antiserum?

Nicoll: We haven't done that. However, as you suggested, that is a highly conjectural question.

Friesen: In our preparations of monkey prolactin derived from incubation media we have also seen two bands which had the same mobility as HGH. We were never convinced that prolactin was confined to one or other of the two components. On starch gel electrophoresis (of culture media from either pregnant or lactating monkey glands) we invariably saw the same two bands, both containing a lot of radioactive protein. After absorption with antibody to growth hormone there were still two radioactive proteins which remained. If we measured prolactin in the eluates from the two bands we always found some prolactin in both. Have you done any incubation studies using radioactive amino acids to see whether you could identify which band contained most of the radioactivity, because that band, I believe, would represent prolactin?

Nicoll: We have not done any incorporation studies, but we are not in disagreement because we feel that the two prominent bands that are present in the crude medium are two forms of prolactin, and this would agree with your results. The upper (slower-migrating) one was isolated by Dr Peckham and when it was run on the acidic system the front-running band was no longer present. This suggests that in the medium the faster-migrating protein is a deamidated form of prolactin.

Friesen: When you ran the electrophoresis at an alkaline pH, did prolactin move ahead of growth hormone?

Nicoll: Yes, and it also does at an acidic pH, but the secondary prolactin band is not apparent in the acidic electrophoretic system.

Friesen: We found an interesting difference there between monkey and human prolactin. It appears that human prolactin moves ahead of growth hormone toward the cathode at an acid pH, but not at an alkaline pH.

Nicoll: Bill Peckham and I have obtained the same results with human prolactin.

Greenwood: During the purification of medium from tissue cultures of foetal pituitaries we find two areas on a gradient of DEAE Sephadex, and we find immunoactivity in both. Professor Li, how easy is it to deamidate a molecule which might exist called human prolactin? Our purifications are done directly on tissue culture media, so there is minimal chemical manipulation. The molecules have been secreted into the medium, and they have been in Medium 199 with insulin.

Nicoll: The medium which we use is stored frozen on dry ice and it is later concentrated by ultrafiltration. It becomes alkaline (pH about 8·5) during ultrafiltration. It is not lyophilized.

Li: For deamidation you need a pH of about 12 or 13.

Sherwood: Lewis and Cheever (1965) obtained clear-cut deamidation of ovine prolactin at pH 8–9 and we have confirmed this for HPL with measurements of ammonia release (Handwerger, Catousse and Sherwood 1970).

Nicoll: During culturing the pH is maintained at 7·4 (by an atmosphere of 95 per cent O_2 and 5 per cent CO_2). Then the medium is removed and the pH rises to 8·5. It is then frozen and during ultrafiltration the pH is around 8·5, so we feel that it is in these alkaline conditions that deamidation occurs.

Li: It is also possible that the second band might be due to polymerization.

Wilhelmi: If it were a polymer it would run more slowly on the gel.

Turkington: We have isolated human prolactin by electrophoresis and have characterized it on polyacrylamide gels. We find that human prolactin migrates at alkaline pH ahead of human growth hormone, at all gel concentrations.

Nicoll: Could it be modified growth hormone?

Turkington: No, because we showed that it had no immunoreactivity against anti-growth hormone antisera and very high biological prolactin activity (by the ^{32}P-casein assay). It couldn't have anything less than 500 times the specific activity of human growth hormone, so it's unreasonable to think that it's a modified growth hormone.

Friesen: What was the source of the material?

Turkington: We have isolated it from serum of a patient with a pituitary tumour and shown that the molecule is identical to the circulating prolactin which is released in a lactating female, by electrophoretic characterization.

Meites: I noticed a couple of apparent discrepancies in the presentations of Dr Pasteels and Dr Nicoll. In his earlier work with human foetal pituitaries, Dr Pasteels observed after a few days almost complete disappearance of growth hormone and a continuation of prolactin secretion. In the work that Dr Nicoll has described here on monkey pituitaries (and there may be a species difference), growth hormone and prolactin release showed parallel patterns.

Secondly, Dr Pasteels reported on the lack of effect of lactation (and I'm not sure whether you mean suckling or just postpartum lactation without considering the suckling interval) on the growth hormone content of the pituitary. Both Grosvenor, Krulich and McCann (1968) and we (Sar and Meites 1969) found that suckling significantly depletes the rat pituitary of growth hormone.

Pasteels: On the first point, I showed our old results of tissue culture experiments to indicate the difference between the two conditions of culture. The first results were obtained in conditions where prolactin cells were proliferating in the cultures and this is probably why they secreted more and more prolactin with time. In those specific conditions, the other pituitary cells were dying and this probably accounts for the significant decrease of growth hormone production. But in *organ* culture, undertaken to preserve the pituitary in a better state, we see evidence of growth hormone secretion by radioimmunoassay and by immunofluorescence, even after two years in culture. I can't say whether prolactin paralleled growth hormone, as Dr Nicoll finds in monkeys. I can only say that in pooled medium some growth hormone was present. In such long experiments we don't follow the secretion from week to week. There were some cells containing growth hormone, as shown by immunofluorescence. This is an important point, because hypothalamic factors were probably present in very low amounts, as we had only 10 per cent normal rabbit serum present. It is interesting to know that in such conditions the cells are able to function rather autonomously. I believe the discrepancy between Dr Nicoll's results and mine is probably due to the technique of tissue culture.

On the other point, in lactating rats submitted to suckling there could be some degranulation of somatotrophs, but it was not important enough to be shown by differential staining. All I can say is that there was a striking degranulation of prolactin cells whereas the somatotrophs appeared unchanged, but that doesn't mean there was absolutely no release of growth hormone. We didn't actually measure growth hormone.

Nicoll: The fact that the prolactin and growth hormone secretion by the adult male and female monkeys was so similar was surprising to us also, because we expected a greater divergence on the basis of our *in vitro* work with rats, rabbits, guinea pigs, hamsters and other species. It appears that the capacity of adult male and female monkeys to secrete prolactin in organ culture is not very unusual in comparison to other mammals but that they have an unusually high capacity to secrete growth hormone over an extended period.

Regarding growth hormone depletion in response to suckling in rats, we looked at this also, using acrylamide electrophoresis to measure prolactin and growth hormone in the adenohypophysis. We obtained a very pronounced and rapid depletion of prolactin from the gland in response to suckling and no change in growth hormone.

Li: What is the specific activity of your isolated monkey prolactin?

Nicoll: I can't give you that figure yet in units per mg protein. We have been interested in getting the preparation as free of growth hormone as possible, and now that we know how to do this we shall try to isolate a highly purified preparation and determine its specific activity.

Li: I would like to comment on Dr Pasteels' heating experiment. You can heat human growth hormone in neutral solution for 20 minutes in a boiling water bath without loss of either lactogenic or growth-promoting activity. However, animal growth hormone loses activity, whereas animal prolactin retains its activity after boiling.

Greenwood: This is what worried us, because we knew that Dr Pasteels had done this and we also knew that heating may or may not modify immunological activity.

Friesen: Dr Nicoll, how much protein is secreted into the media from your pituitaries? We observed that both the pregnant and the lactating monkey pituitary secrete two to three times as much *protein* as pituitaries from males. Prolactin and growth hormone don't account for this major difference. So I think there may be still another pregnancy or lactation-related protein secreted by these pituitaries.

Sherwood: I am not sure one can really say that. In our extensive experience with another endocrine system, the parathyroid gland in organ culture, only one hormone is produced by the gland, yet we find a number of other labelled proteins in the culture medium (Sherwood, Rodman and Lundberg 1970). I have no idea what they are, but I doubt they are hormones. They might be other proteins present in secretory granules, for example.

Friesen: After gel filtration most of these proteins emerge with the same elution volume as growth hormone and prolactin. The radioactive proteins can almost totally be accounted for by the two hormones, but not the proteins, so it seems to be a storage form of protein which isn't turning over rapidly.

Nicoll: I do not have this kind of measurement on the total protein levels which could be equated with the amounts of prolactin and growth hormone.

Greenwood: I want to take up Dr Turkington's point about the tumour material, because if we fractionate, I think we get a family of peptides of similar charge and size but which are not all prolactin, and hence the specific immunoactivity is lower than expected. It took me back to when we extracted ectopic ACTH-producing tumours, using a classical pituitary fractionation. We got a homogeneous peak on the column but there was only 10 per cent biological activity there.

Turkington: Our source of prolactin was serum, which was not treated in any way but was simply electrophoresed. The protein behaves as a homogeneous protein at all gel concentrations and two alkaline pH's (8·2 and 10). But I agree that this doesn't rule out the possibility that there are other forms.

Cotes: Going back to Dr Friesen's point that there may be storage forms of protein in the pituitary gland, it would be attractive to speculate that there is a precursor of prolactin in the gland. This could account for Dr Greenwood's high figures for the amount of prolactin secreted by the normal human subject in response to phenothiazine. In the lactating subject the prolactin content of the gland may be greater than normal, so it is not surprising that immediately after suckling he found that more prolactin appears in the plasma than is present in the normal gland. But in the normal subject, one hour after an oral dose of phenothiazine, Dr Greenwood found that the plasma contained approximately ten times the normal gland content of prolactin. This is rapid synthesis of prolactin.

Greenwood: I would only mention here Professor Herlant's earlier remark that there may well be prolactin cells there that we can't recognize. That leaves it wide open!

Nicoll: The point about the histological appearance of the gland and the possible discrepancies with its hormone content are well illustrated by data of Dr Pasteels on some of the monkey glands we sent to him. Infant glands contain about the same number of identifiable growth hormone cells as the adult glands but they contain substantially more growth hormone, as measured by electrophoresis-densitometry, so that although cell counts and hormonal content are sometimes closely related, this is not always so.

REFERENCES

GROSVENOR, C. E., KRULICH, L. and McCANN, S. M. (1968) *Endocrinology* **82**, 617–619.
HANDWERGER, S., CATOUSSE, S. L. and SHERWOOD, L. M. (1970) *Program of the Endocrine Society, 52nd Meeting*, p. 164 (abst. 255).
HERBERT, D. C. and HAYASHIDA, T. (1970) *Science* **169**, 378–379.
LACY, P. (1970) *Diabetes* **19**, 895–905.
LEWIS, U. J. and CHEEVER, E. V. (1965) *J. Biol. Chem.* **240**, 247–252.
NAKANE, P. K. (1971) In *3rd Karolinska Symposium on Research Methods in Reproductive Endocrinology*: In Vitro *Methods in Reproductive Cell Biology*, ed. Diczfalusy, E., pp. 190–204. Stockholm: Karolinska sjukhuset. Also as Supplement 153 of *Acta Endocrinol. (Copenh.)*.
PASTEELS, J. L. (1963) *Arch. Biol. (Brussels)* **74**, 439–553.
SANO, M. (1962) *J. Cell Biol.* **15**, 85–97.
SAR, M. and MEITES, J. (1969) *Neuroendocrinology* **4**, 25–31.
SHERWOOD, L. M., RODMAN, J. S. and LUNDBERG, W. B. (1970) *Proc. Natl. Acad. Sci. (U.S.A.)* **67**, 1631–1638.

IMMUNOFLUORESCENCE STUDIES ON THE ADENOHYPOPHYSIS IN PREGNANCY

J. SWANSON BECK

*Department of Pathology, The University, Aberdeen**

THE immunofluorescence method is disarmingly simple, at least at the technical level. In theory, proteins of an immune serum are conjugated with a fluorochrome and the resulting conjugate is used as a histological stain: the procedures are technically simple and do not require expensive apparatus (Beck 1971). In practice, however, the interpretation of observations with this technique can be considered valid only when there is very rigorous control of both the serological reaction and the cytological localization of staining. It is therefore pertinent to start by considering the criteria I have adopted for establishing serological specificity and cytological localization, before presenting the results of my immunofluorescence staining experiments on the adenohypophysis in pregnancy.

SEROLOGICAL SPECIFICITY OF IMMUNOFLUORESCENCE STAINING

The immunofluorescence technique can be performed in several different ways. (*a*) *Direct technique*, in which the globulin fraction from the immune serum is combined with a fluorochrome and the conjugate is directly applied to the tissue. (*b*) *Indirect technique*, in which the tissue is treated with the antiserum and the attached immunoglobulin is then stained with the appropriate anti-immunoglobulin conjugate. (*c*) *Anti-complement system*, in which the tissue is treated with heat-inactivated antiserum and fresh guinea-pig serum and later is stained with fluorescein-conjugated antiserum to guinea-pig complement components.

The indirect technique is generally the method of choice since it is versatile and reliable. Monospecific anti-immunoglobulin antisera can be readily prepared for this technique and any non-specific staining properties of the corresponding conjugates can be easily removed: furthermore, it avoids the necessity of conjugating individual anti-hormone antisera with the attendant dangers of loss of titre and introduction of unpredictable

*Present Address: Department of Pathology, The University, Dundee.

non-specific staining effects. Many variations of this technique have been proposed: the methods that I have found satisfactory have been described previously (Beck et al. 1966, 1969). It must be remembered that both stages of the indirect immunofluorescence technique are serological reactions and consequently they must be controlled with equal care.

The immunofluorescence technique is used as a histochemical staining method for cytological localization of hormones on the tacit assumption that it will have the high degree of specificity of the corresponding antibody in the immune serum. It is therefore important to realize the potential fallacies with this method so that precautions can be taken to prevent misinterpretation of results. Apart from tinctorial staining of tissue by free fluorochrome dyes in the conjugate solution, which is readily avoided, irrelevant or non-specific staining can result in two main ways (Beck and Currie 1967). (a) *Irrelevant immunofluorescence staining*. The immunoglobulin fraction of an immune serum will contain specific antibodies against the immunizing antigen and possibly also antibodies against other antigens that contaminate the preparation used in immunization. This fraction will also contain many other antibodies reflecting the previous immunological experience of the animal and various 'natural' antibodies, such as those of the Forssman system. (b) *Non-immunological staining by conjugated serum proteins*. Non-specific staining of this type generally results when the net charge on the labelled protein molecules is opposite to that of a tissue component: this is usually caused by over-labelling the serum proteins and this non-specific property can often be removed from the conjugate by absorption with tissue powders. Rarely it can result from other highly specific, but non-immunological protein–protein interactions.

Selection of antisera for immunofluorescence staining

Ideally, isolated high-avidity monospecific antibody should be used in the immunofluorescence technique, but this is not at present practicable. Instead, the best compromise available to most workers is to use high-titre antisera that appear to be monospecific when tested with other immunological techniques.

However, it must be realized that for many hormones the only preparations available as immunogens are of doubtful purity. When this is so, the usual practice is to remove contaminating antibodies to other antigens by appropriate absorptions before undertaking immunofluorescence staining experiments. As an additional precaution, it is advisable to immunize animals with hormone preparations extracted by alternative methods, when these are available, to demonstrate that the resultant antisera give a similar localization of immunofluorescence staining.

When this stage has been reached it is important to demonstrate that all high-titre antisera stain tissues known to contain the corresponding antigen or any cross-reacting antigen, and that non-immune sera or antisera to totally different antigens will not stain the hormone under study. It is, however, equally important to show that the potent anti-hormone antisera will never stain tissues that do not synthesize or store the hormone under investigation. When a number of antisera are available it will usually be found that the intensity of the immunofluorescence reaction will roughly parallel the titre of the serum in some other immunological system, but high-avidity antisera usually give better immunofluorescence staining than their titre would suggest.

Absorption experiments

When an antiserum is absorbed with the optimum quantity of the corresponding antigen, it will lose completely its staining capacity in the corresponding immunofluorescence system. If it is absorbed with a closely related antigen with which it cross-reacts its staining reaction will be diminished in the specific reaction and completely lost in the cross-reaction. If, however, the antiserum is absorbed with a completely unrelated antigen, there will be no specific absorption of the antibody.

Experiments of this type with similar quantities of the different antigens are essential for establishing the serological specificity of immunofluorescence staining. Since many hormone preparations are impure it is valuable to be able to demonstrate that preparations extracted by alternative methods give similar results in these absorption experiments, in order to eliminate the possibility that immunofluorescence staining has been caused by highly immunogenic contaminants of particular antigen preparations.

Nature of serological factor

It is important to demonstrate that the active factor in the serum is an immunoglobulin, since it is highly improbable that any reaction will be immunological if it is determined by any other serum protein. Furthermore, it is preferable that the specific antibody in the antiserum is predominantly in the $7S\gamma$ (IgG-type) fraction, since such antibodies are more stable and penetrate the tissue sections more readily than the 19S (IgM-type) macroglobulin antibodies.

The 'blocking' test of inhibition of immunofluorescence staining by pretreatment with an unconjugated antiserum of the same specificity is of only limited value in establishing the specificity of immunofluorescence staining. Clearly it is meaningless when performed with unconjugated and conjugated samples of the same serum.

State of antigen in tissue sections

In cryostat sections the tissue antigen is preserved as nearly as possible to its native condition, and so from the serological point of view this would appear to be the best method for preparing tissue for immunofluorescence studies. It must, however, be remembered that antisera are usually prepared by immunizing animals with a purified hormone extract and that this immunogen may be modified or even partly denatured during the extraction process. Thus even under the apparently ideal conditions of cryostat sections, complete concordance need not necessarily be seen between native tissue antigen and the corresponding antibody. Experiments can, however, be performed on unfixed cryostat sections to show that the tissue antigen has solubility and stability characteristics similar to the known properties of the hormone under consideration. Such experiments have a limited value in corroborating the serological specificity of immunofluorescence staining.

Unfixed cryostat sections are not very suitable for cytological localization of immunofluorescence staining (see below) and, consequently, immunofluorescence investigations are often attempted on fixed tissue. The effect of histological fixation on the antigenicity of the polypeptide hormones has not been studied in depth, but it appears probable that it results in loss of some antigenic determinants: it is remotely possible that histological fixation might uncover hidden antigenic determinants or even create new ones.

CYTOLOGICAL LOCALIZATION OF IMMUNOFLUORESCENCE STAINING

The cellular site and distribution of immunofluorescence staining must be compared with that of other established tinctorial staining techniques, so that it can be localized in terms of the known cytology of the tissue. Comparisons of this type are particularly difficult in the adenohypophysis, which is composed of a heterogeneous mixture of different cell types.

The simplest, but least satisfactory, approach is to study an adenoma that is known from conventional histological studies to be composed exclusively or largely of one cell type. Cryostat or fixed sections can be used in these investigations, but the results can be of only limited value since the cells are neoplastic and this method will never be suitable for discriminating cytological investigation.

An alternative method is to mount pairs of consecutive sections on separate slides so that the contiguous surfaces of the sections are uppermost; in this way the majority of cells will be present in both sections, but they will be arranged in a 'mirror-image' distribution. One of the sections is then stained with the immunofluorescence method and the other with a conventional histological method. For technical reasons, this approach can be used only with paraffin sections of fixed tissue. It has the great advantage that the histological staining characteristics of the tissue are not influenced by the immunofluorescence technique, but the method is laborious and interpretation is difficult when a cell is divided unequally between the sections. This is not a very good technique for cytological localization of immunofluorescence staining.

The most satisfactory method in my hands has been to stain fixed sections with the immunofluorescence technique and record the distribution of this staining in recognizable areas photographically. The sections are then stained with the periodic acid-Schiff (PAS)–orange G method to localize the immunofluorescence staining under the light microscope. We have shown with contiguous pairs of 'mirror-image' sections of adult human adenohypophysis that sections prestained with the immunofluorescence method showed slightly less intense staining and somewhat poorer colour contrast than in the directly stained section: however, prestaining with the immunofluorescence method did not alter the nature of the PAS^1_N orange G staining and the final product was perfectly adequate for microscopic examination and cell identification (Beck et al. 1966). This method can be used to establish the cytological localization of immunofluorescence staining in a large number of cells in fixed sections. In our hands, re-staining of unfixed cryostat sections has been unsatisfactory.

Whichever method is used to compare the cytological localization of immunofluorescence staining with that of another staining technique, a substantial number of cells must be studied before valid conclusions can be drawn on the cell types that contain the antigen. Furthermore, it is important to use a sensitive immunofluorescence staining technique so that cells with scanty antigen content will be detected, since it is not enough merely to identify cells with abundant antigen in their cytoplasm. A possible cause of confusion in cytological localization studies on tissue sections of adenohypophysis is immunofluorescence staining in a portion of the cytoplasm of a cell of one type that overlies or is underlain by a portion of the cytoplasm of a cell of another type: statistically this situation would not be expected frequently and an error of interpretation can be avoided only by taking great care in the microscopical examination.

IMMUNOFLUORESCENCE STAINING OF ADENOHYPOPHYSES FROM PREGNANT AND PARTURIENT WOMEN WITH ANTISERA TO HUMAN GROWTH HORMONE AND HUMAN PLACENTAL LACTOGEN

The antisera used in these studies were raised in goats immunized with the Raben preparation of human growth hormone (HGH) or the Friesen preparation of human placental lactogen (HPL). A fluorescein conjugate of the immunoglobulin fraction of rabbit anti-goat-immunoglobulin serum was used in the second stage of the immunofluorescence staining.

In previously reported studies we have assessed the validity of immunofluorescence staining of paraffin sections of formaldehyde-fixed normal adult human adenohypophysis with these antisera on the criteria outlined earlier. We concluded that the anti-HGH antiserum was demonstrating the HGH antigen in the tissue, but in the absence of a purified preparation of human prolactin, we could not eliminate the possibility that this system might be tracing both HGH and human prolactin (Beck et al. 1966). The antiserum to HPL gave immunologically specific staining of HPL in placental syncytiotrophoblast and appeared to cross-react immunologically with HGH in sections of the normal adult human adenohypophysis (Beck et al. 1969).

Dr O. A. Haugen and I have stained paraffin sections of formaldehyde-fixed pituitary glands of six pregnant and parturient women, none of whom had any apparent endocrinological defect, with this immunofluorescence technique (Haugen and Beck 1969). Because of shortage of tissue, we were unable to repeat the specificity control experiments on sections of these pituitary glands and so we were forced to accept that the specificity of immunofluorescence staining with these antisera would be similar to that previously demonstrated on sections of normal adult human adenohypophysis: we do not think that this was an unwarranted assumption. After immunofluorescence staining, the sections were stained with the PAS–orange G method. In the adenohypophyses, little difficulty was encountered in recognizing classical acidophil and mucoid cells: 'pregnancy cells' were distinguished from chromophobe cells mainly on the basis of the larger more open nuclei, often with prominent nucleoli, and the more abundant cytoplasm of the 'pregnancy cells', which also usually contained a few faintly orange G-positive granules in their cytoplasm.

We found little difference between the individual cell counts on adenohypophyses from four pregnant women dying after 31 weeks' gestation and two parturient women who died within 14 hours of delivery at term (Haugen and Beck 1969). Consequently we grouped the counts on all the glands together. The findings in the pregnant and parturient women are compared with previous findings on normal non-pregnant adults in Table I. Bearing in mind the not inconsiderable sampling errors of our technique, it appears that the immunofluorescence staining reactions of orange G-positive acidophil cells, PAS-positive mucoid cells and chromophobe cells

TABLE I

RELATION BETWEEN PAS–ORANGE G AND IMMUNOFLUORESCENCE STAINING IN THE HUMAN ADENOHYPOPHYSIS

Reaction of cells to PAS–orange G staining	Percentage of these cells staining with given antiserum in adenohypophysis from			
	Normal adults		Pregnant and parturient women	
	Anti-HGH	Anti-HPL	Anti-HGH	Anti-HPL
Orange G-positive	97·6	98·4	98·8	98·4
PAS-positive	3·2	2·1	1·6	3·2
Chromophobes	5·0	6·3	11·6	7·8
'Pregnancy cells'	—	—	1·9	3·7

The data are taken from Beck and co-workers (1966, 1969) and Haugen and Beck (1969).

are not materially different from those of the corresponding cell types in the adenohypophyses of adult non-pregnant patients without evidence of endocrine defect. Only a very small proportion of 'pregnancy cells' were stained with either anti-HGH serum or anti-HPL serum (1·9 and 3·7 per cent respectively). It is probable that the differences in the findings with these two antisera are not meaningful. There can be no doubt that the staining reactions of 'pregnancy cells' are completely different from those of orange G-positive acidophil cells.

In view of the previous investigations on the specificity of the immunofluorescence systems, it can be deduced that very few of the 'pregnancy cells' contain the human growth hormone antigen. The significance of the immunofluorescence staining of the orange G-positive acidophil cells in the adenohypophysis of pregnant and parturient women is not so clear, but it is probable that the human growth hormone antigen is being localized in these cells by the direct immunological reaction of the anti-HGH antiserum and an immunological cross-reaction with the anti-HPL antiserum. It is not possible from our experiments to make any deductions on the localization of human prolactin, if such a hormone exists, since we cannot be sure that the anti-HGH antiserum is not contaminated with anti-human prolactin antibody, nor that the anti-HGH and the anti-HPL antibodies do not cross-react with human prolactin.

It has recently been claimed that prolactin can be localized in the carminophil cells that are particularly numerous in the adenohypophysis of the pregnant and lactating monkey (Herbert and Hayashida 1970), but inadequate evidence of the serological specificity of this immunofluorescence staining was given to establish this localization conclusively.

SUMMARY

The criteria for establishing the serological specificity of immunofluorescence staining and for its cytological localization are discussed.

Sections of the adenohypophyses of six pregnant and parturient women were stained by the indirect immunofluorescence technique with characterized antisera to the Raben preparation of human growth hormone and the Friesen preparation of human placental lactogen. Very few of the 'pregnancy cells' were stained with either antiserum. The relative numbers of acidophil, mucoid and chromophobe cells staining with these antisera in the adenohypophyses of pregnant and parturient women were similar to those of the corresponding cells in non-pregnant individuals.

REFERENCES

BECK, J. S. (1971) *Immunofluorescence techniques in immunopathology and histopathology.* Broadsheet 69. London: Association of Clinical Pathologists.
BECK, J. S. and CURRIE, A. R. (1967) *Vitam. & Horm.* **25,** 89–121.
BECK, J. S., ELLIS, S. T., LEGGE, J. S., PORTEOUS, I. B., CURRIE, A. R. and READ, C. H. (1966) *J. Pathol. & Bacteriol.* **91,** 531–538.
BECK, J. S., GORDON, R. L., DONALD, D. and MELVIN, J. M. O. (1969) *J. Pathol.* **97,** 545–555.
HAUGEN, O. A. and BECK, J. S. (1969) *J. Pathol.* **98,** 97–104.
HERBERT, D. C. and HAYASHIDA, T. (1970) *Science* **169,** 378–379.

DISCUSSION

Friesen: Dr Beck, did you suggest that HPL only partially inhibits the immunofluorescence localization of anti-HGH?

Swanson Beck: In these experiments we used comparable weights of antigens to absorb different aliquots of the antisera (Beck *et al.* 1966, 1969). With the anti-HGH antisera, the Raben HGH preparation absorbed the activity completely, whereas the HPL preparation that you gave me did not absorb the activity completely, but it greatly diminished the intensity of staining of acidophil cells. With the anti-HPL serum, HPL absorbed completely, whereas HGH gave partial absorption only.

Friesen: If you increased the amount used for the blocking studies, did HPL completely inhibit the localization of anti-HGH serum?

Swanson Beck: We didn't go beyond 16 mg HPL per ml antiserum.

Meites: Have you made any quantitative or semi-quantitative estimates of growth hormone and prolactin during pregnancy in the six cases?

DISCUSSION

Swanson Beck: We did not attempt to quantify the intracellular content of the antigens. We did not attempt to estimate the fluorescence emission, since there are too many variables in the immunofluorescence technique, and it is not possible to control the ratio of antigen to antibody in this situation. The acidophil cells did not appear to fluoresce more brightly in the adenohypophyses of pregnant women than in the glands of non-pregnant women.

Turkington: I have often wondered what the chemical composition of the secretory granules is in the adenohypophysis. Are they more than pure hormone?

Greenwood: Both Kwa and McShan's groups have isolated and studied granule preparations (Kwa et al. 1965; McShan 1965).

Swanson Beck: I think growth hormone is probably located in the cytoplasm as well as the granules, since immunofluorescence staining gives an almost uniform appearance within the cell cytoplasm and is not usually restricted to the granules, although occasionally it can be granular. I wonder whether the diffuse staining indicates newly synthesized hormone in the endoplasmic reticulum.

Sherwood: In the β-cell it is believed that the enzyme which converts proinsulin to insulin is in the granule (Lacy 1970).

Wilhelmi: Recently we had occasion to assay some bovine pituitary granule preparations from Dr Frank LaBella. The bioassay seemed to suggest that they consisted of as much as 25 per cent of growth hormone. The specific activities were about half a unit per mg of granules (dry weight).

Turkington: If you disaggregated these granules and put them on a gel, would you get a number of bands?

Wilhelmi: We didn't try that, but I suspect you would.

Pasteels: We have done some studies on the localization of gonadotropins and there was a good correlation between the granular content and immunofluorescence, especially in the highly active cells where the granules are located at the apical pole of the cells. Only this part of the cell fluoresced. There was no staining of the whole cytoplasm.

Turkington: Isn't the peroxidase-conjugated anti-hormone antibody localized primarily in the granule?

Pasteels: Nakane (1971) has recently found with electron microscopy that some hormone was present within the cisternae of the endoplasmic reticulum, and also within the Golgi area where the granules are concentrated, and of course in secretory granules, but apparently not in the hyaloplasm.

Friesen: The granules obtained from the posterior lobe definitely consist of more than vasopressin; there are other proteins present as well, such as neurophysin. In catecholamine-containing granules from the adrenal medulla there is a lot of protein also. So at least in two other secretory granules there seem to be other components present in addition to the hormone.

Turkington: But those are all very low molecular weight hormones.

Greenwood: There must be some hormone elsewhere than in the storage granules, even if it is only in passage from the ribosomes to the storage granule. We have been looking for immunoreactive prolactin in subcellular fractions, because presumably at some stage there should be enough there for immunogenic activity, but we haven't been successful. There is a marked non-specific adsorption of prolactin to ribosomes.

Swanson Beck: There is another situation, the syncytiotrophoblast, where substantial quantities of a polypeptide hormone (HPL) are present in the cytoplasm in the absence of granules. In our peroxidase-labelled antibody experiments (D. Sharp, unpublished) we have localized HPL to the cytoplasm between the cisternae of the syncytiotrophoblast.

Wilhelmi: In the work of LaBella and Brown (1959) and of Hymer and McShan (1963) in which the granule fractions were separated from the various other components of the cells, the final high-speed supernatant always contained a significant amount of every single hormone in the gland, especially growth hormone.

Sherwood: Disruption of the granules may account for a lot of free hormone during isolation. The question is whether it is an artifact or not.

Wilhelmi: Yes, it might be an unanswerable question. I've never seen a procedure which failed to reveal some free hormone.

Sherwood: We have been doing this kind of work on the parathyroids and find most of the hormone in the cytosol or in the polysome fraction.

Herlant: Dr Swanson Beck, why do you stain your preparations with PAS-orange G? It is not a technique to be recommended for acidophil cells.

Swanson Beck: Because we felt that we could get reasonably reproducible results in our re-stained preparations. Sections that had been stained with the immunofluorescence method always stained less intensely than control sections that had not been so treated. We could obtain useable preparations with the PAS-orange G method, but unfortunately, in our hands, your elegant tinctorial staining methods were not very reproducible.

REFERENCES

BECK, J. S., ELLIS, S. T., LEGGE, J. S., PORTEOUS, I. B., CURRIE, A. R. and READ, C. H. (1966) *J. Pathol.* **91,** 531.
BECK, J. S., GORDON, R. L., DONALD, D. and MELVIN, J. M. O. (1969) *J. Pathol.* **97,** 545.

HYMER, W. C. and MCSHAN, W. H. (1963) *J. Cell Biol.* **17,** 67–86.
KWA, H. G., VAN DER BENT, E. M., FELTKAMP, C. A., RÜMKE, P. and SLOEMENDAL, H. (1965) *Biochim. & Biophys. Acta* **111,** 447–465.
LABELLA, F. S. and BROWN, J. H. U. (1959) *J. Biophys. & Biochem. Cytol.* **5,** 17–25.
LACY, P. (1970) *Diabetes* **19,** 895–905.
MCSHAN, W. H. (1965) In *Proceedings of the Second International Congress of Endocrinology,* pp. 382–391. International Congress Series no. 83. Amsterdam: Excerpta Medica Foundation.
NAKANE, P. K. (1971) In 3rd *Karolinska Symposium on Research Methods in Reproductive Endocrinology*: In Vitro *Methods on Reproductive Cell Biology,* pp. 190–204, ed. Diczfalusy, E. Stockholm: Karolinska sjukhuset. Also as Supplement 153 of *Acta Endocrinol. (Copenh.).*

ON THE ACTIONS OF PROLACTIN AMONG THE VERTEBRATES: IS THERE A COMMON DENOMINATOR?

CHARLES S. NICOLL AND HOWARD A. BERN

Departments of Physiology-Anatomy and Zoology and Cancer Research Genetics Laboratory, University of California, Berkeley

THE importance of prolactin in vertebrate physiology has become appreciated only within the last few years (see Bern and Nicoll 1968, 1969a, b; Ball 1969; Mazzi 1969). Previously, prolactin was largely regarded as the mammotropic hormone, or as the maternal-behaviour hormone, or even solely as a gonadotropin (luteotropin), by endocrinologists generally. However, since the review by Riddle in 1963, in which he emphasized that prolactin should be regarded as a hormone of metabolic as well as of reproductive significance, it has become abundantly clear that the hormone plays a significant role in various aspects of 'vertebrate function and organization' (Riddle 1963). In essence, prolactin is a 'broad-spectrum' hormone, uniquely versatile.

Unlike the other adenohypophysial hormones, prolactin did not become specialized early in vertebrate phylogeny for the regulation of a single physiological process. As examples, prolactin is necessary for gonadal function in some rodents; it favours an aquatic existence in amphibians; it may be a growth factor in some larval amphibians and adult reptiles; it is essential for egg incubation in birds; and it is of major significance in water and electrolyte metabolism in some teleosts. Thus, prolactin serves to control processes which are peculiar to the physiological ecology of various vertebrate groups (classes, orders, families, and even lesser taxa).

The latest compilation of the numerous actions of prolactin is presented in Table I*. Eighty-two different reported actions are known to us, in May

*Note added in proof:

Two papers on cyclostomes and one on elasmobranchs contain information supplemental to Table I. S. Falkmer and A. J. Matty (*Gen. & Comp. Endocrinol.* 6, 334–346, 1966) report a 'pinkish discoloration of the skin' after administration of prolactin to hagfish. Immunohistochemical studies make it appear unlikely that the cyclostome adenohypophysis secretes prolactin, however (G. Aler, G. Båge and B. Fernholm, *Gen. & Comp. Endocrinol.* 16, 498–503, 1971). P. Payan and J. Maetz (*Gen. & Comp. Endocrinol.* 16, 535–554, 1971) found that prolactin (and ACTH) restored normal branchial permeability to water in hypophysectomized elasmobranchs. According to H. Schultheiss, W. Hanke and J. Maetz (personal communication), prolactin has a similar effect in regard to skin permeability to water in hypophysectomized young *Xenopus* tadpoles. Höcker has studied avian responses to prolactin in detail (cf. W. Höcker, S. Darda, D. Petutschnigk and A. Gramlich, *Endokrinologie* 57, 364–382, 1971).

TABLE I

COMPARATIVE ENDOCRINOLOGY OF PROLACTIN

Unabridged list of the manifold actions claimed for prolactin (May, 1971)

Cyclostomes

C-1 Electrolyte metabolism in hagfish (ACTH-like)

Teleosts

T-1 Osmoregulatory actions including:
  a. Survival of hypophysectomized euryhaline freshwater species
  b. Restoration of water turnover in hypophysectomized *Fundulus kansae*
  c. Restoration of plasma Na$^+$ and Ca^{++} in hypophysectomized eels when given with cortisol
  d. Skin, buccal and gill mucus secretion
  e. Reduced gill Na$^+$ efflux (reduced permeability)
  f. Reduced gill permeability to water
  g. Inhibition of gill Na$^+$, K$^+$-ATPase
  h. Renotropic (glomerular and tubular changes)
  i. Increased urinary water elimination and decreased salt excretion
  j. Stimulation of renal Na$^+$, K$^+$-ATPase
  k. Decreased water absorption and increased Na$^+$ absorption in flounder bladder
  l. Decreased salt and water absorption from eel gut

T-2 Adrenocorticotropic
T-3 Resistance to high temperature stress
T-4 Dispersion of pigment in xanthophores of *Gillichthys mirabilis*, *Arothron hispidus* and *Gobius minutus*
T-5 Melanogenesis and proliferation of melanocytes (synergism with MSH)
T-6 Thyroid (TSH?) stimulation

Reptiles

R-1 Somatic growth
R-2 Tail regeneration
R-3 Hyperphagia
R-4 Epidermal sloughing
R-5 Lipid metabolism
R-6 Anti-gonadotropic
R-7 Restoration of plasma sodium levels in hypophysectomized lizard

Birds

B-1 Production of crop 'milk' (columbids)
B-2 Formation of brood patch
B-3 Stimulation of feather growth
B-4 Somatic growth (including splanchnomegaly)
B-5 Lipid metabolism
B-6 Hyperglycaemic-diabetogenic
B-7 Stimulation of nasal (orbital) salt gland secretion
B-8 Anti-gonadal (anti-gonadotropic)
B-9 Synergism with steroids on female reproductive tract
B-10 Parental behaviour
B-11 Suppression of sexual phase of reproductive cycle (including calling and mating in quail)
B-12 Premigratory restlessness (*Zugunruhe*)

Mammals

M-1 Stimulation of mammary growth and development
M-2 Stimulation of milk secretion

T-7   Lipid metabolism
T-8   Growth and secretion of catfish seminal vesicles
T-9   Reduction of toxic effects of oestrogen
T-10 Parental behaviour (nest building, fin fanning, buccal incubation of eggs)
T-11 Maintenance of brood pouch in male seahorse
T-12 Gonadotropic (increased-3β-ol dehydrogenase in cichlid ovaries)

Amphibians

A-1   Water drive, including skin and tail changes
A-2   Larval growth (especially tail, including collagen synthesis, and gills)
A-3   Somatotropic in postmetamorphic anurans; including tail growth in urodeles
A-4   Lipid metabolism in anurans
A-5   Peripheral thyroxin antagonism (anti-metamorphic)
A-6   Goitrogenic-thyrotropic (TSH release)
A-7   Growth of brain in frog tadpoles
A-8   Limb regeneration
A-9   Decreased urea excretion in some anuran tadpoles
A-10 Increased hepatic arginase activity in other anuran tadpoles secondary to thyroid activation
A-11 Hyperglycaemic-diabetogenic
A-12 Proliferation of melanophores
A-13 Skin yellowing in frogs
A-14 Restoration of plasma Na$^+$ levels in hypophysectomized newts
A-15 Na$^+$ and water transport across toad bladder
A-16 Secretion of oviductal jelly
A-17 Anti-spermatogenic
A-18 Cloacal gland development (urodele)
A-19 Stimulation of ultimobranchial and possible hypercalcaemia (toads)

M-3   Stimulation of sebaceous gland size and activity (including preputial glands)
M-4   Hair maturation
M-5   Stimulation of somatic growth (including splanchnomegaly)
M-6   Lipid metabolism
M-7   Hyperglycemic-diabetogenic; increased insulin secretion
M-8   Erythropoietic
M-9   Renotropic including Na$^+$ retention
M-10 Increased fertility in male and female dwarf mice
M-11 Actions on male reproductive organs including:
   a. Synergism with androgens on male sex accessory glands
   b. Increased androgen binding in human prostate
   c. Increased cholesterol levels in mouse testes
   d. Stimulation of β-glucuronidase activity in rodent testes
M-12 Decreased copulatory activity in male rabbits
M-13 Advancement of puberty in rats
M-14 Progesterone secretion by mouse ovaries: possible synergism in other species (luteotropic)
M-15 Luteolytic action in rats
M-16 Parental behaviour (retrieval of young by laboratory rats)
M-17 Vaginal mucification in rats

1971. Although all cannot be considered to be confirmed functions of the hormone, and some of the different targets may represent components of the same functional response, well over half of them are of apparent physiological significance.

In view of the manifold actions of prolactin among the vertebrates, the question logically arises of whether there is any common denominator underlying its numerous physiological effects, which could allow the emergence of some unifying concept. At present no such common denominator is obvious. However, it is possible to classify the numerous actions into one or more of several different categories.

REPRODUCTION

The most obvious, and historically the best-known, category of prolactin actions consists of effects related to reproduction (Table II). About 38 per cent of the reported actions can be listed in this category. This group includes the commonly known *mammalian* actions of the hormone in promoting mammary development and lactation and in stimulating secretion of progesterone from the ovaries of certain species of rodents. Additional reproduction-related actions of prolactin, less well known to endocrinologists, are also found in mammals, including effects on male gonads and sex accessories.

Among *birds*, parental behaviour, pigeon crop 'milk' production and the formation of the incubation patch are perhaps the most widely known of prolactin's influences on reproductive processes. However, as indicated in Table II, several other reproductive actions are evident in this class.

Among *reptiles*, prolactin has been examined for possible physiological effects only by a few investigators. Consequently, the only action related to reproduction reported so far in this class is a gonadotropin-blocking effect.

Several actions claimed for prolactin in *amphibians* are concerned with reproduction, including the migration of certain salamanders to the aquatic habitat where they reproduce. Many of the physiological and biochemical changes accompanying this water drive are dependent on prolactin. Actions on two reproductive organs, the oviduct and cloacal glands, are also claimed, and there are reports that prolactin can have anti-spermatogenic activity in salamanders.

Among *teleosts*, several actions of prolactin are related to reproduction. In certain cichlids and catfishes, prolactin stimulates secretion of mucus by cutaneous and buccal gland cells, and this may be used to nourish the young. Accordingly, it is referred to as 'milk' in those species which form it in abundance. In the Indian catfish, *Heteropneustes fossilis*, prolactin is

TABLE II
ACTIONS OF PROLACTIN RELATED TO REPRODUCTION

Teleosts

1. Skin mucus secretion (e.g. discus 'milk') (T-1d)
2. Reduction of toxic effects of oestrogen (T-9)
3. Growth and secretion of seminal vesicles (T-8)
4. Parental behaviour (nest building, fin fanning, buccal incubation of eggs) (T-10)
5. Maintenance of brood pouch in male seahorse (T-11)
6. Gonadotropic (T-12)

Amphibians

1. Water drive (prior to reproduction) (A-1)
2. Secretion of oviductal jelly (A-1)
3. Anti-spermatogenic (A-17)
4. Stimulation of cloacal gland development (A-18)

Reptiles

1. Anti-gonadotropic (R-6)

Birds

1. Production of crop 'milk' (B-1)
2. Formation of brood patch (B-2)
3. Anti-gonadal (B-8)
4. Premigratory restlessness (B-12)
5. Parental behaviour (B-10)
6. Synergism with steroids on female reproductive tract (B-9)
7. Suppression of sexual phase of reproductive cycle (B-11)

Mammals

1. Mammary development and lactation (M-1, M-2)
2. Preputial gland size and activity (M-3)
3. Synergism with androgen on male sex accessory glands (M-11a)
4. Luteotropic (M-14)
5. Luteolytic action in rats (M-15)
6. Fertility in dwarf mice (M-10)
7. Increased testis cholesterol (M-11c)
8. Increased androgen binding in human prostate (M-11b)
9. Stimulation of glucuronidase activity in rodent testis (M-11d)
10. Parental behaviour (M-16)
11. Decreased copulatory activity in male rabbits (M-12)
12. Advanced puberty in rats (M-13)
13. Vaginal mucification in rats (M-17)

reported to synergize with androgens in stimulating secretion by the seminal vesicles. Other reproductive actions of prolactin are found among the teleosts, including a gonadotropic effect and influences on parental behaviour and on the brood pouch of the seahorse.

A number of the reproductive actions of prolactin have homologous, or at least analogous actions in different vertebrate classes. For example, the hormone stimulates cutaneous 'milk' secretion in certain teleosts (*Symphysodon discus* and other teleosts), crop 'milk' production in columbids and mammary secretion in mammals: all materials used for nurturing the young. The hormone stimulates development of male sex accessory organs in rodents, catfish and urodeles. It is implicated in parental behaviour in teleosts, birds and mammals, and it has a gonadotropic action on the ovaries of some mammals and a teleost.

OSMOREGULATION

A second category of prolactin actions among vertebrates consists of effects of the hormone on water and electrolyte metabolism (Table III). About 25 per cent of the actions listed in Table I are in this category. The first indications that prolactin may have a function in osmoregulation among vertebrates came from studies on euryhaline *teleosts*. It was found that certain species could not survive in fresh water after hypophysectomy unless they were given prolactin replacement in the form of injections or a pituitary transplant. The osmoregulatory role of prolactin in fish has received considerable attention since these early observations (cf. Ball 1969). The list given in Table III clearly indicates that multiple sites of action are involved in the osmoregulatory function of the hormone in fishes. The numerous effects include changes in gill permeability and enzyme activity, in addition to changes in the structure and function of the gut, kidney and bladder. Prolactin stimulation of mucus secretion at various body sites may also be of significance in the overall turnover of water and ions.

Although prolactin is of osmoregulatory significance in certain teleosts, no reports have appeared on its possible roles in other fish groups (i.e., chondrichthyeans, ganoids, dipnoans, *Latimeria*). The only report on the action of prolactin in cyclostomes indicates a possible involvement in osmoregulation in the hagfish.

Several reports indicate that prolactin may be concerned in osmoregulation in *amphibians*. These include effects on skin mucus gland secretion accompanying water drive and associated blood electrolyte changes. The hormone's ability to restore the low plasma sodium levels of hypophysecto-

TABLE III
ACTIONS OF PROLACTIN AFFECTING WATER AND ELECTROLYTE BALANCE

Cyclostomes
1. Electrolyte metabolism in hagfish (ACTH-like) (C-1)

Teleosts
1. Survival of hypophysectomized euryhaline freshwater species (T-1a)
2. Restoration of water turnover in hypophysectomized *Fundulus kansae* (T-1b)
3. Restoration of plasma Na$^+$ and Ca^{++} in hypophysectomized eels when given with cortisol (T-1c)
4. Skin, buccal and gill mucus secretion (T-1d)
5. Reduced gill Na$^+$ efflux (reduced permeability) (T-1e)
6. Reduced gill permeability to water (T-1f)
7. Inhibition of gill Na$^+$, K$^+$-ATPase (T-1g)
8. Renotropic (increased glomerular size) (T-1h)
9. Increased urinary water elimination and decreased salt excretion (T-1i)
10. Stimulation of renal Na$^+$, K$^+$-ATPase (T-1j)
11. Decreased water absorption and increased Na$^+$ absorption in flounder bladder (T-1k)
12. Decreased salt and water absorption from eel gut (T-1l)

Amphibians
1. Skin and electrolyte changes associated with water drive (A-1)
2. Sodium and water transport across toad bladder (A-15)
3. Restoration of plasma Na$^+$ in hypophysectomized newts (A-14)
4. Possible hypercalcaemia in toads (A-19)

Reptiles
1. Restoration of plasma Na$^+$ levels in hypophysectomized lizard (R-7)

Birds
1. Stimulation of nasal (orbital) salt gland secretion (B-7)

Mammals
1. Lactation (M-2)
2. Increased Na$^+$ retention at renal level (M-9)

mized newts to normal is clearly analogous to its effects on euryhaline teleosts. The stimulatory action of prolactin on sodium transport across the anuran urinary bladder is also analogous to its effect on the flounder bladder.

The possible role of prolactin in osmoregulation in *reptiles*, *birds* and *mammals* has only recently been considered. In the hypophysectomized lizard (*Dipsosaurus dorsalis*), prolactin is reported to lower the elevated plasma sodium levels to normal. In the domestic duck, it stimulates secretion by the nasal or orbital salt gland, which is concerned with salt elimination.

We have included the lactogenic action of prolactin in the osmoregulatory category, because lactation involves the movement of large quantities of water and electrolytes from the extracellular fluid compartment, thus providing an osmoregulatory stress, particularly in small mammals. Prolactin is reported to have a renotropic action in rats and salt-retaining effects in rats and in cats. These last studies suggest a possible osmoregulatory function of prolactin in mammals. Although the hormone may not be of major significance in this regard in non-lactating animals, it may be of considerable importance during lactation when the mechanisms which maintain water and electrolyte balance are severely taxed.

These considerations also raise the possibility that placental prolactin may have an important function in osmoregulation during pregnancy, for both the mother and the developing foetus. Certain disorders of water and electrolyte balance in gestation are conceivably due to excess or deficiency in placental prolactin or to hypersensitivity to the hormone. Human placental lactogen may stimulate aldosterone secretion (Melby *et al.* 1966).

The available data on the osmoregulatory role of prolactin in reptiles, birds and mammals are meagre, and scant information exists on amphibians. Nevertheless, the results at hand suggest that in freshwater teleosts, amphibians and mammals, prolactin is concerned with preventing loss of sodium chloride. Thus, it acts to prevent a reduction in the plasma levels of this electrolyte. In reptiles and birds, it may have the opposite action since it lowers plasma sodium in hypophysectomized *Dipsosaurus* and stimulates secretion by the salt-excreting glands of ducks. It will be interesting to learn whether the homologous salt-excreting glands of reptiles and the rectal salt glands of chondrichthyeans are similarly responsive to prolactin.

GROWTH

Table IV shows the actions of prolactin which involve growth promotion. Fifty-two per cent of prolactin's actions can be listed in this category.

TABLE IV

ACTIONS OF PROLACTIN INVOLVING GROWTH PROMOTION

Stimulation of somatic growth observed in adult reptiles, birds and mammals and in larval and adult amphibians (R-1, B-4, M-4, A-2, A-3)

Actions on specific target cells or tissues:

Teleosts
1. Proliferation of melanocytes (T-5)
2. Growth of seminal vesicles (T-8)
3. Renal glomerular growth, tubule stimulation and proliferation (T-1h)

Amphibians
1. Tail and gill growth (A-2)
2. Limb regeneration (A-5)
3. Proliferation of melanocytes (A-9)
4. Structural changes accompanying water drive (A-1)
5. Brain growth in tadpoles (A-6)
6. Cloacal gland development (A-18)
7. Ultimobranchial stimulation (A-19)

Reptiles
1. Tail regeneration (R-2)
2. Skin sloughing (R-4)

Birds
1. Proliferation of pigeon crop sac mucosa (B-1)
2. Epidermal hyperplasia in brood patch (B-2)
3. Feather growth (B-3)
4. Development of female reproductive tract (B-9)

Mammals
1. Mammary development (M-1)
2. Sebaceous and preputial gland growth (M-3)
3. Hair growth (M-4)
4. Erythropoietic (M-8)
5. Renotropic (M-9)
6. Spermatogenic (M-10)
7. Male sex accessory development (M-11a)
8. Luteotropic (M-14)

Growth hormone-like metabolic actions:

1. Effects on lipid deposition and/or mobilization reported in teleosts, amphibians, reptiles, birds and mammals (T-7, R-5, B-5, M-5, A-4)
2. Hyperglycaemic–diabetogenic action reported in amphibians, birds and mammals (A-1, B-6, M-7)
3. Effects on BUN, nitrogen balance, blood glucose, free fatty acids and calcium metabolism in man similar to HGH

These include stimulation of general body growth (and overgrowth) in reptiles, birds, mammals and amphibians, and actions on several specific target cells and tissues. Prolactin also has a variety of metabolic actions similar to those of growth hormone preparations. These include effects on lipid metabolism and a hyperglycaemic–diabetogenic action in several vertebrate classes. Effects of ovine prolactin on blood urea levels, on nitrogen balance and on glucose, fatty acid and calcium metabolism have been reported in man.

INTEGUMENT

An additional category of prolactin actions consists of effects of the hormone on integumentary or ectodermal structures. These are shown in Table V and comprise about 25 per cent of the actions of the hormone listed in Table I. The hormone has effects on integumentary structures in all classes of vertebrates examined. These include actions on the mammary gland and on the brood patch of birds, epidermal sloughing in reptiles, skin changes associated with water drive in amphibians, and a variety of actions in teleosts involving osmoregulation and mucus production. Effects on cutaneous pigment cells are reported in teleosts and amphibians.

SYNERGISM WITH STEROIDS

A final category of actions of prolactin among the vertebrates includes those which involve synergism with steroid hormones, or actions on organs which are also influenced by steroids. These actions are listed in Table VI. About 36 per cent of the activities of prolactin can be put into this category. One of the most recent reports of prolactin synergism with sex steroids concerns vaginal mucification in rats (Kennedy and Armstrong 1971). Here one can recall the stimulation of mucus cells in teleost ectodermal structures, which does not so far appear to involve steroids. Farnsworth (1970) has pointed out that prolactin increases androgen binding by human prostate tissue *in vitro*—a possible explanation for its synergistic action with steroids. Dorfman (1971) has suggested that the synergism between prolactin and androgen may involve increases in intracellular levels of certain enzymes, such as sulphatases and/or glucuronidase. These enzymes may convert inactive esterified androgens to an active form in the target tissue. The work of Evans (1962), which showed that prolactin activates β-glucuronidase activity in homogenates of rat and mouse testes, supports this proposal.

TABLE V

ACTIONS OF PROLACTIN ON INTEGUMENTARY (ECTODERMAL) STRUCTURES

Teleosts

1. Reduced gill Na⁺ efflux (T-1e)
2. Reduced gill permeability to water (T-1f)
3. Inhibition of gill Na⁺, K⁺-ATPase (T-1g)
4. Restoration of water turnover in hypophysectomized *Fundulus kansae* (T-1b)
5. Skin, buccal and gill mucus secretion (T-1d)
6. Melanogenesis and proliferation of melanocytes (synergism with MSH) (T-5)
7. Dispersal of yellow pigment in cutaneous xanthophores (T-4)
8. Maintenance of brood pouch in male seahorse (T-11)

Amphibians

1. Skin changes associated with water drive (A-1)
2. Proliferation of melanophores (A-12)
3. Effects on toad bladder (ectodermal) (A-15)
4. Skin yellowing in frogs (A-13)

Reptiles

1. Epidermal sloughing (R-4)

Birds

1. Production of crop 'milk' (B-1)
2. Formation of brood patch (B-2)
3. Stimulation of feather growth (B-3)
4. Stimulation of nasal gland secretion (B-7)

Mammals

1. Mammary development and lactation (M-1, M-2)
2. Sebaceous and preputial gland size and activity (M-3)
3. Hair maturation (M-4)

TABLE VI

ACTIONS OF PROLACTIN INVOLVING SYNERGISM WITH STEROID HORMONES OR ON ORGANS WHICH ARE ALSO INFLUENCED BY STEROIDS*

Cyclostomes

1. Electrolyte metabolism in hagfish (C-1)

Teleosts

1. Na^+ retention by gills (corticosteroids) (T-1e)
2. Na^+ retention by kidney (corticosteroids) (T-1i)
3. Salt and water movement in gut (corticosteroids) (T-1f)
4. Synergism with androgens on catfish seminal vesicles (T-8)
5. Dispersal of yellow pigment in xanthophores (corticosteroids) (T-4)
6. Maintenance of brood pouch in male seahorse (corticosteroids) (T-19)

Amphibians

1. Stimulation of oviductal jelly secretion (oestrogens and progestins) (A-16)
2. Na^+ transport across anuran bladder (aldosterone) (A-15)
3. Water drive structural changes (sex steroids) (A-1)
4. Cloacal gland development (androgens)

Reptiles

1. Restoration of plasma Na^+ levels in hypophysectomized lizard (corticosteroids) (R-7)
2. Anti-gonadotropic (sex steroids) ? (R-6)

Birds

1. Formation of brood patch (synergism with ovarian or testicular steroids) (B-2)
2. Parental behaviour (possible progesterone synergism) (B-10)
3. Female reproductive tract (synergism with oestrogens and progestins) (B-9)
4. Stimulation of nasal (orbital) gland secretion (corticosteroids) (B-7)
5. Stimulation of feather growth (sex steroids in some species) (B-3)
6. Anti-gonadotropic (sex steroids) ? (B-8)

Mammals

1. Mammary growth (ovarian steroids) (M-1)
2. Milk secretion (corticosteroids) (M-2)
3. Sebaceous and preputial gland secretion (gonadal and cortical steroids) (M-3)
4. Growth and secretion of male sex accessory glands (androgens) (M-11a, b)
5. Luteotropic action (oestrogens ?) (M-14)
6. Renal Na^+ reabsorption (aldosterone?) and renotropic (androgens) (M-9)
7. Spermatogenesis (androgens) (10)
8. Advanced puberty (gonadal steroids) (M-13)
9. Hair growth (androgens, corticoids) (M-4)
10. Vaginal mucification in rats (oestrogen and progesterone) (M-17)

* Indicated by parentheses.

IS THERE A COMMON THEME UNDERLYING THE ACTIONS OF PROLACTIN?

From this categorization of the effects of prolactin it is evident that the dominant action of the hormone among vertebrates consists of influences related to growth and metabolism. Accordingly, this broad, general influence could be considered as the unifying theme to explain most of the actions of prolactin. However, it should be emphasized that most experimental studies are based on the use of mammalian hormonal preparations (largely ovine or bovine). Hence, the growth-promoting and metabolic effects of mammalian prolactin in many vertebrate species may be of no physiological significance, unless it is established that the animal's own prolactin has similar effects. Available evidence indicates that mammalian growth hormones and prolactins are structurally similar (Bewley and Li 1970; Aloj and Edelhoch 1970; Fellows, Hurley and Brady 1970), and their biological properties show considerable overlap, especially those derived from primate pituitaries. Recent studies have shown that prolactin and growth hormone do exist as separate molecular entities in the pituitaries of teleosts, amphibians, reptiles and birds, as well as mammals (Nicoll and Nichols 1971; Nicoll and Licht 1971; Nicoll, Licht and Clarke 1971). It was found also that the prolactins of several vertebrate species have significant somatotropic activity in the toad growth test of Zipser, Licht and Bern (1969). These include the prolactins of two turtles, of ovine and bovine glands, and of the toad *Bufo marinus*. However, it is not known to what extent the prolactins of any of these animals are somatotropic or of metabolic significance in the species of origin.

Several of the reproductive actions of prolactin have been known for about four decades now. Accordingly, it is not surprising that this category should contain about a third of the total number of actions reported for the hormone. Although reproduction (including parental care) is a significant common theme of many of the actions of prolactin, it cannot be considered the dominant one. The majority of the actions of prolactin are patently unrelated to reproduction.

Effects of the hormone on water and electrolyte balance have only recently been appreciated; osmoregulatory research is the most rapidly expanding area of prolactin physiology. Thus, this category already contains about a quarter of the listed actions of the hormone. Within another decade, the significance of the hormone in osmoregulation may become even greater.

It is noteworthy that of the 82 actions of prolactin listed in Table I, only one of them clearly belongs to all the various groupings listed in Tables II to VI. The stimulation of mammary development and lactation involves growth of an integument-derived, reproductive structure requiring synergism with steroids for both proliferation and secretion; its secretory activity depends on massive fluid movements, producing osmoregulatory problems for the organism. Prolactin exemplifies its full spectrum of activity by its influence on the mammary gland, which provides a focal organ for the study of prolactin action at the cellular and subcellular level (cf. Denamur 1969; Topper 1970; see also Turkington, this volume, p. 111).

If one is interested in a fundamental action of prolactin at the biochemical level which might account ultimately for the wide variety of responses seen, an examination of its role in ion and water movements, and on membrane permeability, could be rewarding. Changes in intracellular ionic concentrations can influence protein synthesis (cf. Hendler 1969) and also enzyme activation. The hormones so vital in initiating and supporting milk secretion —prolactin and corticosteroids—are the same hormones which appear to regulate water and ion movements in a variety of other organs, including the intestine of teleost fishes (Utida and Hirano 1971). It no longer stretches the imagination to propose that mechanisms of significance in hormone-regulated water and salt transport in the gut, gill and urinary bladder of fishes, the bladder of anurans and the salt gland of birds may have much in common with those occurring in the mammary gland of mammals.

Despite this discussion, there is no compelling evidence at this juncture to select any of the several categories of prolactin action as being of primary significance in 'vertebrate function and organization' (Riddle 1963). It seems advisable to reserve judgment on this matter for a later date, when the role of prolactin in the physiology of a wider and more representative variety of species has been demonstrated. We suggest that it is most expedient at the present time simply to consider prolactin as that hormone of the pituitary gland which has been used by vertebrates to regulate a variety of physiological processes which are peculiar to certain groups. Prolactin is thus the 'jack-of-all-trades' of the pituitary gland, a versatile hormone that did not become specialized to subserve any single function or group of functions. Inasmuch as prolactin did not become committed early in vertebrate phylogeny to the regulation of a single physiological process, or even of a single category of processes, it was available for the control of 'emerging' physiological mechanisms important in adaptation to various ecological niches. Accordingly, prolactin may have contributed significantly to the diversification of vertebrates.

SUMMARY

The numerous physiological actions claimed for prolactin among the vertebrates are listed and considered in terms of possible underlying themes which might unify them. No single common denominator is obvious. However, most of the 82 different effects of prolactin can be classified into one or more of five categories of actions. Thus 52 per cent of all of the reported actions are related to growth, or somatotropin-like metabolic effects. About 38 per cent of prolactin actions are related to reproduction. A quarter of the listed actions of the hormone are concerned with water and electrolyte metabolism and 26 per cent of them involve effects on integumentary (ectodermal) structures. A final category concerns actions of prolactin which involve synergism with steroid hormones or effects on organs which are also influenced by steroids. Thirty-six per cent of prolactin's actions can be listed in this category.

The available data provide no compelling arguments for selecting any one of these categories as representing the major underlying theme of prolactin actions among the vertebrates. Additional data on other species may allow meaningful conclusions to be made on this matter in the future. For the present, it seems most profitable to view prolactin as the pituitary hormone which is used to regulate physiological processes that are peculiar to different vertebrate groups.

Acknowledgements

The studies in our laboratories were supported by grants from the National Institutes of Health (AM-11161 and AM-13605 to C.S.N. and CA-05388 to H.A.B.) and the National Science Foundation (GB-23033 and GF-372 to H.A.B.), and by a grant from the Population Council to C.S.N.

BIBLIOGRAPHY

In order to limit the size of the pertinent bibliography we are listing here only those references cited in the text and those papers which have appeared or become known to us since our most recent reviews (Bern and Nicoll 1968, 1969a, b). The latter can be consulted for most of the earlier literature. The references not cited or incompletely cited in the text are identified here as to content by a word or two, to indicate the source of the material listed in the tables.

ALOJ, S. M. and EDELHOCH, H. (1970) *Proc. Natl. Acad. Sci. (U.S.A.)* **66**, 830–836.
APOSTOLAKIS, M. (1968) *Vitam. & Horm.* **26**, 197–235 (review).
ARIMATSU, S. (1971) M.Sc. Thesis in Zoology, University of Tokyo (bird behaviour).
BALL, J. N. (1969) *Gen. & Comp. Endocrinol.* Suppl. 2, 10–25 (review).
BALL, J. N. (1971) Personal communication (teleost pigmentation: yellowing of *Gobius*).
BARTMANN, W.-D. (1968) Ph.D. Thesis in Zoology, J. W. Goethe University, Frankfurt-am-Main (*Z. Tierpsychol.* in press 1971) (teleost behaviour; prolactin cells).
BERN, H. A. and NICOLL, C. S. (1968) *Recent Prog. Horm. Res.* **24**, 681–720.
BERN, H. A. and NICOLL, C. S. (1969a) *C.N.R.S. Coll. Int.* **177**, 193–203.
BERN, H. A. and NICOLL, C. S. (1969b) In *Progress in Endocrinology*, pp. 433–439, ed. Gual, C. Amsterdam: Excerpta Medica Foundation.
BEWLEY, T. A., and LI, C. H. (1970) *Science* **168**, 1361–1362.
BLANC-LIVNI, N. and ABRAHAM, M. (1970) *Gen. & Comp. Endocrinol.* **14**, 184–197 (teleost pituitary histology and prolactin content).
BLATT, L. M., SLICKERS, K. A. and KIM, K. H. (1969) *Endocrinology* **85**, 1213–1215 (amphibian metamorphosis).

BLUM, V. and WEBER, K. M. (1968) *Experientia* **24,** 1259–1260 (gonadotropic in fish).
BOISSEAU, J. P. (1969) *C.N.R.S. Coll. Int.* **177,** 205–214.
BOSCHWITZ, D. (1969) *Israel J. Zool.* **18,** 277–289 (amphibian ultimobranchial).
BOSCHWITZ, D. and BERN, H. A. (1971) *Gen. & Comp. Endocrinol.* in press (amphibian hypercalcaemia).
BREAUX, C. B. and MEIER, A. H. (1971) *Am. Midland Nat.* **85,** 267–271 (amphibian metamorphosis, circadian cycle).
BROWN, P. S. and FRYE, B. E. (1969) *Gen. & Comp. Endocrinol.* **13,** 126–138, 139–145 (amphibian growth and metamorphosis).
BROWNING, H. C. (1969) *Gen. & Comp. Endocrinol.* suppl. **2,** 42–54 (reproduction).
CALLARD, I. P. and ZIEGLER, H., JR (1970) *J. Endocrinol.* **47,** 131–132 (reptile gonad inhibition).
CAMPANTICO, E., OLIVERO, M. and PEYROT, A. (1968) *Ric. Scient.* **38,** 980–985 (thyrotropic action in amphibians).
CHADWICK, A. (1969) *Gen. & Comp. Endocrinol.* suppl. **2,** 63–68 (review: birds and mammals).
CHADWICK, A. (1970) *J. Endocrinol.* **47,** 463–469 (teleost prolactin).
CHADWICK, A. and JORDAN, B. J. (1971) *J. Endocrinol.* **49,** 51–58 (crop sac stimulation).
CHAMBOLLE, P. (1969) *C. R. Hebd. Séance Acad. Sci., Sér. D* **268,** 1215–1217; **269,** 229–232 (teleost osmoregulation and reproduction; prolactin cells).
CHAN, D. K. O., CALLARD, I. P. and CHESTER JONES, I. (1970) *Gen. & Comp. Endocrinol.* **15,** 374–387 (reptile osmoregulation).
CHESTER JONES, I., PHILLIPS, J. G. and BELLAMY, D. (1962) *Gen. & Comp. Endocrinol.* suppl. **1,** 36–47 (hagfish osmoregulation).
COHEN, D., GREENBERG, J. A., LICHT, P., BERN, H. A. and ZIPSER, R. D. (1971) *Gen. & Comp. Endocrinol.* in press (amphibian metamorphosis).
DALTON, T. and SNART, R. S. (1969) *J. Endocrinol.* **43,** vi–vii (amphibian osmoregulation).
DENAMUR, R. (1969) In *Progress in Endocrinology*, pp. 959–972, ed. Gual, C. Amsterdam: Exerpta Medica Foundation.
DERBY, A. (1970) *J. Exp. Zool.* **173,** 319–328 (amphibian metamorphosis).
DERBY, A. and ETKIN, W. (1968) *J. Exp. Zool.* **169,** 1–8 (amphibian metamorphosis).
DHARMAMBA, M. (1970) *Gen. & Comp. Endocrinol.* **14,** 256–269 (teleost osmoregulation).
DONALDSON, E. M., YAMAZAKI, F. and CLARKE, W. C. (1968) *J. Fish. Res. Bd Can.* **25,** 1497–1500 (teleost osmoregulation).
DORFMAN, R. I. (1971) *Proceedings of the Third International Congress on Hormonal Steroids* (Hamburg). Amsterdam: Excerpta Medica. In press.
ENEMAR, A., ESSVIK, B. and KLANG, R. (1968) *Gen. & Comp. Endocrinol.* **11,** 328–331 (amphibian growth).
ENSOR, D. M. and PHILLIPS, J. G. (1970) *J. Endocrinol.* **48,** 167–172 (avian osmoregulation).
ETKIN, W. (1970) *Mem. Soc. Endocrinol.* **18,** 137–155 (amphibian metamorphosis).
ETKIN, W., DERBY, A. and GONA, A. G. (1969) *Gen. & Comp. Endocrinol.* suppl. **2,** 253–259 (amphibian metamorphosis).
EVANS, A. J. (1962) *J. Endocrinol.* **24,** 233–244.
FARNSWORTH, W. E. (1970) *Program of the Endocrine Society, 52nd Meeting*, p. 159 (male mammal sex accessories).
FELLOWS, R. E., JR, HURLEY, T. W. and BRADY, K. L. (1970) *Fed. Proc. Fed. Am. Soc. Exp. Biol.* **29** (2), 579 (abst. 1866).
GONA, A. G. (1968) *Gen. & Comp. Endocrinol.* **11,** 278–283 (amphibian goitrogen).
GONA, A. G. and ETKIN, W. (1970) *Gen. & Comp. Endocrinol.* **14,** 589–591 (amphibian metamorphosis).
GONA, A. G., PEARLMAN, T. and ETKIN, W. (1970) *J. Endocrinol.* **48,** 585–590 (amphibian metamorphosis).

GOURDJI, D. and TIXIER-VIDAL, A. (1969) *C.N.R.S. Coll. Int.* **179**, 231–241 (effects on birds; prolactin cells).
GUARDABASSI, A., OLIVERO, M., CAMPANTICO, E., RINAUDO, M. T., GIUNTA, C. and BRUNO, R. (1970) *Gen. & Comp. Endocrinol.* **14**, 148–151 (amphibian thyrotropin).
HENDLER, R. W. (ed.) (1969) *Protein Biosynthesis and Membrane Biochemistry.* New York: Wiley.
HIRANO, T., JOHNSON, D. W. and BERN, H. A. (1971) *Nature (Lond.)* **230**, 469–471 (teleost osmoregulation).
HOSICK, H. L., STROHMAN, R. C. and BERN, H. A. (1969) *J. Exp. Zool.* **171**, 377–384 (amphibian growth).
JOHNSON, D. W., HIRANO, T. and BERN, H. A. (1970) *Am. Zool.* **10**, 497 (teleost osmoregulation).
JONES, R. E. (1969) *Proc. Soc. Exp. Biol. & Med.* **131**, 172–174 (bird gonads).
JONES, R. E. (1969) *Gen. & Comp. Endocrinol.* **12**, 498–502; **13**, 1–13 (bird incubation patch).
JONES, R. E. (1971) *Biol. Rev.* **49** (in press) (bird incubation patch).
JONES, R. E., KREIDER, J. W. and CRILEY, B. B. (1970) *Gen. & Comp. Endocrinol.* **15**, 398–403 (bird incubation patch).
KENNEDY, T. G. and ARMSTRONG, A. T. (1971) *Fed. Proc. Fed. Am. Soc. Exp. Biol.* **30** (2), 420 (abst. 1249).
KIKUYAMA, S. (1971) Personal communication (urodele cloacal gland, synergism with steroids).
KNIGHT, P. J., INGLETON, P. M., BALL, J. N. and HANCOCK, M. P. (1970) *J. Endocrinol.* **48**, xxix–xxxi (teleost prolactin).
LAHLOU, B. and GIORDAN, A. (1970) *Gen. & Comp. Endocrinol.* **14**, 491–509 (teleost osmoregulation).
LAHLOU, B. and SAWYER, W. H. (1969) *Gen. & Comp. Endocrinol.* **12**, 370–377 (teleost osmoregulation).
LAM, T. J. (1968) *Can. J. Zool.* **46**, 1095–1097; (1969) *Can. J. Zool.* **47**, 865–869; (1969) *Comp. Biochem. & Physiol.* **31**, 909–913 (teleost osmoregulation).
LAM, T. J. and LEATHERLAND, J. F. (1969) *Can. J. Zool.* **47**, 245–250; (1969) *Gen. & Comp. Endocrinol.* **12**, 385–394; (1970) *Comp. Biochem. & Physiol.* **33**, 295–302 (teleost osmoregulation).
LEATHERLAND, J. F. (1970) *Z. Zellforsch. & Mikrosk. Anat.* **104**, 301–317, 318–336, 337–344 (teleost prolactin cells).
LEATHERLAND, J. F. (1970) *J. Endocrinol.* **48**, xxxi–xxxii (teleost prolactin cells).
LEATHERLAND, J. F. and LAM, T. J. (1969) *Can. J. Zool.* **47**, 787–792 (teleost mucus cells), **47**, 989–995 (teleost osmoregulation).
LICHT, P. and HOYER, H. (1968) *Gen. & Comp. Endocrinol.* **11**, 338–346 (reptile growth).
LICHT, P. and HOWE, N. R. (1969) *J. Exp. Zool.* **171**, 75–84 (reptile tail regeneration).
LICHT, P. and NICOLL, C. S. (1969) *Gen. & Comp. Endocrinol.* **12**, 526–535 (reptilian prolactin cells).
LOCKETT, M. F. (1965) *J. Physiol. (Lond.)* **181**, 192–199 (renotropic in cat).
LOCKETT, M. F. and NAIL, B. (1965) *J. Physiol. (Lond.)* **180**, 147–156 (renotropic in rat).
MADERSON, P. F. A. and LICHT, P. (1967) *J. Morphol.* **123**, 157–172 (reptile epidermal sloughing).
MATTEIJ, J. A. M. and SPRANGERS, J. A. P. (1969) *Z. Zellforsch. & Mikrosk. Anat.* **99**, 411–419 (teleost prolactin cells; mucus cells).
MAZZI, V. (1969) *Boll. Zool.* **36**, 1–60.
MEDDA, A. K. and FRIEDEN, E. (1970) *Endocrinology* **87**, 356–365 (amphibian metamorphosis, nitrogen metabolism).
MEIER, A. H. (1969) *Gen. & Comp. Endocrinol.* suppl. 2, 55–62 (vertebrate lipid metabolism, circadian cycle).
MEIER, A. H. (1969) *Gen. & Comp. Endocrinol.* **13**, 222–225 (bird gonad, migration).

MELBY, J. C., DALE, S. L., WILSON, T. E. and NICHOLS, A. S. (1966) *Clin. Res.* **14,** 283.
NAGAHAMA, Y. and YAMAMOTO, K. (1969) *Bull. Fac. Fish. Hokkaido Univ.* **20,** 159-168; (1970) **20,** 293-302; (1970) **21,** 169-177 (teleost prolactin cells).
NICOLL, C. S. and FIORINDO, R. P. (1969) *Gen. & Comp. Endocrinol.* suppl. **2,** 26-31 (hypothalamic control).
NICOLL, C. S. and LICHT, P. (1971) *Gen. & Comp. Endocrinol.* in press (tetrapod somatotropins).
NICOLL, C. S., LICHT, P. and CLARKE, W. C. (1971) In *Growth Hormone (Proceedings of the Second International Symposium)*, ed. Pecile, A. and Müller, E. E. Amsterdam: Excerpta Medica. In press (vertebrate prolactins and somatotropins).
NICOLL, C. S. and NICHOLS, C. W. (1971) *Gen. & Comp. Endocrinol.* **17,** 300-310 (tetrapod prolactins).
OGAWA, M. (1970) *Can. J. Zool.* **48,** 501-503 (teleost mucus cells).
OGAWA, M. and YAGASAKI, M. (1971) Ms in preparation (teleost osmoregulation: branchial water influx).
OLIVEREAU, M. (1969) *Gen. & Comp. Endocrinol.* suppl. 2, 32-41 (teleost prolactin cells).
OLIVEREAU, M. (1969) *C.N.R.S. Coll. Int.* **177,** 225-230 (teleost TRF activity).
OLIVEREAU, M. and BALL, J. N. (1970) *Mem. Soc. Endocrinol.* **18,** 57-85 (teleost osmoregulation).
OLIVEREAU, M. and LEMOINE, A.-M. (1968) *Z. Zellforsch. & Mikrosk. Anat.* **88,** 576-590; (1969) **95,** 361-376 (teleost kidney).
OLIVEREAU, M. and OLIVEREAU, J. (1970) *Z. Vgl. Physiol.* **68,** 429-445 (teleost adrenocorticotropic).
OLIVERO, M., LATTES, M. G., CAMPANTICO, E. and TOPPINO, G. (1967) *Ric. Scient.* **37,** 997-1004 (amphibian growth and metamorphosis).
OLSSON, R. (1968) In *Current Problems in Lower Vertebrate Phylogeny*, pp. 455-472, ed. Orrig, T. Stockholm: Almqvist and Wiksell (teleost prolactin cells).
OSEWOLD, T. and FIEDLER, K. (1968) *Z. Zellforsch. & Mikrosk. Anat.* **91,** 617-632 (teleost TRF activity).
PEAKER, M., PHILLIPS, J. G. and WRIGHT, A. (1970) *J. Endocrinol.* **47,** 123-127 (bird salt gland).
PEYROT, A., MAZZI, V., VELLANO, C. and LODI, G. (1969) *J. Endocrinol.* **45,** 525-530 (amphibian neural control).
PEYROT, A., VELLANO, C., ANDREOLETTI, G. E., PONS, G. and BICIOTTI, M. (1971) *Gen. & Comp. Endocrinol.* **16,** 524-534 (amphibian thyrotropic).
PICKFORD, G. E., GRIFFITH, R. W., TORRETTI, J., HENDLEZ, E. and EPSTEIN, F. H. (1970) *Nature (Lond.)* **228,** 378-379 (teleost osmoregulation: Na^+, K^+-ATPase).
POTTS, W. T. W. and FLEMING, W. R. (1970) *J. Exp. Biol.* **53,** 317-327 (teleost osmoregulation).
REDDI, A. H. (1969) *Gen. & Comp. Endocrinol.* suppl. **2,** 81-86 (male mammal sex accessories).
RIDDLE, O. (1963) *J. Natl. Cancer Inst.* **31,** 1039-1110.
SAGE, M. (1970) *J. Exp. Zool.* **173,** 121-128 (teleost pigmentation).
SAGE, M. and BERN, H. A. (1970) *Am. Zool.* **10,** 499 (teleost pigmentation).
SAGE, M. and BERN, H. (1971) *Int. Rev. Cytology* in press (teleost prolactin cells).
SAMPIETRO, P. and VERCELLI, L. (1968) *Boll. Zool.* **35,** 419 (amphibian osmoregulation).
STANLEY, J. G. and O'CONNELL, J. K. (1970) *Am. Zool.* **10,** 298 (teleost osmoregulation and mucus cells).
STETSON, M. H. and ERICKSON, J. E. (1970) *Gen. & Comp. Endocrinol.* **15,** 484-487 (bird gonad).
SUNDARARAJ, B. (1962) *Proc. Natl. Inst. Sci. India, B* **28,** 193-200 (teleost 'milk').
SUNDARARAJ, B. I. and NAYYAR, S. K. (1969) *Gen. & Comp. Endocrinol.* suppl. 2, 69-80 (male teleost sex accessories; hypothalamic control).

TASSARA, R. A. (1969) *J. Exp. Zool.* **170**, 33–54; **171**, 451–458 (amphibian regeneration).
TOPPER, Y. J. (1970) *Recent Prog. Horm. Res.* **26**, 287–308.
UTIDA, S., HATAI, S., HIRANO, T. and KAMEMOTO, F. I. (1971) *Gen. & Comp. Endocrinol.* **16**, 566–573 (teleost osmoregulation).
UTIDA, S. and HIRANO, T. (1971) In *Responses of Fish to Environmental Changes*, ed. Chavin, W. and Egami, N. in press. (teleost osmoregulation).
VELLANO, C. and LODI, G. (1968) *Boll. Zool.* **35**, 149–156 (amphibian thyroid and gonads).
VELLANO, C., LODI, G., BANI, G., SACERDOTE, M. and MAZZI, V. (1970) *Monitore Zool. Ital.* **4**, 115–146 (amphibian integument).
VELLANO, C., MAZZI, V. and SACERDOTE, M. (1970) *Gen. & Comp. Endocrinol.* **14**, 535–541 (amphibian growth).
VELLANO, C., PEYROT, A. and MAZZI, V. (1969) *Gen. & Comp. Endocrinol.* **13**, 537 (amphibian TRF stimulation).
VELLANO, C., PEYROT, A., MAZZI, V. and BICIOTTI, M. (1967) *Ric. Scient.* **37**, 260–261 (amphibian thyrotropin).
YOSHIZATO, K. and YASUMASU, I. (1970) *Dev. Growth & Diff.* **11**, 305–317; **12**, 265–272 (amphibian larval growth).
ZIPSER, R. D., LICHT, P. and BERN, H. A. (1969) *Gen. & Comp. Endocrinol.* **13**, 382–391 (amphibian growth).

DISCUSSION

Sherwood: Most of us have had clinical experience with men with prostatic carcinoma receiving large doses of oestrogen or occasionally women with breast cancer who developed sodium retention and oedema. It would be interesting to know whether these effects were directly due to the oestrogen or were perhaps secondary, through an effect of prolactin.

Greenwood: Among the 84 effects of prolactin, can one distinguish between primary and secondary actions? By analogy, insulin causes a release of growth hormone, which itself has a number of biological effects. If we didn't know of the existence of growth hormone, we would say that insulin had these effects. I wonder if the number of effects of prolactin is artificially inflated in this way? I can't think of any hormone that acts in a vacuum when injected at high levels, without itself producing some reverberations in the endocrine system.

Nicoll: This hasn't been analysed in most cases. Some of the effects, like those in the hagfish on sodium balance, seem to involve adrenal cortical tissue. Prolactin is reported to have a corticotropic effect in teleosts and amphibians, as detected histologically. However, these and many of the other reported actions could be indirect.

Denamur: Do you know any experiments that show an effect of prolactin on sodium reabsorption in mammalian tissue?

Nicoll: Lockett (1965) and Lockett and Nail (1965) have reported that prolactin reduces urinary sodium excretion in rats and cats.

Friesen: The only suggestion we have had that prolactin has a role in sodium regulation is that among patients with renal failure, a surprising number have very high prolactin levels. We haven't studied the effects of hypotonicity or hypertonicity in great detail, but in some patients studied following renal dialysis, a rise in prolactin occurs.

Nicoll: What is the situation in eclampsia with regard to HPL?

Spellacy: Watanabe and co-workers (1965) demonstrated that aldosterone secretory rates were decreased in severe toxaemia. We have found HPL levels to be low in severe toxaemia too. This raises the question of whether HPL and aldosterone secretion are related. In 1966, Melby and co-workers reported that HPL increased aldosterone secretion 30 per cent.

Greenwood: This might be relevant to that problem. When measuring plasma prolactin in goats after intravenous hypertonic saline infusion, Dr N. E. Palumbo, our collaborator, made the serendipitous observation that in stress situations, the goat shows increases in serum glutamic-oxaloacetic transaminase (SGOT). From the literature he then found that SGOT concentrations rise in psychic and physical stress in monkeys. We had been trying to monitor the stress element in our goat studies by plasma cortisol values, without much success. The SGOT and prolactin levels correlate beautifully in a psychically stressed goat. However, we can also get modest rises in SGOT without rises in prolactin, after isotonic saline infusions. In other words, the SGOT level appears to be a more sensitive indicator of stress than prolactin or cortisol.

Frantz: I think this cannot be the case in man. SGOT is used as one of the prime indicators of myocardial infarction, and people with severe chest pain who are presumably exposed to all the stress of a threatened infarction either have rises if they have an infarction, or don't if they don't, but the stress may be quite similar psychically.

Turkington: It seems to me that in biology, chemical interactions are of fundamental importance, and what really counts are covalent bonds, hydrogen bonding, hydrophobic bonding, and so on. So one can look at the spectrum of hormonal effects not only in terms of one of the reactants, the hormone, but also in terms of whatever molecules it interacts with. Since we don't know all the species that might have prolactin or what that molecule might be, one of the other important variables is the retention through all the vertebrate phyla of molecules which can react with prolactin—that is, a prolactin receptor which could be modified in the course of evolution to have a greater or lesser affinity for prolactin, and also all the different kinds of differentiation of these reactive cells. One would expect a variety of responses of these various kinds of cells, depending on how reactive their prolactin receptors are and what kind of cell differentiation they have, so that they can respond to some permutation of that receptor.

Nicoll: Prolactin has widespread effects on water and electrolyte shifts across epithelial membranes. Do you feel that this may give a clue to a possible common mechanism of action? Changes in intracellular electrolytes can certainly alter protein synthesis. Do you feel that when prolactin interacts with its receptor, one of the effects may be to change permeability to, or intracellular concentrations of, ions?

Turkington: That's one of our working hypotheses, since we haven't identified anything more specific about the cell membrane. One thing prolactin may do is to alter many general properties of the membrane in terms of permeability. That idea has to be substantiated by experiments.

Short: It's interesting that in our lifetime we have witnessed the elimination of one of the actions of prolactin. Dr Nicoll referred to the important effect of prolactin on broodiness in chickens. The commercial poultry breeder has succeeded in eliminating broodiness completely by intensive selection. If one is interested in the mechanisms of action of prolactin, it would be fascinating to compare these strains of commercial laying hens with primitive, unimproved birds that continue to go broody.

Nicoll: There are indications that the non-brooding breeds of chickens have a lower prolactin concentration in their pituitary glands than the strains which do show this behaviour. In addition, it appears that the former may be refractory to prolactin injections. Accordingly, there may have been a dual change in the animals, through selection—both in the hormone levels and in the receptors.

Forsyth: Drs J. Shani, B. J. A. Furr, B. S. Thomas and I did some experiments on cocks of a commercial non-brooding strain of fowl, trying to repeat the experiments of Nalbandov (1945). He found that prolactin was anti-gonadal and induced broody behaviour in the male. We could not demonstrate any effect of even very high doses (up to 80 i.u. per day for 20 days) of ovine prolactin on the histological appearance of the testis, on plasma testosterone levels or on comb size, so it may be that disappearance of the response of the testis to prolactin and loss of broody behaviour are genetically linked.

Herlant: As a new example of the effects of prolactin, we have observed in the male mole that during sexual activity, which is short but particularly intense in the mole, the prolactin cells are stimulated and hypertrophied, but afterwards they involute completely, when sexual activity reduces and ceases.

Nicoll: This would go along with the results of studies on rodents and other species which indicate that prolactin is involved in the function of male reproductive organs (see p. 304).

Cowie: What is the evidence for an effect of prolactin on maternal behaviour? I have noticed in rats hypophysectomized during lactation that even when milk secretion is inhibited, the maternal behaviour seems quite normal; for example, these rats still retrieve their pups.

Nicoll: This is a very controversial area which I believe remains unsettled.

Meites: Dr M. X. Zarrow reported an interesting study in rabbits (Workshop on Prolactin, National Institutes of Health, January 1971). Rabbits characteristically at the end of pregnancy or pseudopregnancy build a nest; they pull out their hair and use it to make a downy lining. Zarrow showed that this behaviour could be induced in non-pregnant rabbits by giving a combination of oestrogen, progesterone and prolactin. He also showed that if you give ergocornine at the end of pregnancy or pseudopregnancy, this effect could be prevented in the rabbit. This indicates a role for prolactin in maternal behaviour.

Dr Nicoll and Dr Bern deserve a great deal of credit for their very fine comparative studies on prolactin, but I wouldn't underestimate the role of prolactin in mammary growth and lactation. After all, the existence of most mammalian species for a critical period after birth depends on milk secretion, and that's not an unimportant function.

Most of the comparative studies were made with mammalian prolactins, and they appear to have all these effects throughout the vertebrates, but the question is, do fish, amphibia and reptiles produce anything comparable to mammalian prolactin? Will fish prolactin induce mammary growth or lactation in mammals? What is the chemical relationship, if any, between fish prolactin and mammalian prolactin? In other words, is there any justification for calling all these substances 'prolactin'?

Nicoll: The answer to this question depends on how one defines a hormone. Prolactin obtained its various names on the basis of its observed actions in mammals and pigeons. If the first work on the hormone had been done in certain teleosts, we would probably be calling it by a name which denotes its osmoregulatory actions.

There is good evidence that the pituitary glands of every group of jawed vertebrates contain hormones which are homologous with virtually all the pituitary principles of mammals. The question then is, how do we establish homology? Cells with the same tinctorial properties as mammalian lactotrophs (carminophils or erythrosinophils) have been identified in the adenohypophyses of all classes of the jawed vertebrates. It has been demonstrated by Dr E. Emmart and her colleagues by immunofluorescent procedures that the prolactin cells of a teleost (*Fundus heteroclitis*) bind rabbit antiserum to ovine prolactin which is tagged with fluorescein. This suggests that the teleost prolactin has immunological determinants similar to the ovine hormone. It has also been demonstrated that fish pituitaries contain principles with some of the biological activities of mammalian prolactins (e.g. osmoregulatory action in euryhaline species, eft water drive, cutaneous mucus secretion). Accordingly, there are anatomical, immunochemical and physiological reasons for homologizing a fish pituitary principle with mammalian prolactin.

If one chooses to adopt a restricted definition of prolactin, and maintain that such a hormone must be active in mammals, then the adenohypophyses of vertebrates other than tetrapods would not have prolactins. However, this would be akin to asserting that fish pituitaries do not contain a growth hormone because teleostean pituitary preparations do not promote growth in the rat. This ignores the fact that it is possible to isolate a protein from the pituitary of teleosts which does promote growth in teleosts. If one wished to be even more restrictive in defining growth hormones, one could assert that growth hormones are those pituitary principles which promote growth in humans. According to this definition, the ovine and bovine pituitary factors which promote growth in the rat would not be growth hormones but some preparations of ovine prolactins would be. Such restricted definitions are obviously inadequate. Clearly, when several criteria of homology are satisfied, as is the case for vertebrate prolactins, we are justified in considering the hormonal principles as homologous. The question which you raised, Dr Meites, reflects the inadequacy of the term 'prolactin' and the various other names which are applied to this hormone.

Meites: Has any work been done on the extraction of prolactins from vertebrates such as fishes, and what effects do they produce in mammals? You say they don't produce mammary growth or lactation. They don't produce epithelial growth in pigeon crop sacs and they have no luteotropic function. What effects do they have in mammals?

Nicoll: As far as I know, fish prolactins have no typical effect in mammals.

Meites: Are we justified in calling them prolactins?

Denamur: According to Chadwick (1966), fish prolactin has a mammotropic effect.

Nicoll: It's not a typical mammotropic effect.

Wilhelmi: We tried to isolate prolactin from fish pituitaries. The difficulty was that the fish from which the glands were taken (hake and pollack) are marine teleosts living in conditions in which they don't need prolactin very much. (Prolactin is apparently required for adaptation to fresh water, but not by fish that live in the sea all the time.) The concentration of hormone in the gland, so far as we could tell from some rather crude quantitative estimates using anti-ovine prolactin, by Dr Emily Emmart, is very low. I never had a fraction that gave good stimulation of the crop glands in the pigeon, so we couldn't say that we could demonstrate any effect of prolactin. I would prefer to reserve opinion on this until we have a source of fish pituitaries that would enable us to isolate a good prolactin preparation. Incidentally, Dr Nicoll, have you vetted your various effects for the possibility that some of the preparations used may not have been good prolactins at all?

Nicoll: Almost all the effects that I listed have been reported with the recently available NIH preparations of prolactin. However, in most cases the ovine hormone was used.

Meites: Dr Li has worked on this problem from a chemical point of view. There are chemical similarities, aren't there, between the different growth hormones, for example?

Li: You are really asking for a definition of a hormone. In my view, a 'growth hormone' is for promoting body growth—whether of rat or fish or human growth. It promotes an increase in protein, and new tissue.

Wilhelmi: If a substance produces growth in any animal it might be described as a growth hormone, but there are paradoxical effects and you must define the test animal. For example, if you inject prolactin in the plateau intact female rat, it produces growth.

Li: Compared to the same dose of growth hormone, prolactin causes very little growth in plateau intact rats. Prolactin increases appetite and so does growth hormone, so Reisfeld and his colleagues (1961) said that prolactin promotes growth in hypophysectomized rats, but this is wrong.

Wilhelmi: We have been unable to repeat this. In plateaued female rats it has been known for many years that without dietary restriction, in the presence of their own growth hormone, if you administer prolactin there is a good growth response in 15 days and you can even develop a dose-response curve. This has been shown by a number of people. The odd thing is that if you hypophysectomize the rat, you get no growth with prolactin, even with large doses (4 mg/day), using homogeneous protein free of growth hormone. We've done this many times. I have never tested rats on restricted food intake, and I know that a critical way of differentiating the effects is just this. I would agree with you that an animal on a restricted food intake, say a young adult male rat, won't gain weight. If you give growth hormone he will gain weight, providing he has enough body fat to afford it.

Beck: We can abolish the effects of growth hormone in man by restricting nitrogen intake, just as Dr Jane Russell showed she could abolish some of the actions of growth hormone in the rat by reducing nitrogen intake.

Wilhelmi: You have to restrict it very severely.

Li: This goes back in history, to the disagreement between Oscar Riddle and H. M. Evans. Riddle insisted that prolactin is a growth hormone.

We have tested this many times. On a restricted diet prolactin does not promote body growth in rats.

MacLeod: But are we talking about increases in weight or in length? These are two entirely different things. Rats bearing pituitary tumours which secrete very large amounts of prolactin may or may not gain weight, but they certainly don't increase their bone length, so in those terms prolactin is not a growth hormone.

Meites: Dr Beck probably has the best evidence for the growth-promoting effects of one type of prolactin (ovine), but I would tend to agree with Dr Li on the whole, because the type of growth obtained with prolactin in most mammalian species is extremely limited. Even by the most sensitive biological test for growth hormone, the tibia test, you can get only a very limited increase in tibial-cartilage width with prolactin, 60 μm or less, whereas you see tremendous increases with pituitary growth hormone.

Cowie: Antibody production might be limiting the growth response when heterologous prolactins are used. Most studies have been done with sheep prolactin and if the recipient animal makes antibodies to it, these could neutralize the growth-promoting effect. We came up against this problem in using sheep prolactin as replacement therapy in rabbits hypophysectomized during lactation; antibodies are formed and the exogenous prolactin becomes less effective in maintaining milk secretion (Cowie, Hartmann and Turvey 1969).

Greenwood: Dr Nicoll, do you regard these prolactin tissue receptors as both primitive and stable in an evolutionary sense? As far as I can see, the ovine prolactin molecule binds to all receptors, from fish to mammals. Therefore there must be a sequence in the prolactin molecule of the recipient species which is similar to ovine prolactin; this may explain why anti-ovine prolactin serum is useful across the species board.

Nicoll: The recent data on the primary structure of prolactins and growth hormones indicate that they are very similar. This suggests that they probably arose from a common ancestral molecule. If we speculate on this precursor molecule, I think that it was most probably a prolactin, rather than a growth hormone. The available evidence indicates that foetal and neonatal mammals do not need a growth hormone to grow up to a certain point. There are other cases of vertebrates which do not need the adenohypophysis to grow normally, such as the guinea pig, or *Ambystoma tigrinum*. Accordingly, the need for a growth hormone in early vertebrates is open to question. However, they probably needed a hormone like prolactin for osmoregulation because they were aquatic. Thus, prolactin was probably functioning as an osmoregulatory hormone in early vertebrate history.

To speculate further, the primitive prolactin may have given rise to an isohormone which was slightly different, but still able to combine with the appropriate receptors and regulate the same physiological processes. By further modification, the isohormones became more different. In order for this isoprolactin to become an effective growth hormone, the receptors of these primitive prolactins in some tissues would also have to undergo some alteration to develop a higher affinity to the isoprolactin, and they must have become coupled with processes concerned with growth regulation. Assuming that extant prolactins and growth hormones arose by such a mechanism, the receptors, like the hormonal principles, are probably very similar in many cases.

Since the prolactins and somatotropins of mammals are able to react with the receptors of fish and other 'lower' vertebrates, this indicates a certain degree of stability in both the receptors and the hormones. However, fish prolactins are not fully active in the mammary gland or pigeon crop sac and fish growth hormone is essentially inactive in the rat (although it may be active in another mammalian species). Accordingly, some of the receptors and some of the hormones have changed with evolution.

Greenwood: The human ought to have receptors which distinguish human prolactin from human growth hormone.

Nicoll: Yes, but the evidence is that the human receptors aren't very good discriminators. However, if the receptor can discriminate adequately between both endogenous hormones within the normal physiological range, that is all that is required to achieve differential regulation of physiological processes.

REFERENCES

CHADWICK, A. (1966) *J. Endocrinol.* **35,** 77–81.
COWIE, A. T., HARTMANN, P. E. and TURVEY, A. (1969) *J. Endocrinol.* **43,** 651–662.
LOCKETT, M. F. (1965) *J. Physiol. (Lond.)* **181,** 192–199.
LOCKETT, M. F. and NAIL, B. (1965) *J. Physiol. (Lond.)* **180,** 147–156.
MELBY, J. C., DALE, S. L., WILSON, T. E. and NICHOLS, A. S. (1966) *Clin. Res.* **14,** 283.
NALBANDOV, A. V. (1945) *Endocrinology* **36,** 251–258.
REISFELD, R. A., TONG, G. L., RICKES, E. L. and BRINK, N. G. (1961) *J. Am. Chem. Soc.* **83,** 3717–3719.
WATANABE, M., MEEKER, E. I., GRAY, M. J., SIMS, E. A. H. and SOLOMON, S. (1965) *J. Clin. Endocrinol. & Metab.* **25,** 1665.

HYPOTHALAMIC CONTROL OF PROLACTIN SECRETION

Joseph Meites

Department of Physiology, Michigan State University, East Lansing, Michigan

HYPOTHALAMIC INHIBITION

UNDER most conditions, the mammalian hypothalamus has an inhibitory influence on the synthesis and release of prolactin. An early experiment by Everett (1954) showed that when the pituitary of female rats was removed from hypothalamic control by transplanting it under the kidney capsule, functional corpora lutea (indicative of prolactin action) were maintained for at least three months. The amount of prolactin released by the transplant was judged to be comparable to that produced during pseudopregnancy in the rat. More recent work by our laboratory (Chen *et al.* 1970) showed that when a single pituitary from a mature female rat was transplanted under the kidney capsule of a hypophysectomized mature female rat, the serum prolactin concentration, as measured by radioimmunoassay, was about as high as in oestrous rats for the first two weeks after transplantation, and then fell to levels slightly higher than in dioestrous rats for the remaining eight weeks of the experiment. Transplantation of two or four pituitaries resulted in correspondingly higher levels of serum prolactin (Fig. 1).

Bilateral lesions placed in the median eminence or anterior or posterior hypothalamus of ovariectomized rats, produced significant elevations in serum prolactin as compared to sham-lesioned controls (Chen *et al.* 1970) (Fig. 2). More recently, bilateral lesions placed in the median eminence of ovariectomized rats were observed to result in about a ten-fold rise in serum prolactin by the end of 30 minutes, and the concentration remained elevated for at least six months thereafter (Welsch *et al.* 1971). Presumably such lesions destroy areas of the hypothalamus which inhibit the release of prolactin. Similar lesions placed in these areas of the hypothalamus were found to initiate lactation in rabbits (Haun and Sawyer 1960) and rats (DeVoe, Ramirez and McCann 1966), and to induce pseudopregnancy in rats (Flerko and Bardos 1959).

The culture *in vitro* of pituitary tissue from a variety of mammalian species demonstrated that the pituitary can synthesize and release prolactin autonomously for prolonged periods of time (Meites, Kahn and Nicoll 1961; Pasteels 1961; Meites, Nicoll and Talwalker 1963). Anterior pituitary tissue from rats, mice, rabbits, guinea pigs and monkeys released ten to 16 times more prolactin during a six-day period of culture than was

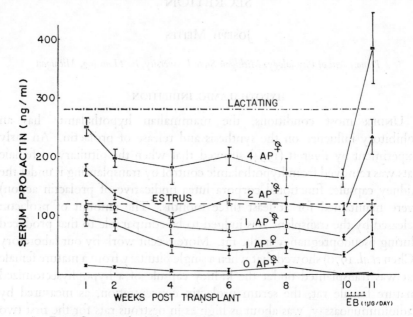

FIG. 1. Serum prolactin concentrations, measured by radioimmunoassay, in hypophysectomized rats with 0, 1, 2 or 4 anterior pituitary grafts under the kidney capsule. All rats were ovariectomized except one group with one pituitary graft. These are compared with intact rats on the day of oestrus and with postpartum lactating rats immediately after being suckled by their young for 30 minutes. Oestradiol benzoate injections (1 μg/day per rat) were begun at the end of the tenth week to demonstrate the direct effects of oestrogen on prolactin release. Vertical bars represent standard errors of the mean. (After Chen *et al.* 1970.)

originally present in the fresh pituitary tissue (Nicoll and Meites 1962a). These and related studies provide ample evidence that the mammalian hypothalamus chronically depresses prolactin secretion, and that more prolactin is released when the pituitary is separated from the hypothalamus.

Prolactin-inhibiting activity was demonstrated in hypothalamic extracts from a variety of species, including the rat, sheep, cow, pig and man (Meites 1966; Pasteels 1961; Talwalker, Ratner and Meites 1963; Schally *et al.* 1965). The effects of rat hypothalamic extract on serum prolactin in rats are shown in Fig. 3. The presumed neurohormone was named 'prolactin inhibiting factor' (PIF), since early evidence indicated that it inhibited both synthesis and release of prolactin (Meites, Kahn and Nicoll

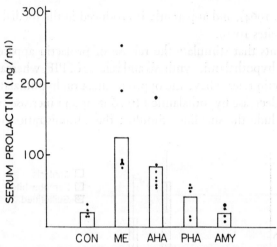

FIG. 2. Serum prolactin concentrations measured by radioimmunoassay two weeks after placement of bilateral hypothalamic lesions in ovariectomized rats. CON=sham-operated controls, ME=median eminence lesions, AHA=anterior hypothalamic lesions, PHA= posterior hypothalamic lesions, AMY=amygdaloid lesions. Note that the AMY lesions had no effect on serum prolactin levels. (After Chen et al.1970.)

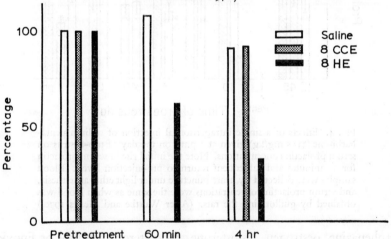

FIG. 3. Effects of a single intraperitoneal injection of the equivalent of eight hypothalami extracted (HE) from rats compared with the effects of a similar amount of cerebral cortical extract (8 CCE). The extracts were injected at 10 a.m. on the morning of pro-oestrus and blood samples were removed 60 minutes and four hours later. Only the hypothalamic extract significantly reduced serum prolactin levels. (After Amenomori and Meites 1970.)

1961; Talwalker, Ratner and Meites 1963). Chemical characterization of PIF has not yet been reported, although it appears to be a small molecule. It has been separated from luteinizing hormone releasing factor (LRF)

(Schally et al. 1964), and apparently is produced in the medial basal hypothalamus (Meites 1966).

Many agents that stimulate the release of prolactin apparently act by reducing the hypothalamic synthesis and release of PIF, whereas agents that inhibit prolactin release have the opposite effect on hypothalamic activity. Agents that decrease hypothalamic PIF content and increase the release of prolactin include the suckling stimulus; the administration of reserpine,

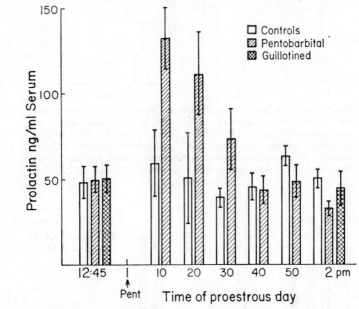

FIG. 4. Effects of a single intraperitoneal injection of sodium pentabarbitone (31·5 mg/kg) given at 1 p.m. on the day of pro-oestrus on serum prolactin concentrations. Note the initial rise in serum prolactin for 30 minutes and subsequent return to pre-injection levels. Blood samples were collected by heart puncture under light ether anaesthesia, and serum prolactin concentrations were the same as when blood was obtained by guillotining the rats. (After Wuttke and Meites 1970.)

perphenazine, oestrogen, testosterone, progesterone, cortisol or Enovid; and several stresses and non-specific agents (see Meites and Nicoll 1966). Repeated injections of adrenaline and acetylcholine also reduce hypothalamic PIF content (Mittler and Meites 1967), but these effects may be non-specific in nature. A single injection of sodium pentobarbitone increased prolactin concentrations in serum in rats during the first 30 minutes after injection by reducing hypothalamic PIF activity, and subsequently depressed prolactin release by directly inhibiting the pituitary (Fig. 4) (Wuttke and Meites 1970).

Agents that elicit significant increases in hypothalamic PIF activity, and decrease serum prolactin, include prolactin itself (Chen, Minaguchi and Meites 1967; Clemens and Meites 1968), ergot drugs (Wuttke, Cassell and Meites 1971), L-dopa and several monoamine oxidase inhibitors (Lu and Meites 1971). A single injection of dopamine into the third ventricle of male rats was also shown to increase PIF activity in the hypothalamo-pituitary portal blood (Kamberi, Mical and Porter 1970).

HYPOTHALAMIC STIMULATION

The possibility that the hypothalamus may promote the release of prolactin under some conditions was suggested by reports that oxytocin can induce prolactin release (Benson and Folley 1956). Since oxytocin is produced in hypothalamic nuclei and the nerve tracts carrying oxytocin to the posterior pituitary traverse the median eminence, it appeared reasonable to assume that some oxytocin may be released into the portal circulation and evoke prolactin release. The work of Benson and Folley (1956) was based on the demonstrated ability of injected oxytocin to retard involution of the mammary glands of postpartum mother rats after removal of their litters, a property also shared by prolactin. No actual blood or pituitary prolactin values were reported. Subsequent studies by other investigators failed to show that oxytocin increased prolactin release (Meites, Nicoll and Talwalker 1963; Pasteels 1968). More recently F. C. Greenwood reported (at the Prolactin Workshop, National Institutes of Health, Bethesda, January 11–12, 1971) that oxytocin injections did not raise blood prolactin levels in sheep. Also, the effects of oxytocin on the mammary glands of postpartum unsuckled rats were found not to duplicate the effects of prolactin, although both hormones retarded mammary involution. It appears therefore that oxytocin is not a releaser of prolactin.

The first attempt to demonstrate prolactin-releasing activity in hypothalamic extracts was reported by our laboratory (Meites, Talwalker and Nicoll 1960). Injections of neutralized acid extracts of rat hypothalamus into oestrogen-primed rats initiated mammary secretion. However, initiation of lactation in rats cannot be considered an exclusive property of prolactin alone, since adrenal cortical hormones are also essential. Also, many non-specific or stressful agents can initiate mammary secretion in oestrogen-primed rats (Meites, Nicoll and Talwalker 1963). Several years later Mishkinsky, Khazen and Sulman (1968) confirmed our results and concluded that this suggested the presence of prolactin-releasing activity in hypothalamic extracts. However, it still remains to be proved that rat hypothalamic extracts contain a prolactin-releasing factor.

In vitro studies by Nicoll and co-workers (1970) suggested that crude rat hypothalamic extracts contained both prolactin-inhibiting and prolactin-releasing activities. Incubation of rat pituitary with hypothalamic extract inhibited prolactin release for the first four hours, but stimulated prolactin release for the following four to eight hours. Under somewhat different *in vitro* conditions, our laboratory (Meites 1970) failed to duplicate the results of Nicoll and co-workers (1970) and we observed only inhibition of prolactin release during an eight-hour period of incubation.

Experiments with drugs have also suggested the possibility of the presence of prolactin-releasing activity in the hypothalamus. Schneider and Midgley (1970) reported that an injection of L-threodihydroxyphenylserine (DOPS) resulted in an increase in blood prolactin in rats. They also found that injections of blood from the DOPS-treated rats into untreated recipient rats increased serum prolactin concentrations. However, the possibility was not excluded that some of the drug remained in the blood injected into the recipient rats, or that they were merely eliminating PIF activity in the blood. In our laboratory, Lu (unpublished data) recently found that a single intraperitoneal injection of chlorpromazine into hypophysectomized rats with a single pituitary transplant under the kidney capsule increased serum prolactin although the increase was less than after an injection of chlorpromazine into intact rats with no pituitary transplant. Incubation of untreated pituitary tissue with blood from the chlorpromazine-treated rats resulted in increased prolactin release. However, it appears that this increase was due to removal of PIF activity, since the systemic blood of hypophysectomized rats was previously shown to contain PIF activity (Meites, Nicoll and Talwalker 1963; Sud, Clemens and Meites 1970). Unequivocal evidence that the mammalian hypothalamus contains prolactin-releasing activity requires the demonstration that extracts of hypothalamic tissue can induce release of prolactin both *in vivo* and *in vitro*. A point of interest is the recent report by Tashjian, Barowsky and Jensen (1971) that synthetic thyrotropin-releasing factor can induce increased release of prolactin *in vitro* by pituitary cells obtained from rats with pituitary tumours. This raises the interesting possibility that factors other than PIF in the hypothalamus may help to regulate prolactin secretion.

In contrast to mammals, there is general agreement that the avian hypothalamus contains prolactin-releasing rather than prolactin-inhibiting activity (see Meites and Nicoll 1966). This has been demonstrated in hypothalamic extracts from the pigeon, tricoloured blackbird, Japanese quail, chicken and duck (Meites and Nicoll 1966) and most recently in the turkey (Chen *et al.* 1968). An explanation for the apparent difference in hypothalamic regulation of prolactin secretion in mammals and birds is not at hand. Agents such as oestrogens, androgens or reserpine stimulate prolactin release when injected into mammals but have no effect on prolactin release in pigeons (unpublished observations).

THE ROLE OF CATECHOLAMINES

Catecholamines are present in relatively higher concentrations in the hypothalamus than in the rest of the brain (Vogt 1954), and there is considerable evidence that they are involved in the release of several anterior pituitary hormones (Coppola 1968; Fuxe and Hökfelt 1969; Wurtman 1970). An adrenergic tonus is believed to stimulate LH and FSH release, but to inhibit prolactin release. Early experiments indicated that administration of adrenergic drugs induced LH release and ovulation, whereas anti-adrenergic drugs blocked these actions (Sawyer 1969). The injection of adrenaline and noradrenaline into rats and rabbits with developed mammary glands induced lactation, but anti-adrenergic drugs and non-specific agents also initiated lactation (see Meites, Nicoll and Talwalker 1963). Blood prolactin concentrations could not be measured during these earlier years because of the unavailability of radioimmunoassays. Since lactation is not a specific reaction to prolactin alone but also requires the action of adrenal cortical hormones, these earlier observations on the effects of administered catecholamines are of uncertain significance.

Recent work indicates that drugs that decrease hypothalamic catecholamines produce an increase in serum prolactin, whereas drugs that increase hypothalamic catecholamines elicit a reduction in serum prolactin (Lu et al. 1970; Lu and Meites 1971). A single intraperitoneal injection into female rats of drugs that reduce brain catecholamine levels, including reserpine, chlorpromazine, α-methyl-p-tyrosine, α-methyl-m-tyrosine, methyldopa or D-amphetamine, produced many-fold increases in serum prolactin concentration by 30 minutes to four hours after injection (Fig. 5). Reserpine depletes the brain of catecholamines and prevents their re-uptake by nerve endings, chlorpromazine interferes with the action of catecholamines at receptor sites, the tyrosine analogues inhibit the conversion of tyrosine to L-dopa, and methyldopa inhibits the transformation of dopa into dopamine (Koelle 1970; Innes and Nickerson 1970). Amphetamine may interfere with receptor-site actions of catecholamines.

A single intraperitoneal injection of drugs that increase hypothalamic catecholamines, including L-dopa, pargyline, iproniazid or Lilly compound 15641, significantly decreased serum prolactin concentrations by 30 minutes to two hours later (Fig. 6). L-Dopa is the immediate precursor of dopamine and the other three drugs inhibit the action of monoamine oxidases, thereby depressing the metabolism of catecholamines and causing them to accumulate in the brain (Koelle 1970; Innes and Nickerson 1970). Each of these drugs increased hypothalamic PIF content (Lu and Meites, unpublished observations).

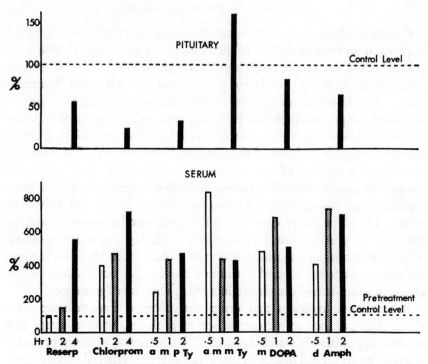

FIG. 5. Effects of single intraperitoneal injections of drugs that decrease hypothalamic catecholamines on serum prolactin concentrations. All injections were made at 10 a.m. on the morning of pro-oestrus and blood samples were collected for assay thereafter at the indicated time intervals. All drugs increased serum prolactin significantly. When the rats were killed at the end of blood collections, pituitary prolactin concentrations were significantly reduced except in the rats given α-methyl-*m*-tyrosine, reflecting the pronounced release of prolactin from the pituitary by the drugs. (After Lu *et al.* 1970; Lu and Meites 1971.)

The catecholamines apparently act as neurotransmitters to effect the release of hypothalamic hypophysiotropic hormones (Coppola 1968; Fuxe and Hökfelt 1969; Kamberi, Mical and Porter 1970; Kamberi, Schneider and McCann 1970). The latter two groups of workers provided evidence that dopamine increases the release by the hypothalamus of LRF amd FSH-releasing factor (FSH-RF). Kamberi, Mical and Porter (1970) also reported that a single injection of dopamine into the third ventricle of rats resulted in increased PIF activity in the portal blood and decreased prolactin in the systemic blood. Reserpine depresses hypothalamic catecholamine levels, decreases hypothalamic PIF content and increases blood prolactin concentration (Ratner, Talwalker and Meites 1965). The concept that hypothalamic catecholamines act to reduce the release of prolactin

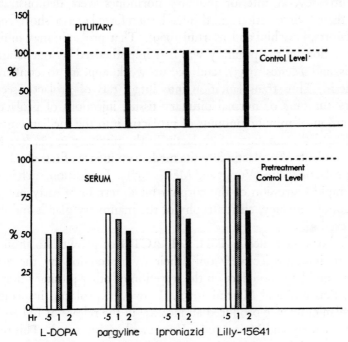

FIG. 6. Effects of single intraperitoneal injections of drugs that increase hypothalamic catecholamine levels on serum prolactin values. All drugs reduced serum prolactin significantly but had little or no effect on pituitary prolactin concentration. (After Lu and Meites 1971.)

by increasing the release of PIF therefore appears to be correct. Whether dopamine is the principal catecholamine involved in this process, as indicated by Kamberi, Mical and Porter (1970) and Kamberi, Schneider and McCann (1970), or whether all catecholamines act similarly, remains to be determined.

Some doses of catecholamines were reported to inhibit directly the release of prolactin by rat pituitary tissue *in vitro* (MacLeod 1969; Birge *et al.* 1970). However, smaller doses of catecholamines were subsequently shown to increase prolactin release by the rat pituitary *in vitro* (Koch, Lu and Meites 1970). Catecholamines could not be detected in portal blood (Wurtman 1970), and failed to alter prolactin release *in vivo* when infused into portal vessels (Kamberi, Mical and Porter 1970). It appears doubtful, therefore, that the direct effects of catecholamines on the release of pituitary prolactin observed *in vitro* are of physiological significance.

INHIBITION OF PROLACTIN SECRETION BY PROLACTIN

Recently several anterior pituitary hormones were demonstrated to inhibit their own secretion and this has been referred to as a 'short feedback loop' (Motta, Fraschini and Martini 1969). That prolactin may inhibit its own secretion by the pituitary was first hypothesized by our laboratory (Sgouris and Meites 1953), and recent work appears to confirm this hypothesis. Thus transplantation into intact rats of prolactin-secreting pituitary tumours or normal pituitary tissue, injections of prolactin, or implantation of minute amounts of prolactin into the median eminence, all decreased the amounts of prolactin in the pituitary *in situ* (see Meites 1970). Implants of prolactin into the median eminence of rats also lowered serum prolactin values (Voogt and Meites 1971). In addition, such implants induce rapid regression of the corpora lutea, terminate early pregnancy and pseudopregnancy, elicit atrophy of the mammary glands and depress milk secretion.

Other hormones such as FSH, LH, ACTH or growth hormone each appear to inhibit selectively only their own secretion by the anterior pituitary, and have no effect on the secretion of other pituitary hormones (Motta, Fraschini and Martini 1969). However, prolactin implants into the median eminence produced an increase in the release of LH and FSH (Voogt, Clemens and Meites 1969; Voogt and Meites 1971). This resulted in the resumption of cycling in early pregnancy, in pseudopregnant or in postpartum lactating rats. In prepubertal female rats, injections of prolactin or implantation of minute amounts of prolactin into the median eminence at 21 days of age hastened the time of onset of puberty by about one week, from an average of about 38 to 31 days after birth (Clemens *et al.* 1969; Voogt, Clemens and Meites 1969).

The mechanisms of these actions by prolactin and their physiological significance remain to be determined. It has been shown that the hypothalamic content of PIF is increased in rats carrying pituitary tumour transplants or implants of prolactin in the median eminence (see Meites 1970). This suggests that the inhibiting action of prolactin on prolactin secretion is exerted via the hypothalamus, although a possible direct action of prolactin on the pituitary cannot be completely ruled out. Fuxe and Hökfelt (1969) presented evidence that prolactin injections markedly activate the tubero-infundibular *dopaminergic* neurons in rats. This activation of dopamine in the hypothalamus could explain why PIF is increased and prolactin release is decreased after implantation of prolactin in the median eminence. This would also explain why LH and FSH release are increased by the implant, since dopamine has been reported to increase the release into the portal vessels of LRF and FSH-RF (Kamberi, Mical and Porter 1970). Evidence has already been presented that administration of L-dopa results in a decrease in serum prolactin. More recently, we have shown that L-dopa increases PIF activity in the hypothalamus (Lu and Meites, unpublished).

Whether prolactin really has a role in inhibiting its own secretion, or in stimulating the release of LH and FSH in normal physiological states, remains to be demonstrated. It has been reported that an implant of prolactin in the median eminence can inhibit the stimulatory action of small doses of oestrogen (Welsch et al. 1968) or of the suckling stimulus (Clemens and Meites 1968) on prolactin release. These and other observations suggest that prolactin probably does have a role in regulating its own secretion and perhaps that of FSH and LH as well.

DIRECT ACTION OF SOME HORMONES AND DRUGS ON PITUITARY PROLACTIN RELEASE

Although the hypothalamus appears to be the principal regulator of prolactin secretion, there is evidence that some agents can act directly on the pituitary to influence prolactin release. Oestrogen can stimulate the release of prolactin by the pituitary *in vitro* (Nicoll and Meites 1962b). This steroid therefore acts both via the hypothalamus, by decreasing PIF activity, and directly on the pituitary to promote prolactin release. Thyroxine and tri-iodothyronine also can increase prolactin release by the pituitary *in vitro* (Nicoll and Meites 1963). The injection of thyroxine into rats does not change the hypothalamic PIF content (Chen and Meites 1970), suggesting that thyroid hormones promote the release of prolactin only by direct stimulation of the pituitary.

Ergocornine increases hypothalamic PIF activity (Wuttke, Cassell and Meites 1971), but also can act directly on the pituitary to inhibit prolactin release (Lu and Meites 1971). Incubation of rat pituitary with ergocornine resulted in a significantly reduced release of prolactin but in increased accumulation of prolactin in the pituitary. Incubation of the pituitary with oestradiol increased prolactin release, but this effect was inhibited by ergocornine (Fig. 7). Injection of ergocornine into rats *in vivo* also inhibited the stimulatory action of oestrogen on prolactin release and prevented enlargement of the pituitary (Lu, Koch and Meites 1971).

In addition to the above, sodium pentobarbitone was shown to directly inhibit pituitary prolactin release *in vitro* (Wuttke, Gelato and Meites 1971), and it is probable that this is the principal mechanism by which this drug inhibits the release of prolactin. Vasopressin, oxytocin and bradykinin had no effect on prolactin release *in vitro* (Meites, Nicoll and Talwalker 1963). The possible direct effects of other drugs on prolactin release remain to be tested.

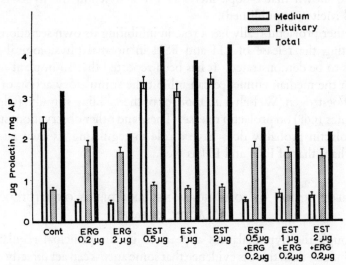

FIG. 7. Effects of incubating individual anterior pituitary halves from mature male rats in 2 ml of Medium 199 with ergocornine (ERG) or oestradiol (EST) or both for 12 hours. Prolactin was measured by radioimmunoassay. Note that both doses of ergocornine markedly inhibited prolactin release into the medium and increased prolactin concentration in the pituitary. All three doses of oestrogen significantly increased prolactin release with little effect on pituitary concentration. Ergocornine completely suppressed the stimulatory action of oestrogen on prolactin release. (After Lu, Koch and Meites 1971.)

SUMMARY

The mammalian hypothalamus exerts an inhibitory influence on the synthesis and release of prolactin under most conditions. Hypothalamic extracts from many species depress prolactin release and are believed to contain a prolactin-inhibiting factor (PIF). Removal of hypothalamic connexions to the pituitary results in enhanced prolactin release. Stimuli that promote prolactin release decrease hypothalamic PIF activity, whereas agents that inhibit prolactin release increase hypothalamic PIF activity. Hypothalamic catecholamines are important in regulating prolactin secretion; an increase in adrenergic tonus is associated with increased PIF activity and reduced prolactin release, whereas a decrease in adrenergic tonus results in lower PIF activity and greater prolactin release. A prolactin-releasing factor may also be present in the hypothalamus, but this remains to be established. Although most stimuli act through the hypothalamus to influence prolactin secretion, some hormones and drugs can act directly on the pituitary to alter the release of prolactin.

Acknowledgements

The author was aided in part by research grants AM 04784 and CA 10771 from the National Institutes of Health, United States Public Health Service.

REFERENCES

AMENOMORI, Y. and MEITES, J. (1970) *Proc. Soc. Exp. Biol. & Med.* **134**, 492–495.
BENSON, G. K. and FOLLEY, S. J. (1956) *Nature (Lond.)* **177**, 700–701.
BIRGE, C. A., JACOBS, L. S., HAMMER, C. T. and DAUGHADAY, W. H. (1970) *Endocrinology* **86**, 120–130.
CHEN, C. L., AMENOMORI, Y., LU, K. H., VOOGT, J. L. and MEITES, J. (1970) *Neuroendocrinology* **6**, 220–227.
CHEN, C. L., BIXLER, E. J., WEBER, A. I. and MEITES, J. (1968) *Gen. & Comp. Endocrinol.* **11**, 489–494.
CHEN, C. L. and MEITES, J. (1970) *Proc. Soc. Exp. Biol. & Med.* **131**, 576–578.
CHEN, C. L., MINAGUCHI, H. and MEITES, J. (1967) *Proc. Soc. Exp. Biol. & Med.* **126**, 317–320.
CLEMENS, J. A. and MEITES, J. (1968) *Endocrinology* **82**, 878–881.
CLEMENS, J. A., MINAGUCHI, H., STOREY, R., VOOGT, J. L. and MEITES, J. (1969) *Neuroendocrinology* **4**, 150–156.
COPPOLA, J. A. (1968) *J. Reprod. & Fert.* suppl. 4, 35–44.
DEVOE, W. F., RAMIREZ, V. D. and MCCANN, S. M. (1966) *Endocrinology* **78**, 158–164.
EVERETT, J. W. (1954) *Endocrinology* **54**, 685–690.
FLERKO, B. and BARDOS, V. (1959) *Acta Neuroveg.* **20**, 248–262.
FUXE, K. and HÖKFELT, T. (1969) In *Frontiers in Neuroendocrinology*, 1969, pp. 47–96, ed. Ganong, W. F. and Martini, L. New York: Oxford University Press.
HAUN, C. K. and SAWYER, C. H. (1960) *Endocrinology* **67**, 270–272.
INNES, I. R. and NICKERSON, M. (1970) In *The Pharmacological Basis of Therapeutics*, pp. 478–523, ed. Goodman, L. S. and Gilman, A. New York: Macmillan.
KAMBERI, I. A., MICAL, R. S. and PORTER, J. C. (1970) *Fed. Proc. Fed. Am. Soc. Exp. Biol.* **29 (2)**, 378(abst. 751).
KAMBERI, I. A., SCHNEIDER, H. P. G. and MCCANN, S. M. (1970) *Endocrinology* **86**, 278–284.
KOCH, Y., LU, K. H. and MEITES, J. (1970) *Endocrinology* **87**, 673–675.
KOELLE, G. B. (1970) In *The Pharmacological Basis of Therapeutics*, pp. 422–432, ed. Goodman, L. S. and Gilman, A. New York: Macmillan.
LU, K. H., AMENOMORI, Y., CHEN, C. L. and MEITES, J. (1970) *Endocrinology* **87**, 667–672.
LU, K. H., KOCH, Y. and MEITES, J. (1971) *Endocrinology* **89**, 229–233.
LU, K. H. and MEITES, J. (1971) *Proc. Soc. Exp. Biol. & Med.* **137**, 480–483.
MACLEOD, R. M. (1969) *Endocrinology* **85**, 916–923.
MEITES, J. (1966) In *Neuroendocrinology*, vol. I, pp. 669–707, ed. Martini, L. and Ganong, W. F. New York: Academic Press.
MEITES, J. (1970) In *Hypophysiotropic Hormones of the Hypothalamus: Assay and Chemistry*, pp. 261–281, ed. Meites, J. Baltimore: Williams and Wilkins.
MEITES, J., KAHN, R. H. and NICOLL, C. S. (1961) *Proc. Soc. Exp. Biol. & Med.* **108**, 440–443.
MEITES, J. and NICOLL, C. S. (1966) *A. Rev. Physiol.* **28**, 57–88.

MEITES, J., NICOLL, C. S. and TALWALKER, P. K. (1963) In *Advances in Neuroendocrinology*, pp. 238–279, ed. Nalbandov, A. V. Urbana: University of Illinois Press.

MEITES, J., TALWALKER, P. K. and NICOLL, C. S. (1960) *Proc. Soc. Exp. Biol. & Med.* **103**, 298–300.

MISHKINSKY, J., KHAZEN, K. and SULMAN, F. G. (1968) *Endocrinology* **82**, 611–613.

MITTLER, J. C. and MEITES, J. (1967) *Proc. Soc. Exp. Biol. & Med.* **124**, 310–311.

MOTTA, M., FRASCHINI, F. and MARTINI, L. (1969) In *Frontiers in Neuroendocrinology, 1969*, pp. 211–253, ed. Ganong, W. F. and Martini, L. New York: Oxford University Press.

NICOLL, C. S., FIORINDO, R. P., MCKENNEE, C. T. and PARSONS, J. A. (1970) In *Hypophysiotropic Hormones of the Hypothalamus: Assay and Chemistry*, pp. 115–144, ed. Meites, J. Baltimore: Williams and Wilkins.

NICOLL, C. S. and MEITES, J. (1962a) *Nature (Lond.)* **195**, 606–607.

NICOLL, C. S. and MEITES, J. (1962b) *Endocrinology* **70**, 272–277.

NICOLL, C. S. and MEITES, J. (1963) *Endocrinology* **72**, 544–551.

PASTEELS, J. L. (1961) *C. R. Séance Soc. Biol.* **253**, 2140–2142.

PASTEELS, J. L. (1968) In *La Physiologie de la reproduction chez les mammifères*, pp. 530–544, ed. Jost, A. Paris: Centre National de la Recherche Scientifique.

RATNER, A., TALWALKER, P. K. and MEITES, J. (1965) *Endocrinology* **75**, 377–382.

SAWYER, C. H. (1969) In *The Hypothalamus*, pp. 289–430, ed. Haymaker, W., Anderson, E. and Nauta, W. J. H. Springfield, Ill.: Thomas.

SCHALLY, A. V., KUROSHIMA, A., ISHIDA, Y., REDDING, T. W. and BOWERS, C. Y. (1965) *Proc. Soc. Exp. Biol. & Med.* **118**, 350–352.

SCHALLY, A. V., MEITES, J., BOWERS, C. Y. and RATNER, A. (1964) *Proc. Soc. Exp. Biol. & Med.* **117**, 252–254.

SCHNEIDER, H. P. G. and MIDGLEY, A. R. (1970) *Abstracts, Third Annual Meeting of the Society for the Study of Reproduction*, p. 12. New York: Academic Press.

SGOURIS, J. T. and MEITES, J. (1953) *Am. J. Physiol.* **175**, 319–321.

SUD, S. C., CLEMENS, J. A. and MEITES, J. (1970) *Indian J. Exp. Biol.* **8**, 81–83.

TALWALKER, P. K., RATNER, A. and MEITES, J. (1963) *Am. J. Physiol.* **205**, 213–218.

TASHJIAN, A. H., BAROWSKY, N. J. and JENSEN, D. K. (1971) *Biochem. & Biophys. Res. Commun.* **143**, 516–523.

VOGT, M. (1954) *J. Physiol. (Lond.)* **123**, 451–481.

VOOGT, J. L. and MEITES, J. (1971) *Endocrinology* **88**, 286–292.

VOOGT, J. L., CLEMENS, J. A. and MEITES, J. (1969) *Neuroendocrinology* **4**, 157–163.

WELSCH, C. W., SAR, M., CLEMENS, J. A. and MEITES, J. (1968) *Proc. Soc. Exp. Biol. & Med.* **129**, 817–820.

WELSCH, C. W., SQUIERS, M. D., CASSELL, E., CHEN, C. L. and MEITES, J. (1971) *Am. J. Physiol.* In press.

WURTMAN, R. J. (1970) In *Hypophysiotropic Hormones of the Hypothalamus: Assay and Chemistry*, pp. 184–194, ed. Meites, J. Baltimore: Williams and Wilkins.

WUTTKE, W. and MEITES, J. (1970) *Proc. Soc. Exp. Biol. & Med.* **135**, 648–652.

WUTTKE, W., CASSELL, E. and MEITES, J. (1971) *Endocrinology* **88**, 737–741.

WUTTKE, W., GELATO, M. and MEITES, J. (1971) *Endocrinology* in press.

DISCUSSION

Sherwood: What is the state of purity of prolactin-inhibiting factor and what other information is available on its chemistry?

Meites: We know very little about its chemistry. It appears to be a small molecule, like the other hypothalamic factors; it goes through a semipermeable membrane. It is not destroyed by boiling. Beyond that, I can't tell you very much.

Greenwood: On the question of auto-feedback, when you transplanted one to four pituitaries under the kidney capsule of a hypophysectomized rat, there seemed to be a dose–response relationship, with *increasing* prolactin production, whereas one would expect an auto-feedback effect, if prolactin targets the hypothalamus, to cause an increase in PIF in the circulation?

Meites: We have speculated that the decline in serum prolactin levels that occurs in the first two weeks after pituitary transplantation may be due precisely to this mechanism, to the prolactin from the transplanted pituitary acting back on the hypothalamus to increase release of hypothalamic PIF and to decrease secretion of prolactin. When several pituitaries are transplanted, autonomous prolactin production by the transplants appears to take precedence over any hypothalamic inhibition.

Greenwood: Would you also extend that interpretation to the oestrogen treatment of hypophysectomized rats with transplanted pituitaries, where you found an increase in prolactin? Is the oestrogen supposed to be acting on the isolated transplanted pituitary, or is it acting on the hypothalamus?

Meites: When Dr Nicoll was with us (Nicoll and Meites 1962) he showed that when rat anterior pituitary cultures were incubated with very small doses of oestradiol there was a significant increase in prolactin release, and we have recently confirmed this in short-term (12-hour) incubations (Lu, Koch and Meites 1971). So I don't think there is any doubt that oestrogen can act directly on the pituitary. However, oestrogen has also been shown to decrease hypothalamic PIF activity (Ratner and Meites 1964), and we believe therefore that oestrogen increases prolactin by a dual mechanism, by direct stimulation of the pituitary and by decreasing hypothalamic PIF activity.

Greenwood: Does oestrogen act on PIF via hypothalamic catecholamines?

Meites: The reported effects of oestrogens on catecholamines are contradictory. Some people have found that they increased catecholamine activity, but others have noted a decrease (see Martini and Meites 1971). The methods for measuring catecholamines are still far from perfect.

Pasteels: We also have evidence for a dual action of oestradiol on the control of prolactin secretion. Labelled oestradiol is taken up simultaneously by the hypothalamus, by two different areas, as shown by Stumpf (1968), and by anterior pituitary cells. We also showed that the implantation of oestradiol into the hypothalamus can stimulate prolactin secretion, even when the implants are located in the anterior part of the hypothalamus, quite remote from the portal vessels (Pasteels and Ectors 1971). But I would like to ask Dr Meites a question. He knows that I have good reason for wanting to believe in the existence of a specific PIF! However, I am puzzled by Dr MacLeod's recent findings, and I would like your opinion on your earlier experiments (Ratner, Talwalker and Meites 1965) on the influence of reserpine on the PIF of hypothalamic extracts, where you described it as an effect on PIF content. You showed an influence on PIF *activity*, certainly, but can you speak of PIF *content*? Was this a crude hypothalamic extract, and what is known about its possible catecholamine content? In fact, could the effect on PIF be a result of catecholamine?

Meites: We are only just beginning to try to measure catecholamines in the hypothalamus. This is not easy to do, because they are present in nanogramme quantities per gramme of tissue. So we don't know how much catecholamine is present in the hypothalamic extract. Only one laboratory, so far as I am aware, has tried to measure catecholamines in the portal circulation (Porter, Goldman and Wilber 1970) and they were unable to detect catecholamines. Porter also injected catecholamines directly into a single portal vessel and observed no effect on prolactin release, which led him to conclude that the catecholamines had no direct effect on the pituitary. On the other hand, there is no question that when catecholamines are incubated *in vitro* with pituitary tissue in sufficiently large doses, they inhibit prolactin release, as Dr MacLeod and also Birge and co-workers (1970) have shown and as we have confirmed. But we also reported that much smaller doses of catecholamines incubated *in vitro* can stimulate prolactin *release*. For these reasons I believe that these are pharmacological rather than physiological effects. The evidence is much better for an action of catecholamines at the level of the hypothalamus, where they increase production and release of PIF. Of course we can't rule out completely a direct action of catecholamines on the pituitary.

Pasteels: For this reason I don't entirely agree that you are measuring the influence of hypothalamic extracts on PIF content; it must be considered as an effect on prolactin-inhibitory *activity* so long as the substance PIF has not been extracted and purified.

Meites: We remove the hypothalamus from rats, extract it with hydrochloric acid, neutralize it and measure its effects *in vitro* on release of prolactin. We are measuring activity in the hypothalamic extract.

MacLeod: This is then not a measure of the content of PIF. 'Content' implies an amount of a definite substance.

Meites: You would like the biochemical evidence that there is a specific

PIF molecule. There was the same sort of debate many years ago about the anterior pituitary hormones, which were originally called factors or substances, and not until they were well differentiated chemically were some people prepared to agree that they were separate hormones. The same history is taking place with the hypothalamic factors, which I believe can properly be considered as hormones. But the ultimate proof will unquestionably be the demonstration that these exist as separate structural entities, like TRF (TRH).

Frantz: Dr Meites, your experiments on the transplanted pituitary seem to show that this is not a very good way of studying prolactin control, because the pituitary continues to be responsive to hypothalamic influences even at a distant site, and the feedback mechanism evidently continues to operate. But this is very interesting because in this respect, prolactin appears to be different from the other pituitary hormones which more or less stop being secreted from the transplanted pituitary, evidently because the releasing factors don't get out into the peripheral circulation in high enough concentrations to be effective. Might this mean that PIF is secreted in relatively larger amounts than the releasing factors for the other hormones?

Meites: We see this PIF activity only in *hypophysectomized* rats (with or without pituitary transplants); the releasing factors are also found only in hypophysectomized rats. When we destroy the median eminence of hypophysectomized rats (Sud, Clemens and Meites 1970) by extensive lesions, we see increased release of prolactin by the transplanted pituitary, presumably because PIF is no longer being released into the systemic circulation.

Frantz: Have you evidence that in the hypophysectomized rat with a transplanted pituitary the feedback mechanism for gonadotropins continues to operate?

Meites: Gonadotropins are secreted by the transplanted pituitary in the hypophysectomized rat in extremely small amounts. This is probably due to the presence of LRF and FSH-RF in the systemic circulation, but there is no evidence that the gonadotropins feed back on to the hypothalamus. If we remove the rat pituitary from its normal site and place it in a culture medium, in a few days release of all the hormones ceases, as far as we can determine by bioassay, with the exception of prolactin. This is believed to be due to lack of hypothalamic influence on the cultured pituitary.

Frantz: But Dr Greenwood's studies (pp. 207–217) showed continued secretion of the other hormones (FSH, LH, TSH, ACTH-MSH and GH) by human foetal pituitaries.

Greenwood: Our results simply confirm Dr Pasteels' extensive observations (cf. Pasteels 1969).

Pasteels: It depends on the culture method. Petrovic (1963) showed that the electron microscopic appearance of a guinea pig pituitary cultured away from the hypothalamus was unchanged with regard to the various categories of pituitary cells; probably they were less active, but from this there appears to be some autonomous activity of the pituitary cells, subjected to further control by hypothalamic factors.

Friesen: In partial reply to Dr Frantz's comment about the detection of releasing factors in the systemic circulation, Dr Reichlin's group has reported that LRF is present around the period of ovulation in the systemic circulation in women (Malacara, Seyler and Reichlin 1971).

Denamur: We found differences in prolactin secretion between the sheep and the rat. For example, milking or suckling provoke, in the sheep as in the rat, significant increases in blood prolactin from a basal level of 20 ng/ml to several hundred ng/ml already three to five minutes after suckling has begun. But unlike the rat, the pituitary prolactin content does not change and even increases in the sheep. So in the ewe a few minutes after suckling has started there is an increase in the release and synthesis of prolactin (Kann, Sotliaroff and Denamur 1971).

Cowie: Dr Meites, have you studied pituitary transplants under the kidney capsule in other species and also catecholamines in other species?

Meites: Our studies have been confined mainly to the rat, although we did some studies in rabbits with tranquillizers.

MacLeod: We have done some studies in the mouse and obtained very similar results to those reported by Dr Meites.

Cowie: I ask in connexion with pituitary transplants because we have tried implanting pieces of goat pituitary under the kidney capsule of the goat, and the tissue is simply absorbed, even pieces of the animal's own pituitary. Also, my colleagues Shani, Knaggs and Tindal (1971) have been doing some preliminary studies on the effects of catecholamines on lactation in rabbits and their results do not quite fit in with your story in the rat in that lactation was not greatly depressed, so I wonder if species differences are involved.

Meites: On the other hand, I was very interested to learn from Dr Friesen (p. 107) that L-dopa decreases serum prolactin levels in human subjects, just as we've shown in the rat. Apparently L-dopa increases brain catecholamine levels and increases PIF activity in both species.

Li: It is very difficult to do a complete hypophysectomy in the rat. In our experience, a small amount of gonadotropic hormone activity continues after 60–100 days, because of small remnants of pituitary tissue.

That is why you still find a slight amount of gonadotropic activity. Dr I. I. Geschwind (unpublished work, 1958) in our laboratory transplanted an animal's pituitary under the kidney capsule, and he found no evidence of any gonadotropic hormones. However, growth hormone activity continued; the rate of growth matched very well with the normal animal. Presumably growth hormone releasing factor was not operating in this transplant, so apparently growth hormone can be released without the hypothalamus. This was also shown by Hertz (1959).

Meites: We have also reported on this (Meites and Kragt 1964).

Li: Do you agree that growth hormone release in a rat has dual control, by itself and by the hypothalamus?

Meites: Yes, to some extent. The hormone that is secreted in largest amounts next to prolactin by the isolated pituitary appears to be growth hormone, and we reported that a single pituitary transplant in a young hypophysectomized rat elicits about 45 per cent of normal growth (Meites and Kragt 1964). So apparently growth hormone continues to be produced by the transplanted pituitary. On the other hand, when we incubated rat pituitary gland *in vitro* we could not detect growth hormone by the rat tibia test after the first three or four days of culture, suggesting that *in vivo* there is still some growth hormone releasing factor acting on the transplanted pituitary whereas it isn't present in the culture system. After eight days of culture, when no growth hormone detectable by the tibia test is being released, if we add a crude hypothalamic extract, large amounts of growth hormone are again released and apparently are being synthesized (Deuben and Meites 1965).

Li: Releasing factors are evidently still operating in the hypophysectomized animal and in transplanted pituitary tissue. Why then are the releasing factors for the gonadotropic hormones less effective?

Meites: I'm not sure I know. They have been detected in hypophysectomized rats, and there are indications that they may be more effective on the gonads in male than in female rats.

Nicoll: I should like to say something on behalf of a prolactin-releasing factor (PRF). In addition to the data we have published already, suggesting the existence of a PRF in rat hypothalami, evidence was reported at the Workshop on Prolactin (National Institutes of Health, January 1971) by several groups including R. Midgley's, and by W. Wuttke and his co-workers, for a PRF in rats. The most significant report was that of Valverde and colleagues (see Valverde and Chieffo 1971) from Dr S. Reichlin's laboratory. They have isolated a fraction from pig hypothalami which has prolactin-releasing activity *in vivo*.

Dr Meites mentioned the effects of several drugs which reduce brain catecholamines and cause pronounced increases in serum levels of prolactin and *decreases* in pituitary content. However, one drug, α-methyl-*m*-tyrosine, caused an increase in serum levels and an *increase* in pituitary levels. I wonder how that can be explained on the basis of one factor?

Meites: This drug may be more effective in removing PIF activity. If the pituitary can synthesize and release prolactin autonomously I think that this in itself may be sufficient explanation.

Nicoll: I don't! If you are eliminating PIF effects and getting release, all the drugs should have worked about the same way, if the result was entirely due to a change in PIF.

MacLeod: There is an explanation for that. When you release the pituitary from the inhibitory influence of the hypothalamus, you have a simultaneous release of the preformed prolactin and increase in the rate of previously suppressed prolactin synthesis. Although I was surprised to see an increase in the prolactin content after only four hours, it could be due to an increase in synthesis of prolactin.

Nicoll: I don't agree. If you remove hypothalamic influence from the rat adenohypophysis by making hypothalamic lesions, transplanting the pituitary or placing the gland *in vitro*, the gland content falls and remains low. In all these circumstances, prolactin turnover is high. The gland appears to require the hypothalamic influence to maintain a high content of prolactin. Accordingly, removal of PIF control from the gland should not cause a simultaneous increase in prolactin concentration in the tissue and in the serum.

We have data on the intrapituitary turnover of rat prolactin *in vivo* which are difficult to explain on the basis of one regulating factor, either stimulatory or inhibitory. These studies were done in my laboratory by Dr K. C. Swearingen. A report on some of them has already appeared (Nicoll and Swearingen 1970) and the remainder of the work is in press (Swearingen and Nicoll 1971).

For these studies, rats were given an intravenous injection of ^{3}H-labelled leucine. Their adenohypophyses were removed at different times after the leucine injection and the prolactin and growth hormone were separated by disc electrophoresis. The amounts of both hormones in the gel columns were estimated by densitometry and the radioactivity in both bands was measured by liquid scintillation techniques.

The decline in specific activity of the labelled prolactin, in general, followed a double exponential function. At this juncture, we believe that the slow component of this decline is more closely related to the secretory activity of the gland than is the initial fast component. The data on the prolactin concentration in the glands of several experimental groups, and on the rate constants of the decline in specific activity of prolactin in these groups, are shown in Table I.

TABLE I (Nicoll)
PROLACTIN CONCENTRATIONS AND RATE CONSTANTS OF ITS TURNOVER IN RAT ADENOHYPOPHYSES *IN VIVO*

Experimental group	Prolactin concentration µg/mg wet weight	Rate constant k
Untreated males	2·0	0·01
Oestrogen-treated females	5·0	0·05
Reserpine-treated females	8·2	0·03
Oestrogen and reserpine-treated females	6·2	0·1
Renal capsule transplants of female adenohypophyses in intact females	2·4	0·1

The results show that the male glands have a small pool of prolactin which is turning over very slowly. The transplanted glands also have a small pool but it is turning over about ten times faster than the prolactin of the male rat. The adenohypophyses of the oestrogen- or reserpine-treated animals have large pools of prolactin which are turning over faster than those of the male glands but not as fast as in the transplants. If either treatment had simply removed PIF control from the adenohypophyses, the turnover rate should have been the same as in the transplanted glands and the size of the pools should have been small. A high turnover rate in the glands *in situ* equivalent to that observed in the transplants was obtained only with animals treated with oestrogen plus reserpine. This was accompanied by a relatively large pool of prolactin.

These results are more amenable to interpretation by a prolactin-stimulating factor *and* a prolactin-inhibiting factor than by a single regulatory factor. If the prolactin-stimulating factor regulates the pool size of prolactin and the inhibitory factor controls its turnover rate (by controlling release), the results can be readily explained. The oestrogen presumably increased the pool size of prolactin by a direct effect on the adenohypophysis but it failed to raise the turnover rate to a level equivalent to that of a gland removed from control by the inhibitory factor (i.e. the transplanted adenohypophyses). Treatment with reserpine increased the content of prolactin in the gland while increasing the turnover rate only slightly. This may reflect an increase in secretion of a prolactin-stimulating factor with little change in the effect of the inhibitory factor on the adenohypophysis.

Meites: These are very interesting data, but the final proof for the existence of a PRF has to come from studies on the hypothalamus itself. It would be necessary to show that after you give reserpine, PRF activity is produced in hypothalamic extracts. We have given reserpine, chlorpromazine and several other tranquillizing drugs to rats. We can eliminate PIF activity in hypothalamic extracts but so far haven't been able to demonstrate PRF activity in the extracts. If you postulate the existence of a PRF in the hypothalamus, you have to show that it is present in the hypothalamus.

Reichlin and his collaborators reported at the Prolactin Workshop (National Institutes of Health, January 1971) that porcine hypothalamic extract increased prolactin release *in vivo* in rats, as you said. But others, including Dr Schally and his co-workers (1967), have found that porcine hypothalamic extracts show PIF activity *in vitro*. We tested porcine hypothalamic extract in an *in vitro* study and found only PIF activity (unpublished results).

What is the evidence that the hypothalamus is necessary for stimulation of prolactin release? In the pig has anyone sectioned the pituitary stalk or transplanted the pituitary or made median eminence lesions, to show that the pituitary requires hypothalamic stimulation, unlike every other mammalian species that has been studied?

Greenwood: I think it may be a poor refuge to think in terms of species specificity for the releasing hormones. What about toxicity?

Nicoll: The earlier work showing only inhibition was done with crude hypothalamic extracts, but Valverde and his associates are using fractions purified by Sephadex chromatography.

Greenwood: Dr Meites, in your median eminence lesions when you get a rapid five or ten minute rise in prolactin, couldn't this be a stress stimulus *per se* to the hypothalamus?

Meites: Perhaps, but how then do you account for the observation that serum prolactin remains elevated for six months thereafter?

Greenwood: I agree that stress cannot explain the long-term effects.

Nicoll: Dr W. Wuttke also reported evidence for PRF activity in rat hypothalami at the Prolactin Workshop.

Meites: Dr Wuttke is interested in the question of PRF but I don't think he would claim that its existence in mammals is proved. We *are* looking for this, of course!

Denamur: Do catecholamines have the same inhibitory action on prolactin secretion in birds?

Meites: We haven't tried catecholamines in birds. We found that reserpine has no effect on the release of prolactin in pigeons (Kragt and Meites 1965).

Greenwood: I am confused about the catecholamine story. Presumably *blood* catecholamines are not playing a role because they don't cross the blood–brain barrier. What is the evidence that adrenaline is present and synthesized in the hypothalamus? According to Dr Nicoll and Dr MacLeod there is no evidence that it is present or synthesized in the hypothalamus, so it can only be noradrenaline; is that right?

Meites: There is *some* adrenaline in the hypothalamus, but more noradrenaline is found. The hypothalamus contains very little dopamine, but dopaminergic nerve endings are present in the median eminence.

Greenwood: I was under the impression that the brain does not synthesize adrenaline.

Sherwood: L-Dopa is converted to dopamine and eventually to catecholamines.

MacLeod: We find about 2·5 times as much noradrenaline as dopamine in the hypothalamus, but no demonstrable adrenaline.

REFERENCES

BIRGE, C. A., JACOBS, L. S., HAMMER, C. T. and DAUGHADAY, W. H. (1970) *Endocrinology* **86,** 120–130.
DEUBEN, R. and MEITES, J. (1965) *Proc. Soc. Exp. Biol. & Med.* **118,** 409–412.
HERTZ, R. (1959) *Endocrinology* **65,** 926–931.
KANN, G., SOTLIAROFF, M. and DENAMUR, R. (1971) *Neuroendocrinology* in press.
KRAGT, C. L. and MEITES, J. (1965) *Endocrinology* **76,** 1169–1176.
LU, K. H., KOCH, Y., and MEITES, J. (1971) *Endocrinology* **89,** 229–233.
MALACARA, J. M., SEYLER, L. E. and REICHLIN, S. (1971) *Clin. Res.* **19,** 376.
MARTINI, L. and MEITES, J. (ed.) (1971) *Neurochemical Aspects of Hypothalamic Function.* New York: Academic Press.
MEITES, J. and KRAGT, C. L. (1964) *Endocrinology* **75,** 565–570.
NICOLL, C. S. and MEITES, J. (1962) *Endocrinology* **70,** 272–277.
NICOLL, C. S. and SWEARINGEN, K. C. (1970) In *The Hypothalamus,* pp. 449–462, ed. Martini, L., Motta, M. and Fraschini, F. New York: Academic Press.
PASTEELS, J. L. (1969) *Mém. Acad. R. Méd. Belg.* **7,** 1–45.
PASTEELS, J. L. and ECTORS, F. (1971) In *Basic Actions of Sex Steroids on Target Organs,* pp. 200–207, ed. Hubinont, P. O., Leroy, F. and Galand, P. Basel and New York: Karger.
PETROVIC, A. (1963) In *Coll. Int. 128 La cytologie de l'adénohypophyse,* pp. 121–136. Paris: CNRS.
PORTER, J. C., GOLDMAN, B. D. and WILBER, J. F. (1970) In *Hypophysiotropic Hormones of the Hypothalamus: Assay and Biochemistry,* pp. 282–297, ed. Meites, J. Baltimore: Williams and Wilkins.
RATNER, A. and MEITES, J. (1964) *Endocrinology* **75,** 377–382.
RATNER, A., TALWALKER, P. K. and MEITES, J. (1965) *Endocrinology* **77,** 315–319.
SCHALLY, A. V., KASTIN, A. J., LOCKE, W. and BOWERS, C. Y. (1967) In *Hormones in Blood,* pp. 491–525, ed. Gray, C. H. and Bacharach, A. L. London: Academic Press.
SHANI (MISHKINSKY), J., KNAGGS, G. S. and TINDAL, J. S. (1971) *J. Endocrinol.* **50,** 543.
STUMPF, W. E. (1968) *Science* **162,** 1001–1003.
SUD, S. C. CLEMENS, J. A. and MEITES, J. (1970) *Indian J. Exp. Biol.* **8,** 81–83.
SWEARINGEN, K. C. and NICOLL, C. S. (1971) *J. Endocrinol.* submitted.
VALVERDE, C. and CHIEFFO, V. (1971) *Program of the Endocrine Society, 53rd Meeting,* p. 84 (abst. 83).

PLASMA PROLACTIN ACTIVITY IN PATIENTS WITH GALACTORRHOEA AFTER TREATMENT WITH PSYCHOTROPIC DRUGS

M. APOSTOLAKIS, S. KAPETANAKIS, G. LAZOS AND A. MADENA-PYRGAKI

Department of Physiology and Department of Neurology and Psychiatry, University of Thessaloniki, Greece

IT is well documented that galactorrhoea can appear as a side-effect in patients treated with psychotropic drugs (Hooper, Welch and Shackelford 1961; Kirchgraber 1963; Taubert, Haskins and Moszkowski 1966; Rees 1967; Lazos 1971). However, no systematic study has yet been made covering a large number of drugs on the one hand and investigating the nature of the secretion produced on the other. This paper describes preliminary data of this type obtained in the course of longitudinal studies on the effects of psychotropic drug administration generally on the endocrine status of male and female subjects.

Two hundred female patients were treated by the neuroleptic agents (major tranquillizers) chlorpromazine (Largactil), levopromazine (Nozinan), promazine (Sparine), thioridazine (Melleril), trifluoperazine (Stelazine), perphenazine (Trilafon), fluphenazine (Sevinol), thiopropazate (Dartalan), chlorprothixene (Truxal) and haloperidol (Aloperidin), and by the thymoleptic agents amitriptyline (Tryptizol) and imipramine (Tofranil), or by a combination of two of these. The number of patients treated with each drug and the daily dosage administered are shown in Table I.

It can be seen that out of a total of 200 cases under treatment galactorrhoea appeared in 101, or just over 50 per cent. No cases of galactorrhoea appeared after the administration of promazine or amitriptyline alone. Imipramine was not given alone. The combination of trifluoperazine and thioridazine appeared to be most active in this connexion, galactorrhoea being obtained in twenty out of the twenty-four cases treated, or over 80 per cent. It should be stressed that copious galactorrhoea did not occur in all cases, but that mammary secretion appeared when the gland was 'milked' by the investigating doctor. The amount secreted ranged generally from 0·5 to 6·0 ml. Most of the patients were on long-term treatment

TABLE I

THE EFFECTS OF ADMINISTERING PSYCHOTROPIC DRUGS TO FEMALE SUBJECTS

Drug administered	Daily dose mg	Number of cases treated	Cases presenting galactorrhoea
Chorpromazine	100–250	22	8
Levopromazine	75–100	13	3
Promazine	100–150	4	0
Thioridazine	100–250	28	12
Trifluoperazine	8–18	12	8
Perphenazine	16–24	6	4
Fluphenazine	2–4	8	4
Thiopropazate	30–50	8	4
Chlorprothixene	200–400	16	6
Haloperidol	6–14	17	5
Amitriptyline	75–100	3	0
Trifluoperazine+thioridazine	(3–6)+(75–100)	24	20
Thioridazine+haloperidol	(75–125)+(3–6)	15	11
Haloperidol+chlorprothixene	(3–6)+(100–200)	10	7
Thioridazine+imipramine	(75–125)+(75–100)	14	9
Totals		200	101

and in some of them the resulting galactorrhoea had been going on for over 18 months.

Generally speaking, larger daily dosages of the drugs were more apt to produce galactorrhoea than small ones; nevertheless in a few cases, especially when chlorprothixene or thioridazine was administered, even a minimal dose caused substantial secretion.

Fig. 1 shows the distribution of the patients with and without galactorrhoea according to age. The greatest incidence occurs in the 35 to 40 age group. It should be stressed that Fig. 1 is a cumulative representation which includes all types of treatment and that detailed investigation of the data obtained with each drug separately may show a somewhat different age-distribution pattern. Nevertheless, there appears to be little doubt that galactorrhoea in women during treatment with psychotropic drugs is a phenomenon occurring mainly in the reproductive period, that is, between 20 and 45 years. A further interesting finding was that (Fig. 2) when all the subjects of reproductive age (137 individuals) were taken together, galactorrhoea appeared about twice as often (76 per cent) in women who had never lactated before as in those who had lactated previously, after a pregnancy (40 per cent). A number of other correlations investigated, such as the influence of the presence of ovulatory cycle disturbances on the appearance of galactorrhoea, will be discussed in a further publication.

In the same study 60 male patients were also treated with these drugs. Galactorrhoea appeared in six cases, or 10 per cent of the total. The treatment given to these six subjects is shown in Table II. As far as we are

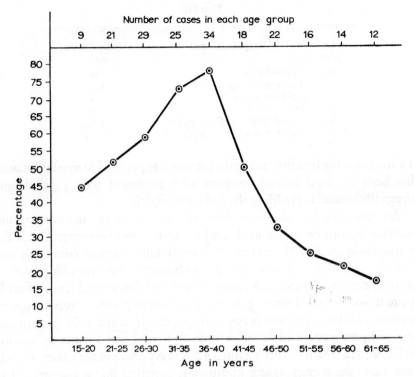

FIG. 1. Percentage distribution according to age group of cases of galactorrhoea among 200 women treated with psychotropic drugs.

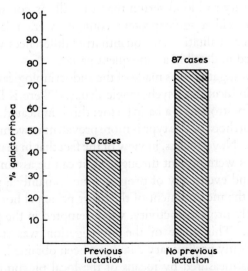

FIG. 2. Correlation of galactorrhoea following psychotropic drugs and previous lactation in women of reproductive age (137 cases).

Table II

TREATMENT OF SIX MALE SUBJECTS IN WHOM GALACTORRHOEA DEVELOPED

Age years	Drug administered	Daily dose mg
20	Thioridazine	150
22	Chlorpromazine	100
40	Trifluoperazine	18
34	Haloperidol	10
23	Thioridazine + trifluoperazine	75 + 12
24	Chlorpromazine + haloperidol	75 + 4

aware, this is the first time that galactorrhoea, as opposed to gynaecomastia, has been reported in male patients after treatment with psychotropic drugs (Robinson 1957; Margolis and Gross 1967).

An extensive immunoelectrophoretic analysis of the mammary fluid secreted by our cases was made and the results were compared with the immunoelectrophoretic pattern of normal milk, normal colostrum and blood serum. Fig. 3 shows, from top to bottom, a normal milk pattern, a normal colostrum pattern, a pattern of galactorrhoeic fluid from a female patient and a normal serum pattern. The galactorrhoeic secretion appears to take a midway position between the relatively simple milk pattern and the complex pattern of blood serum. A typical and apparently specific milk protein which has been named α_{2x} is also present in the galactorrhoeic secretion; the corresponding immunoprecipitation line is also present in the secretion obtained from male patients (Fig. 4), whereas it appears to be lacking in normal serum. On the other hand the galactorrhoeic fluid showed a greater similarity to blood serum than to milk or colostrum in that its transferrin and acid α_1-glycoprotein contents were relatively high. A number of other qualitative and quantitative differences were found and will be described in detail in a subsequent paper.

An initial investigation was made of the endocrinological background of galactorrhoea induced by psychotropic drugs. There is little doubt that the presence of oestrogens is a basic factor; this is indicated by the observation that galactorrhoea of this type is more prevalent in women of reproductive age (Fig. 1). Nevertheless, in view of the fact that in many cases normal ovulatory cycles were present throughout, it can be assumed that the role of oestrogens and eventually of progesterone is mainly a permissive one. For this reason the measurement of the other principal hormonal factor in lactation, namely prolactin activity, was attempted in the plasma of some of the patients. This phase of the investigation was started relatively recently and so only preliminary data have been obtained.

Prolactin was measured by means of the local pigeon crop assay procedure including the so-called 'corticosteroid shielding' (Apostolakis

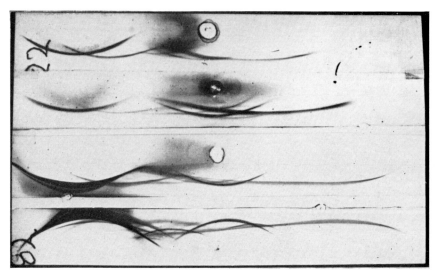

FIG. 3. Immunoelectropherograms of (from top to bottom) normal milk, normal colostrum, galactorrhoeic fluid and normal serum, in all cases against an anti-human serum rabbit antiserum.

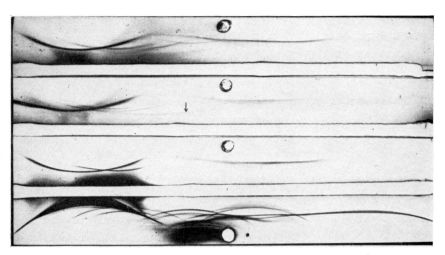

FIG. 4. Immunoelectropherograms of (top three) samples of galactorrhoea fluid from male subjects and (bottom) normal human serum, in all cases against an anti-human serum rabbit antiserum. The arrow indicates the α_{2x} globulin immunoprecipitation line.

(*facing page 352*)

1968) which to a large extent prevents unspecific reactions. In some subjects blood plasma was extracted by the procedure described by Canfield and Bates (1965); in these cases the results obtained were multiplied by a factor of 2·5 because in our experience this method extracts only about 40 per cent of the prolactin activity present in plasma. The data obtained are shown in Table III. A total of 17 determinations were made in 13

TABLE III
PLASMA PROLACTIN ACTIVITY IN PATIENTS TREATED WITH PSYCHOTROPIC DRUGS

Name and sex	Age years	Treatment mg/day	Prolactin activity i.u./100 ml
Ba.(f)	30	Thioridazine (150)	3·0
			3·4
		Thioridazine (150) + chlordiazepoxide (100)	21·8
Ke.(f)	45	Levopromazine (75)	4·9
			2·4
Za.(m)	21	Haloperidol (15)	5·8
		Haloperidol (15) + chlordiazepoxide (100)	25·0
Fr.(f)	36	Levopromazine (50)	1·9
De.(f)	48	Haloperidol (10)	6·6
Karag.(f)	58	Levopromazine (75) + haloperidol (15)	13·6
Sy.(f)	42	Thioproperazine (1)	<0·5
Mi.(f)	17	Chlorpromazine (200) + haloperidol (5)	31·2
Karak.(f)	25	Chlorpromazine (300)	15·6
As.(f)	36	Thioproperazine (20)	<0·5
Gi.(f)	41	Haloperidol (10)	37·5
Ta.(f)	23	Haloperidol (10)	2·6
Ch.(f)	39	Trifluoperazine (30)	2·7

* Treatment started at least one month previously.

patients. In all except two cases, measurable amounts of prolactin activity were found. It should be stressed that generally in non-pregnant and non-lactating women prolactin values above 0·5 i.u./100 ml are not found with the procedures employed by us. It is of especial interest to note that in cases Ba. and Za., whereas initially during treatment with thioridazine and haloperidol respectively a relatively low level of prolactin activity was present, later when chlordiazepoxide was added a substantial increase appeared, which was accompanied by an increase in the rate of mammary secretion. The significance of this observation is not clear at present. In the two cases with no measurable blood prolactin the dosage administered was also low; nevertheless galactorrhoea was present in both. No acromegalic changes were observed in any of our patients, male or female.

CONCLUSIONS AND SUMMARY

The main conclusions to be drawn from these preliminary data are as follows:

(1) The administration of psychotropic drugs, especially the major tranquillizers (such as phenothiazine derivatives), appears to cause galactorrhoea as a side-effect in an unexpectedly high percentage of patients.

(2) This effect occurs most often in female patients of reproductive age who have never previously lactated.

(3) The immunoelectrophoretic pattern of the fluid secreted in these cases has characteristics in common with those of both blood serum and milk or colostrum.

(4) Abnormally high blood prolactin activity was found in most of the cases studied. A causal relationship between these high levels and the appearance of the galactorrhoea seems probable.

REFERENCES

APOSTOLAKIS, M. (1968) *Vitam. & Horm.* **26,** 197.
CANFIELD, C. J. and BATES, R. W. (1965) *New Engl. J. Med.* **273,** 897.
HOOPER, J. H., JR, WELCH, V. C. and SHACKELFORD, R. T. (1961) *J. Am. Med. Ass.* **178,** 506.
KIRCHGRABER, D. (1963) *Schweiz. Arch. Neurol. & Psychiat.* **91,** 412.
LAZOS, G. (1971) M.D. Thesis, University of Thessaloniki, Greece.
MARGOLIS, I. B. and GROSS, C. G. (1967) *J. Am. Med. Ass.* **199,** 942.
REES, W. D. (1967) *Practitioner* **198,** 835.
ROBINSON, B. (1957) *Med. J. Austr.* **2,** 239.
TAUBERT, H. D., HASKINS, A. L. and MOSZKOWSKI, E. F. (1966) *Southern Med. J.* **59,** 1301.

DISCUSSION*

Spellacy: We have been interested in the syndrome of lactation in women using oral contraceptives, and in order to put this into proper perspective we looked at the incidence of lactation in women using mechanical contraception too. Breast secretion was present in 32·7 per cent of 52 women who were multiparous and more than one year *post partum* and using an intrauterine device (Shevach and Spellacy 1971). This is clearly a common condition. Dr Apostolakis described lactation in 50 per cent, but one may have to discount 30–35 per cent as the normal occurrence and thus consider only the remaining 15–20 per cent as being drug induced. This then relates to the 10 per cent incidence he found in his male subjects taking similar drugs. The other point here is that when progestins are administered alone there is no change in the incidence of lactation although when an oestrogen plus a progestin is administered, as in the combination-type oral contraceptives, there appears to be a decreased incidence (19 per cent). There is also a decrease in the amount of secretion.

* Dr Apostolakis was unable to be present at the meeting. See p. 357 for his comments on this discussion.

Now that we have the assays to measure prolactin, we must redefine the condition of galactorrhoea or it will be difficult to interpret the data.

Beck: One very crucial point about Dr Apostolakis' paper is that the incidence of galactorrhoea in the group of women before initiation of therapy is unknown.

Sherwood: Dr Spellacy, would there be any value, in women who have difficulty lactating after pregnancy, in giving tranquillizers for a couple of days? Admittedly the infants won't be able to nurse during this time since the drugs are secreted in the milk.

Spellacy: The answer to that would presuppose that women who have difficulty in nursing are deficient in prolactin, and I don't know that that is established. I would hesitate to go to therapy before I knew the pathophysiology.

Cowie: Dr J. Shani, Dr W. H. Broster and I have given these tranquillizers to lactating cows and have observed a marked depression in milk yields!

MacLeod: Is that because they stop eating?

Cowie: Partly.

Greenwood: I think it's also because cows are already hyperprolactinaemic animals, selectively bred for that trait.

Cowie: These tranquillizers also depress milk yields in goats and in rabbits (J. Shani and A. T. Cowie, unpublished).

Greenwood: Dr Frantz, were the values obtained by Dr Apostolakis in the pigeon crop assay for patients on tranquillizers comparable to those you obtain?

Frantz: Dr Apostolakis' values are all quite a bit higher, about ten times, and this parallels other results reported with the pigeon crop sac assay after extraction, which tend to be somewhat higher than ours.

Cotes: Dr Apostolakis' figures were interesting in that, using a conversion factor of 30 i.u. by bioassay per mg of human prolactin, they were in good agreement with Dr Greenwood's immunoassay figures for subjects after chlorpromazine.

Greenwood: I would not like to do those calculations myself, Dr Cotes! Our figures may be gross overestimates. Our standard may be only 10–15 per cent prolactin and I don't know the specific bioactivity of human prolactin.

Turkington: We have measured serum prolactin in a series of a hundred patients treated with various psychotropic drugs, including many of those used by Dr Apostolakis, and in this group we did not find a single patient without elevated prolactin, in the range of 25–500 ng per ml. A low percentage of the female patients, about 20 per cent, were aware of galactorrhoea. Whereas galactorrhoea may or may not occur, in our experience prolactin is uniformly elevated in patients receiving therapeutic amounts of the tranquillizing drugs. The other thing that impressed us was the duration of the effect. We have seen a number of patients who had been chronically treated and then stopped, who have maintained elevated prolactin levels for as long as ten days or two weeks after the final dose.

Friesen: We have studied prolactin levels in blood in about 20 patients on long-term therapy with tranquillizers and we found roughly the same as you did, except that the men generally had normal basal levels but showed an increase with their morning dose, whereas most of the patients with chronically elevated levels of blood prolactin were women.

Meites: It has been demonstrated in laboratory animals (rats, mice and rabbits) that tranquillizers influence the secretion not only of prolactin but of other pituitary hormones as well (Meites 1970). Reserpine and chlorpromazine depress FSH and LH secretion; in fact, the evidence is quite good that reserpine depresses all the anterior pituitary hormones except prolactin. In respect to the long-acting effect of some of these drugs, a single injection of reserpine into an oestrogen-primed rabbit initiates lactation a week or ten days later and this lactation persists for many weeks (Meites 1957), as it persists in other animals and in man.

The other point is the interesting observation of Dr Apostolakis that lactation occurs more frequently in women of reproductive age than in postmenopausal women taking these psychotropic drugs. There is support for this in our work, in that when chlorpromazine or reserpine or other tranquillizers are given to intact rats, they produce a much more pronounced elevation of serum prolactin than when given to ovariectomized rats. We do get a rise in prolactin in the latter but it's much less than in the intact rat or in ovariectomized rats primed with oestrogen (Lu and Meites, unpublished).

MacLeod: Is it known whether patients undergoing this therapy have a higher incidence of mammary carcinoma?

Greenwood: I have been trying to find that information by asking breast cancer physicians if they have seen an exacerbation of breast carcinoma by phenothiazines. Since no one seems to have any data on it at all, I think that a statement should be made that there are tranquillizing drugs which apparently do not cause release of prolactin and these would be the drugs of choice, rather than prescribing those known to increase prolactin.

Beck: I have put this question to several cancer registries, and they do not have the information.

Sherwood: Is there any information on two commonly used drugs, digitalis and spironolactone, which are associated with gynaecomastia? Do they increase prolactin secretion?

Friesen: Digitalis had no effect on serum prolactin, nor did aldactone or Aldomet (methyldopa) in six patients we studied. Some of these patients had severe gynaecomastia yet had no measurable serum prolactin.

Frantz: That's not what we found with Aldomet, in one person who had galactorrhoea, apparently as a consequence of taking this drug, and who had an elevated prolactin concentration of 1·7 m i.u./ml.

Meites: What happens to the menstrual cycle of women treated with tranquillizers?

Beck: Dr Apostolakis reports no change in menstrual cyclicity in his group, but in my experience the cycle becomes very variable.

Greenwood: In how many? I thought that amenorrhoea was a notable side-effect of tranquillizers.

Beck: Amenorrhoea is a side-effect of psychosomatic disease, which is a difficult thing to evaluate when you add pharmacological agents to this.

Wilhelmi: Dr Apostolakis found normal ovulatory cycles throughout 'in many cases', but the control is lacking because he did not have past histories.

Short: While we are discussing inappropriate lactation it's worth putting on record the fact that very occasionally normal fertile male goats will lactate; I know of one male stud goat that lactated quite heavily.

Cowie: I have never seen this in male goats in our herd. They occasionally have large teats but if milked only a few drops of secretion are obtained.

Cotes: While we are considering the various methods for assaying prolactin, I wonder whether it would help to make available a pool of plasma from subjects treated with phenothiazine drugs for distribution as an interim reference preparation. Measurements made in different laboratories could then be put on a directly comparable basis.*

Comment added by Professor M. Apostolakis:

In reply to Dr Spellacy's remarks, it should be stressed that out of our total of 137 women in the reproductive period, 87 had had no previous pregnancies leading to lactation. Of these, 66 (75 per cent) presented psychotropic-drug galactorrhoea whereas of the remaining 50, who had had a previous lactation, psychotropic-drug galactorrhoea occurred in 20 (40 per cent). It would appear therefore that actually psychotropic drug-induced galactorrhoea is more likely to occur in patients who have had no previous lactation. On the other hand I must admit that we do not have complete data on the appearance of galactorrhoea in the pretreatment period in many of our patients as most of them were chronic mental cases

* See General Discussion, pp. 396–400.

who had been on and off psychotropic drugs for a long time. On the whole I would consider that Dr Spellacy's figure of 30 per cent galactorrhoea as a normal, spontaneous occurrence is far too high for our group of patients, which includes 63 postmenopausal subjects; I would suggest about 15 per cent as a fair figure and I would thus attribute 30–35 per cent of the galactorrhoea to the drugs. I further agree that we should try to measure the amount of galactorrhoea; we attempted to do so and in some cases succeeded, but in most cases the difficulties were insurmountable, the mental illness of the patients being an added complication.

As far as menstrual activity is concerned, out of the total of 137 women of reproductive age, normal menstruation was present in 80 (58 per cent) and more specifically in 46 out of the 87 patients presenting galactorrhoea and in 34 out of the 50 patients without galactorrhoea. Menstrual irregularities and galactorrhoea were therefore not always associated but the two conditions very often coexisted. I would stress further that our group of female patients was relatively inhomogeneous, insofar as a variety of drugs were administered in different dosages and for unequal time periods. I indicated that this report was only a preliminary one and that more detailed studies under more controlled conditions are being planned.

Finally, as to our absolute prolactin levels in plasma, I agree that they are somewhat higher than those reported by some other investigators. Nevertheless a direct comparison of such data is in my opinion not possible as long as a common assaying technique is not being used. What we are measuring is 'pigeon crop-stimulating activity' which perhaps, by definition, should not be considered equivalent to 'prolactin' as such; however I am not quite sure that anyone can claim to be measuring pure prolactin. I am sure we all agree that the various types of pigeon crop tests on the one hand and the prolactin radioimmunoassays on the other are not measuring exactly the same thing. In these circumstances and in view of the fact that similar discrepancies have been found when comparing gonadotropin bioassays and radioimmunoassays, I think the important point about the findings of the various groups at present is the parallelism of the effects seen—that irrespective of whether the pigeon crop assay or radioimmunoassay is used, increases or decreases in the level of whatever is being measured occur concordantly. This suggests that we are all attacking the same problem, namely the measurement of prolactin, fairly successfully. I would agree that the final method of measurement which will give us 'absolute' values will very probably be a radioimmunological one. This will come only when we are more sure of the chemical identity of human prolactin and it will therefore be possible for the prolactin radioimmunoassay to grow out of its present *Kinderkrankheiten*.

REFERENCES

Meites, J. (1970) In *Hypophysiotropic Hormones of the Hypothalamus: Assay and Biochemistry*, pp. 261–281, ed. Meites, J. Baltimore: Williams and Wilkins.
Meites, J. (1957) *Proc. Soc. Exp. Biol. & Med.* **96**, 728–730.
Shevach, A. B. and Spellacy, W. N. (1971) *Obstet. & Gynecol. (N.Y.)* **38**, 286.

BIOLOGICAL EFFECTS OF NON-PRIMATE PROLACTIN AND HUMAN PLACENTAL LACTOGEN

E. E. McGARRY AND J. C. BECK

Division of Endocrinology and Metabolism, McGill University Clinic, Royal Victoria Hospital, Montreal

THE multiplicity of metabolic effects associated with the administration of human growth hormone (HGH) in man has been reported from numerous centres (Beck *et al.* 1957; Beck *et al.* 1960; Henneman *et al.* 1960; Raben 1959; Elsair *et al.* 1964; Medical Research Council 1959; Luft *et al.* 1958) and extensive reviews of the subject have been made (Raben 1962; Knobil and Hotchkiss 1964; Daughaday and Parker 1965; Najjar and Blizzard 1966), the most recent being by McGarry and Beck (1971). Because of the limited supply of HGH extensive investigation of its possible therapeutic usefulness in conditions other than growth failure has not been possible, and this task remains largely unaccomplished some 15 years after the effects of primate growth hormones in monkey and man were first described.

Thus, we and others (Blizzard *et al.* 1966; Sonenberg *et al.* 1968) have sought a substitute pituitary preparation from a source other than primates. Such a substitute must, of course, produce at least some of the crucial physiological actions in man that human somatotropin produces. The apparent difficulty in separating prolactin from growth hormone in chemical fractionation of human pituitaries (Li 1962; Wilhelmi 1961) and the demonstration in animals of overlapping biological activity of these hormones (Barrett, Friesen and Astwood 1962; Butt 1962; Chadwick, Folley and Gemzell 1961; Ferguson and Wallace 1961; Lyons, Li and Johnson 1961) prompted our first studies on the metabolic effects in man of various preparations of ovine prolactin. Data obtained by others (Bergenstall and Lipsett 1958; McCalister and Welbourn 1962; Knobil and Greep 1959; Sinkoff and DeBodo 1953; Reisfeld *et al.* 1961) suggested that sheep prolactin exerted some of the physiological actions of HGH in both animals and humans. We have reported that an ovine prolactin preparation has most of the metabolic actions of HGH in human subjects (McGarry and Beck 1962; Beck *et al.* 1964, 1965; McGarry, Rubinstein and Beck 1968). Blizzard and his associates (1966) have also compared the effect of animal prolactins and HGH in hypopituitary subjects.

Ehrhardt first demonstrated prolactin-like activity in extracts of human placenta in 1936. In 1961, Fukushima prepared a placental extract which exhibited a growth hormone-like effect. Almost simultaneously other Japanese workers obtained evidence for prolactin-like activity in placental tissue (Kurosaki 1961; Ito and Higashi 1961; Higashi 1961). However, it was the report of Josimovich and MacLaren (1962) that aroused the most widespread interest. Their observation that the placenta and retroplacental blood contained a substance (placental lactogen, HPL) which reacted immunochemically with antiserum to HGH was rapidly confirmed by Kaplan and Grumbach (1964), who demonstrated its origin in the syncytiotrophoblast and designated the material 'human chorionic growth hormone-prolactin'. The physiological role of this protein remains uncertain, and variable biological effects have been described both in animals and man. HPL clearly has somatotropic, luteotropic and lactogenic activities (Friesen 1965; Josimovich and Mintz 1968). In addition, HGH and HPL show similarities in molecular weight and amino acid composition (Friesen 1965; Catt, Moffat and Niall 1967). These qualities led to our interest in the metabolic effects of this protein in man.

We wish to present here some additional studies on the metabolic effects of ovine prolactins and bovine prolactin as well as some studies with HPL.

MATERIALS AND METHODS

Studies with prolactin have been made in 37 patients. Data on the postpubertal hypopituitary patients and on the four patients with hypoparathyroidism have already been published, as have some results on two of the diabetic patients and five of the prepubertal hypopituitary patients, and for the most part these data will not be included in this paper. Fifteen patients with prepubertal hypopituitarism received ovine prolactin (PLR) prepared by a modification of the method of Reisfeld and co-workers (1961). Two of these patients received NIH ovine prolactin (NIH-PL) and two patients received ovine prolactin (PLC) prepared by a modification of the method of Wallace and Ferguson (1961) for the extraction of growth hormone from human pituitaries. One patient received bovine prolactin and enzymically degraded bovine prolactin. All patients had, or subsequently, received HGH while on balance study. Four patients with diabetic retinopathy, three of them having diabetes of the juvenile type, received PLR; three patients received it before and four after hypophysectomy. Four patients with bone marrow depression received PLR while on balance study for a 3–6 week period. Two patients with gonadal dysgenesis, one patient with achondroplasia, one boy with growth retardation attributed to a chromosomal defect (XY/XO), one patient with malabsorption syndrome and two asthmatic patients on steroid therapy also received PLR while on balance study.

Five patients, ranging in age from 12 to 18 years, received human placental lactogen (Purified Placental Protein—lot no. 716049 PPP-H, HG). Three had idiopathic hypopituitarism, one patient had hypopituitary dwarfism following the removal of a craniopharyngioma, and a sixth patient had gonadal dysgenesis. All had previously responded well to HGH as determined by a fall in the blood urea nitrogen (BUN) and hypercalciuria.

All subjects were studied in the Clinical Research Centre at the Royal Victoria Hospital. Measurements were made of urine and stool nitrogen, calcium and phosphorus and urinary sodium and potassium. In some studies free fatty acids and triglycerides were measured at intervals throughout the study as well as urinary aldosterone excretion, 17-ketosteroids and 17-hydroxycorticosteroids. In the four cases with bone marrow depression, plasma erythropoietin, incorporation of ^{59}Fe into erythrocytes and $T_{1/2}$ clearance time of ^{59}Fe, red cell mass and plasma volume were measured. The effect of the prolactin preparations and HPL on carbohydrate tolerance was studied before and at intervals after a breakfast containing at least 50 g of carbohydrate. The analytical techniques have been described elsewhere (Beck et al. 1960).

In the balance studies an equilibrium period on the final diet of at least five days preceded the beginning of the study. The control period which followed lasted in most patients for 12 days.

Two plans of study were employed for HPL. The purpose of the first experimental design was to examine whether HPL would potentiate the action of HGH, since a potentiation of growth hormone effects had been described in animals (Josimovich 1965; Florini et al. 1966). The patients received 5 mg of HGH per day for a period of 12 days, the last six days of which they also received 50 mg of HPL. In the second design two individuals, both with hypopituitarism, received larger amounts of HPL. The first subject received 750 mg of HPL per day in divided doses for three days and the second patient received 500 mg of HPL for three days.

RESULTS

Metabolic effects of prolactin in prepubertal hypopituitary patients

Fifteen hypopituitary patients received both HGH and PLR. In 12 of these patients adequate balance studies were achieved. Two patients showed no response to either HGH or PLR. One of these patients has shown accelerated growth with no evidence of HGH in the serum as measured by radioimmunoassay. The second of these patients, a 15-year-old male with all the clinical and laboratory characteristics of hypopituitary dwarfism, also failed to respond to HGH and PLR. Clinical data on the remaining ten patients are shown in Table I. The dose of HGH given was

TABLE I

HYPOPITUITARY PATIENTS RECEIVING HGH AND PLR

Patient	Sex	Chronological age years	Height cm	Weight kg	Height age	Bone age	Aetiology
J.D.	M	10·5	118·4	24·6	6	*	C[1]
J.Br.	F	21	151	49·6	12	17·5†	C
G.M.	F	5·5	96·4	11·8	3	5·5	C
L.L.	M	13	126	23·9	7	6	I[2]
G.C.	M	14	132·4	32·7	8	8·5	I
N.M.	F	9	124	29·9	8	7·5	C
R.S.	M	19	132	27·2	8	14·5**	I
F.H.	M	18	125	27·2	7	11·5	I
A.G.	M	16	165	61·8	14	—	A[3]
R.C.	M	10	113	21.4	5	5·5	C

[1] C = Craniopharyngioma.
[2] I = Idiopathic.
[3] A = Chromophobe adenoma.
* Bone age 9 at age 12.
** Previous treatment with testosterone.
† Previous treatment with oestrogens.

for the most part 10 mg a day for four days. The dose of prolactin given was 40 mg a day except where otherwise indicated. In some patients HGH or PLR was continued for a longer period and where this was done only the first four days have been used for the comparison of BUN, urine nitrogen and urine calcium. The interval between the discontinuation of one hormone and the beginning of the administration of the second hormone ranged from 14 days to two months except for patient A.G., in whom the interval was only eight days. In two patients administration of NIH-PL was immediately followed by administration of PLR. Data on this aspect of the study are not included in the comparison of HGH and PLR but are presented in detail subsequently. Two patients received PLC between the studies with HGH and PLR (G.C., R.C.). One patient (R.S.) received NIH-PL between the administration of HGH and PLR.

The effect of HGH and PLR on the BUN is shown in Fig. 1. It will be seen that comparable falls in BUN occurred with the two hormones. The mean control BUN for the HGH studies was 16·8, with a mean BUN on HGH of 10·2. The mean control BUN for the PLR studies was 15·9, whereas the mean BUN on PLR was 10·3.

The effect of HGH and PLR on urine nitrogen and nitrogen balance is shown in Table II. Six patients received 10 mg a day of HGH for four days,

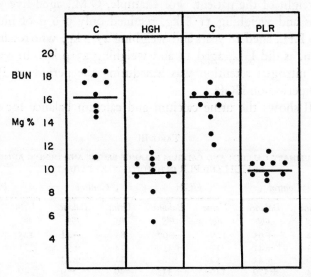

FIG. 1. The effect of giving HGH and PLR on BUN in hypopituitary dwarfs. Each dot represents the mean of several determinations in the control period (C) and the value obtained the fourth day of the HGH or PLR period.

TABLE II

MEAN DAILY URINE NITROGEN AND NITROGEN BALANCE BEFORE AND DURING ADMINISTRATION OF HGH AND PLR IN HYPOPITUITARY PATIENTS

Patient	Control		HGH		Control		PLR	
	Urine g	Balance g	Urine g	Balance g	Urine g	Balance g	Urine g	Balance g
J.D.	8·2	+0·7	7·3	+1·3	8·3	+0·2	6·9	+2·0
J.Br.	5·3	+6·3	4·8	+6·7	5·4	+6·3	5·3	+6·2
G.M.	4·9	+1·7	4·3	+2·3	5·4	+0·7	4·5	+1·7
L.L.	7·8	+0·9	6·6	+1·9	8·4	+0·1	7·5	+1·0
G.C.	9·6	−0·4	7·0	+1·9	9·5	−0·4	8·2	*
N.M.	8·3	−0·1	6·4	+0·3	8·7	−0·3	8·1	*
R.S.	8·4	+0·5	6·8	+2·3	8·6	+0·8	7·7	+0·9
F.H.	11·2	+0·9	9·0	+2·9	11·0	+1·7	9·6	±2·2
A.G.	12·1	−3·8	11·1	−1·2	10·4	−0·9	9·8	−0·24
R.C.	7·5	±0·0	7·0	+0·5	6·7	+0·6	6·3	−0·3**

* Incomplete stool collection.
** PLR 10 mg per day.

one patient (R.C.) received 10 mg a day of HGH for three days, whereas three patients (J.Br., L.L. and G.C.) received 5 mg a day for four days. The control values for both HGH and PLR are for the most part comparable, with the exception of A.G. in whom the interval between the administration of the two hormones was only eight days. With the exception of J.Br. the absolute amount of nitrogen retained tended to vary with the age and weight of the patient. For example, G.M., aged five years and six months and weighing 11·8 kg, retained only 0·6 g of nitrogen in contrast to F.H., aged 18 years and weighing 27·2 kg, who retained 2·2 g of nitrogen, as did L.L., aged 13 and weighing 23·9 kg. In general the degree of nitrogen retention was less during the period on PLR than during the period on HGH.

Table III shows the urine calcium and calcium balance for the same

TABLE III

MEAN DAILY URINARY CALCIUM AND CALCIUM BALANCE BEFORE AND DURING ADMINISTRATION OF HGH AND PLR IN HYPOPITUITARY PATIENTS

Patient	Control		HGH		Control		PLR	
	Urine mg	Balance mg	Urine mg	Balance mg	Urine mg	Balance mg	Urine mg	Balance mg
J.D.	82	+95	127	+97	85	−44	125	+263
J.Br.	13	+41	17	+130	14	−96	31	−202
G.M.	26	+41	52	+24	11	+43	30	+154
L.L.	54	+170	140	+315	29	−9	40	−26
G.C.	183	+192	276	−157	161	+50	207	+366
N.M.	143	+165	307	+119	90	+257	212	*
R.S.	106	+183	137	+179	74	+275	76	+140
F.H.	299	−105	381	+149	151	+206	219	+115
A.G.	247	−473	300	+184	358	−251	474	−671
R.C.	46	+74	55	±0	41	−214	53	**
Mean	120		179		101		147	

* Incomplete stool collection.
** PLR 10 mg per day.

periods in these patients. In six patients the control value of urine calcium for the PLR period was lower than it was for the HGH period. In A.G. the HGH control was higher, perhaps again attributable to the fact that there was only an eight-day interval between the end of the HGH period and the first day of PLR. In two of three patients in whom the control values for urine calcium were comparable, the hypercalciuria was greater with PLR than with HGH.

When the results of the changes in urine nitrogen and urine calcium are expressed as a percentage of control (Fig. 2), six of ten patients showed a greater degree of hypercalciuria with PLR, whereas only two of ten patients showed a greater degree of nitrogen retention with PLR.

The effect on urine phosphorus and phosphorus balance was more marked with HGH than with PLR (Table IV). Only one patient, G.M., showed an increase in urine phosphorus while receiving HGH, one patient, J.D., showed no change, and the other eight patients all showed modest or marked decreases in urine phosphorus while receiving HGH. In contrast, four patients showed an increase in urine phosphorus while

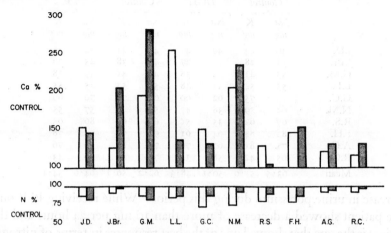

FIG. 2. Effect of HGH and PLR on urine nitrogen and calcium; 100 per cent represents the mean value during the respective control periods. Open bars, HGH treatment period; shaded bars, PLR period.

TABLE IV

MEAN DAILY URINE PHOSPHORUS AND PHOSPHORUS BALANCE BEFORE AND DURING ADMINISTRATION OF HGH AND PLR IN HYPOPITUITARY PATIENTS

Patient	Control		HGH		Control		PLR	
	Urine mg	Balance mg	Urine mg	Balance mg	Urine mg	Balance mg	Urine mg	Balance mg
J.D.	509	+251	506	+357	507	+216	500	+294
J.Br.	217	+217	191	+242	238	+396	199	+373
G.M.	278	+160	306	+159	297	+343	279	+359
L.L.	420	+218	407	+197	486	+43	443	+88
G.C.	647	+170	462	+270	656	+103	404	+295
N.M.	871	−148	505	+230	604	+136	633	*
R.S.	661	+157	531	+314	665	−13	724	+4
F.H.	938	+99	810	+286	935	+77	897	+205
A.G.	1037	−580	959	+82	991	−185	1001	−297
R.C.**	450	+102	389	+354	443	+128	463	*
Mean	603		507		581		454	

* Incomplete stool collection.
** PLR, 10 mg per day.

receiving PLR. Only one patient showed a marked decrease in urine phosphorus on PLR compared to four patients with HGH.

The changes in sodium and potassium excretion were more variable (Table V). Only four patients showed a decrease in urine potassium greater than 10 mEq/day while receiving HGH, though none of them showed an

Table V
MEAN URINARY SODIUM AND POTASSIUM EXCRETION BEFORE AND DURING HGH AND PLR ADMINISTRATION IN HYPOPITUITARY PATIENTS

	Control		HGH		Control		PLR	
	Na mg	K mg	Na mg	K mg	Na mg	K mg	Na mg	K mg
J.D.	49	44	49	41	45	44	42	33
J.Br.	54	28	42	27	45	38	45	29
G.M.	55	41	52	38	49	41	49	38
L.L.	55	54	34	48	50	54	45	55
G.C.	75	75	62	60	68	71	70	69
N.M.	62	59	38	47	56	47	57	55
R.S.	67	62	35	51	66	61	86	65
F.H.	82	84	64	61	96	79	92	78
A.G.	71	81	67	76	76	81	90	76
R.C.	65	42	58	34	71	45	70	43
Mean	63·5	57·0	50·1	48·3	62·2	56·1	64·6	54·1

increase in urine potassium during this period. While receiving PLR only one patient showed a decrease of more than 10 mg per 24 hours and this patient is the one that showed one of the best responses in terms of nitrogen retention. Two patients had a modest increase in urine potassium while receiving PLR while the remainder showed little or no change. Six patients showed a very definite decrease in urine sodium while receiving HGH while in the remainder there was little or no change. No patient showed a significant decrease in urine sodium while receiving PLR. Two patients (R.S., A.G.) showed a definite increase in urine sodium during PLR administration.

Carbohydrate tolerance

Carbohydrate tolerance was measured at intervals following a breakfast containing at least 50 g of carbohydrate. Of 11 patients who had carbohydrate tolerance tests done while on HGH and also on PLR, ten showed varying degrees of carbohydrate intolerance with HGH. One patient showed no change. The degree of change in carbohydrate intolerance was not related to the anabolic response as determined by the fall in BUN and decrease in urine nitrogen. Six of the patients showed changes in carbohydrate tolerance with administration of prolactin. These changes for the most part were modest but two patients showed more marked changes with PLR than with HGH.

The effect of PLR in hypophysectomized diabetic patients

Three patients were studied before and four patients after hypophysectomy. We had previously reported a marked rise in blood and urine sugar occurring in two hypophysectomized juvenile diabetic patients with

PLR (Beck et al. 1964; McGarry et al. 1968). The results in patients with juvenile type diabetes were similar. Fig. 3 shows the effect of PLR before and after hypophysectomy on the fasting blood sugar, free fatty acids and triglycerides in one of these patients. Before hypophysectomy there was

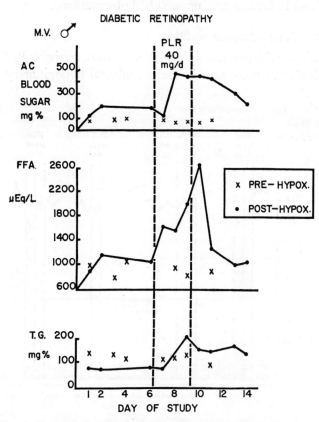

FIG. 3. Changes induced by PLR in a patient with juvenile diabetes before and after hypophysectomy.

little or no effect of PLR but after hypophysectomy there was a sharp rise in fasting blood sugar, free fatty acids and triglycerides. Similar but less marked changes occurred in a third patient with maturity onset diabetes.

Effect of PLR in patients with bone marrow hypoplasia

Four female patients with varying degrees of bone marrow hypoplasia received 40 mg of prolactin per day for three to six weeks. Two patients with chronic panhypoplasia showed an increase in incorporation of ^{59}Fe in red cells; one patient showed an increase in the red cell mass of 140 ml while the second patient showed an increase in plasma iron turnover. The two patients with acute panhypoplasia showed no response to prolactin. There was no increase in the excretion of erythropoietin in the urine of any of these patients. None of these patients showed any evidence of a metabolic effect of PLR as determined by balance study (J. H. Jepson, E. E. McGarry and L. Lowenstein, unpublished observations).

Effect of PLR in miscellaneous conditions

Two patients with gonadal dysgenesis who were receiving PLR showed changes similar to those seen in the prepubertal hypopituitary subjects.

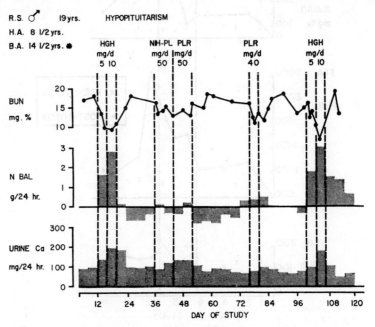

FIG. 4. Effects of NIH-PL and PLR compared to HGH in three hypopituitary patients. Balance data are plotted from a zero line which represents the mean balance for the initial 12-day control period. Mean values of four-day periods are plotted for both nitrogen balance and urine calcium.

Two patients with severe asthma were studied while receiving corticosteroid therapy. One was a postmenopausal female and the second was a prepubertal 18-year-old with growth retardation attributed to corticosteroid therapy. Studies on these patients will be reported in detail elsewhere. Both patients showed a fall in BUN and a modest nitrogen retention and hypercalciuria. No metabolic changes as determined by complete balance studies occurred in one patient with achondroplasia, one boy with growth retardation attributed to a chromosomal defect and one patient with adult onset malabsorption syndrome who received PLR.

Comparison of other prolactin preparations with HGH and PLR

The results of studies made while hypopituitary patients were receiving PLC and NIH-PL are shown in Figs. 4, 5 and 6.

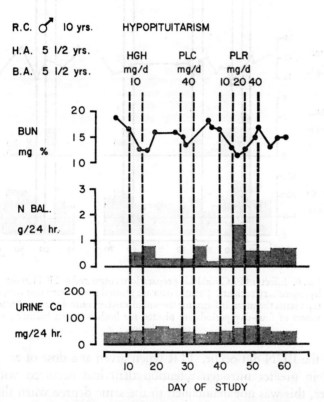

FIG. 5. Effects of PLC and PLR preparations compared to HGH in three hypopituitary patients. Balance data are plotted from a zero line which represents the mean balance for the initial 12-day control period. Mean values of four-day periods are plotted for both nitrogen balance and urine calcium.

In R.S. (Fig. 4) NIH-PL produced a modest fall in the BUN when this preparation was given from days 37 to 40 inclusive. This fall in BUN was maintained when PLR was given from days 41 to 44 inclusive. The changes that occurred, however, were not as great as those induced by HGH given at both the beginning and end of the study. When PLR was given from days 76 to 78, however, a more marked fall in the BUN and a definite, though modest, nitrogen retention occurred. In this individual the hypercalciuria was greater with HGH than with any of the prolactin preparations. In subject R.C. (Fig. 5) only modest hypercalciuria occurred with HGH and PLC. PLC produced no change in nitrogen balance but

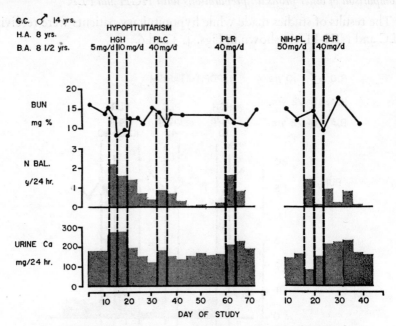

FIG. 6. Effects of PLC and PLR preparations compared to HGH in three hypopituitary patients. Balance data are plotted from a zero line which represents the mean balance for the initial 12-day control period. Mean values of four-day periods are plotted for both nitrogen balance and urine calcium.

a fall in the BUN did occur. PLR when given at a dose of 20 mg a day resulted in greater nitrogen retention than had occurred with HGH. However, this was not maintained to the same degree when the dose of PLR was increased to 40 mg a day. In G.C. (Fig. 6) PLC produced hypocalciuria, which was, however, less than with HGH. A fall in the BUN

TABLE VI

CARBOHYDRATE TOLERANCE TESTS IN R.S.

	Control	HGH*	Post control	NIH-PL*	PLR(1)*	PLR(2)*	Post control	HGH
A.c.	68	83	69	76	73	66	78	85
45 minutes	123	165	137	105	136	160	117	157
1½ hours	94	101	98	103	87	102	124	—
2 hours	91	106	93	94	85	90	96	110
3 hours	73	103	80	79	72	86	86	98
4 hours	68	89	65	75	72	97	74	98

* Done the last day of hormone administration.

occurred with PLC and there was also modest nitrogen retention. NIH-PL produced nitrogen retention and a fall in the BUN comparable to that which had occurred with PLR, and the degree of hypercalciuria was similar with both preparations. Carbohydrate tolerance tests in R.S. throughout

TABLE VII

COMPARISON OF THE METABOLIC EFFECT OF HGH WITH THE EFFECT OF HGH PLUS HPL

		BUN mg%	N balance g/24 hours	Ca balance mg/24 hours	Urine Ca mg/24 hours
C.T.	Control	14	+2·24	+435	46
	HGH*	8	+4·00	+583	112
	HPL**+HGH*	6	+2·82	+501	184
A.T.	Control	16	+0·07	+140	82
	HGH* days 1–4	13	+2·04	+49	160
	days 5–8	—	+2·67	+31	178
	HPL**+HGH*	12·5	+2·02	+29	204
J.B.	Control	16	+0·74	+106	124
	HGH*	7	+1·53	−601	186
	HPL**+HGH*	14	+1·28	+109	221
N.McL.	Control	17	−1·04	+189	59
	HGH*	11	+0·71	+355	98
	HPL**+HGH*	11	+0·55	+171	12
S.S.	Control	11	+3·05	+411	110
	HGH*	7	+3·50	+264	169
	HPL**+HGH*	10	+3·10	+443	198

* HGH, 5 mg per day.
** HPL, 50 mg per day.

TABLE VIII

COMPARISON OF EFFECT OF HGH WITH EFFECT OF HGH PLUS HPL ON SODIUM AND ALDOSTERONE EXCRETION

		Na balance mEq/24 hours	Aldosterone excretion μg/24 hours
C.T.	Control	+18	17
	HGH*	+37	23
	HPL**+HGH*	+5	5·6
A.T.	Control	−0·2	9
	HGH*	+39	3·6
	HPL**+HGH*	+40	7·4
J.B.	Control	+4·2	4·4
	HGH*	+7·4	6·8
	HPL**+HGH*	+2·9	6·8
N.McL.	Control	+0·1	8·4
	HGH*	+10	14·0
	HPL**+HGH*	+2	25·5
S.S.	Control	+10	3·8
	HGH*	+19	2·0
	HPL**+HGH*	+10	4·6

* HGH, 5 mg per day.
** HPL, 50 mg per day.

the study are tabulated in Table VI. Only HGH and PLR given after a control period produced any degree of intolerance.

Studies with human placental lactogen

Results of the study designed to determine whether the metabolic

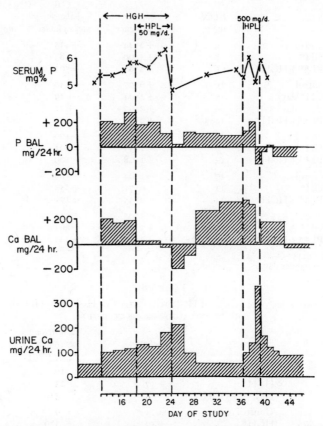

FIG. 7. Changes in serum phosphorus, phosphorus balance, calcium balance and urine calcium in a hypopituitary dwarf while receiving HGH (5 mg per day) and HPL in the amount indicated. Balance data are plotted from a zero line that represents the mean balance for the initial 12-day control period.

effects of HGH would be potentiated by the addition of HPL are shown in Table VII. All individuals showed nitrogen retention and a fall in the BUN as well as hypercalciuria when receiving HGH alone. When HPL was added the fall in the BUN was maintained in four patients but there was no enhancement of nitrogen retention. Urine calcium excretion tended to be greater on HGH plus HPL than on HGH alone. Changes in sodium balance and aldosterone excretion (Table VIII) were greater during the period with HGH alone. In only one patient did aldosterone excretion increase while on HGH, and a further increase occurred on HPL.

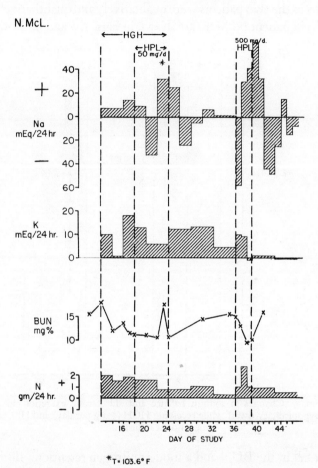

FIG. 8. Changes in BUN, nitrogen and electrolyte balance in a hypopituitary dwarf while receiving HGH (5 mg per day) and HPL in the amount indicated. Balance data are plotted from a zero line that represents the mean balance for the initial 12-day control period.

Two individuals showed a slight rise in serum phosphorus progressively throughout the treatment while the remainder showed no change. Three individuals who showed a definite decrease in carbohydrate tolerance with HGH alone showed a slight enhancement of this intolerance on HGH plus HPL. There was no significant change in either fasting free fatty acids or triglycerides. In these five individuals there was no local or generalized reaction to the injection of HPL although there was a low-grade fever. The patients were asymptomatic, however.

Two patients received HPL alone in high dose (N. McL., F.H.). One received 500 mg of HPL a day, the second received 750 mg of HPL a day. The results in the two patients were qualitatively and quantitatively similar. The results in patient N. McL. are shown in Figs. 7, 8 and 9. Both patients

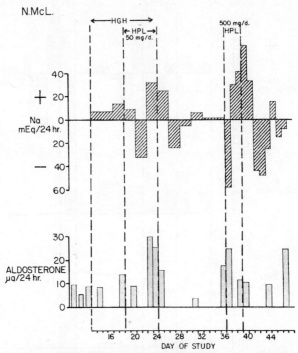

FIG. 9. Changes in sodium balance and aldosterone excretion in a hypopituitary dwarf while receiving HGH (5 mg per day) and HPL in the amount indicated.

showed a fall in the BUN and a modest nitrogen retention. Both patients developed hypercalciuria and both showed a marked increase in urine sodium on the first day of treatment followed by marked sodium retention on the second and third day. N.McL. showed an increase in aldosterone excretion the day before receiving HPL and this was maintained. The second patient showed no change in aldosterone excretion throughout the study. No change occurred in the urinary excretion of 17-ketosteroids or 17-hydroxycorticosteroids in either patient. There was no change in fasting free fatty acids or triglycerides. F.H. showed carbohydrate intolerance on the third day of HPL; N.McL. showed no change in carbohydrate tolerance.

Both patients complained of severe pain at the site of injection of HPL. There was no systemic reaction other than low-grade fever and no local redness or swelling. Similar reactions with this preparation of HPL have been reported by others (Shultz and Blizzard 1966; Grumbach et al. 1968).

In one patient the effects of bovine prolactin (U-19988) and two enzymically degraded preparations of bovine prolactin (N-0103 and N-0102) were compared with the metabolic response to HGH in the same study (Table IX). The enzymically degraded preparation had a small effect on

TABLE IX

EFFECT OF BOVINE PROLACTIN AND DEGRADED BOVINE PROLACTINS IN A HYPOPITUITARY DWARF

	BUN mg%	N	Ca	P	Na	K
Control	16	8·9*	123**	756**	73**	65**
HGH	12·5	6·7	139	632	62	46
Control	12	10·4	116	869	97	62
U-19988, 25 mg twice daily	14	10·5	142	719	101	68
Control	12·5	8·9	93	683	92	62
N-0103, 12·5 mg twice daily	10·5	9·1	132	794	101	79
Control	14	10·1	130	881	71	66
N-0102, 12·5 mg twice daily	11	9·7	137	682	83	64

* g per 24 hours.
** mg per 24 hours.

the BUN but there was no appreciable decrease in urine nitrogen. Administration of bovine prolactin resulted in no change in the BUN nor the urine nitrogen. All three preparations resulted in a modest increase in urine calcium.

DISCUSSION

The results presented here confirm our previously published reports (McGarry and Beck 1962; Beck et al. 1964, 1965; McGarry, Rubinstein and Beck 1968) that many of the metabolic actions of PLR are qualitatively similar to those induced by HGH in man. There are quantitative differences, in that while the fall in BUN was comparable, nitrogen retention was greater with HGH. With regard to the hypercalciuric effect, a comparison is more difficult since, in the majority of instances, the subjects excreted less calcium in the control period for PLR than in the control period for HGH. In all but one patient (G.M.) the HGH period preceded the PLR period. In our experience urine calcium is lower after HGH than in the control period and may remain decreased for an extended period. The intervals of continuous balance study between previous hormone administration and the administration of PLR were 24 days in R.S., G.C. and F.H. and 20 days in L.L., all of whom showed a lower calcium excretion in the eight-day period just before PLR than in the period before HGH. It is of interest that when R.S. received HGH again, 20 days after PLR, the urine calcium excretion in the eight-day period preceding the second HGH period was similar to that seen before PLR administration and the hypercalciuria induced was more modest than with the initial course of HGH (Fig. 5). Changes in phosphorus, sodium and potassium excretion induced by HGH are not as consistent as are the changes in nitrogen. In the studies reported here, however, retention of phosphorus occurred in seven patients with HGH, while only two of the eight patients in whom balance data are available for the PLR period showed definite phosphorus retention. HGH has a variable effect on sodium and potassium excretion, as has PLR.

The effect of PLR on carbohydrate metabolism is most evident in the hypophysectomized subject, a finding also reported with HGH (Luft et al. 1958; Pearson et al. 1960). Elevation of plasma free fatty acids and triglycerides occurred promptly in the hypophysectomized diabetic patients with PLR. Only minimal changes in plasma free fatty acids occurred in non-diabetic subjects with a dose of PLR which evoked the other metabolic responses (Beck et al. 1964).

PLR also increases inulin and PAH clearance in the hypopituitary patient to an extent similar to that seen when human growth hormone is given to such individuals (Beck et al. 1964).

When PLR was given to patients with a variety of disorders, only patients with gonadal dysgenesis and those with allergic disorders on corticosteroid therapy showed any metabolic response, whereas endocrinologically normal individuals showed no metabolic response, paralleling the response to HGH in similar patients (Beck et al. 1965; McGarry and Beck 1969).

The metabolic effects of other prolactin preparations (PLC, NIH-PL) were not as marked as those seen when PLR was given. Blizzard and co-workers (1966) reported minimal metabolic response to NIH-PL, although the results of these authors are not strictly comparable to those reported here since the interval between periods of hormone administration was only six days. Metabolic activity may well be influenced by the method of preparation of prolactin. Hamid and co-workers (1965) found that PLR and PLC were similar to bovine growth hormone with respect to the effect on carbohydrate and fat metabolism *in vitro*. In contrast, a third prolactin preparation was insulin-like in the same system. Winegrad and co-workers (1959) also found an NIH preparation of ovine prolactin to have insulin-like activity *in vitro*. It is conceivable that various methods of preparation may mask, expose or destroy biological determinants of a protein hormone. Although little metabolic effect, except for slight hypercalciuria, was apparent when bovine prolactin or degraded bovine preparations were given to a hypopituitary patient, tryptic digests of bovine growth hormone given to human subjects were found to have some of the metabolic properties of HGH (Sonenberg et al. 1968).

In the studies with HPL, the addition of 50 mg a day of HPL to the existing dose of HGH did not enhance the anabolic effect. Although the urine calcium excretion tended to be greater on HGH plus HPL than on HGH alone, this could be entirely attributed to HGH since this occurs when HGH is continued beyond a four-day period. Similarly, the slight rise in serum phosphorus shown in two individuals as well as the decrease in carbohydrate tolerance that occurred during the HPL period could also have occurred with continuation of HGH alone. It is not surprising, perhaps, that no metabolic effect was discernible with this small dose of HPL since the rate of production of HPL in pregnant women has been estimated to be 1 g per day (Solomon and Friesen 1968; Kaplan et al. 1968), in contrast to the estimated secretion rate of HGH which is in the order of 1 mg per day (Frantz, Killian and Holub 1969; Kowarski, Thompson and Blizzard 1970).

When larger doses of HPL were used alone there was evidence of an anabolic effect, as determined by a fall in the BUN and a modest nitrogen retention. There was also some hypercalciuria. While the changes were not as marked as those seen when such patients received 5 mg of HGH a day, the dose used in these two individuals may be the equivalent of less than 1 mg of HGH a day. Similar results have been reported by Grumbach and co-workers (1968).

Shultz and Blizzard (1966) found no nitrogen retention or hypercalciuria with 200 mg HPL per day. These authors also found that a similar dose of HPL did not augment the metabolic effects of HGH.

Grumbach and co-workers (1968) reported a decrease in the rate of glucose disposal in a patient receiving 400 mg a day of HPL. These authors also found an enhanced insulin response to HPL which was greater than that seen with HGH given at a dose of 2 mg. In addition, HPL has been found to induce increased glycosuria and hyperglycaemia in two hypophysectomized diabetic patients (Samaan et al. 1968). These authors also found an increased insulinogenic response in a stable diabetic following hypophysectomy. Beck and Daughaday (1967) also demonstrated impairment of glucose tolerance with 12-hour infusions of HPL in physiological amounts in humans.

Although our patients show no increase in fasting free fatty acids, HPL has been shown to cause a release in free fatty acids in the fasting individual (Grumbach et al. 1966). HPL in a dose of 100 mg daily for four days has been reported to stimulate aldosterone secretion in man (Melby et al. 1966). Changes in aldosterone excretion with HGH appear to be dose-related, in that increased aldosterone excretion occurred in patients receiving 10 mg per day of HGH but not consistently in patients receiving 5 mg of HGH a day (Beck et al. 1965). That only one patient showed an increment in aldosterone excretion may also be a function of the relatively small dose of HPL.

The immunological similarity of HGH and PLR is well established. Immunological similarities between HGH and enzymically degraded bovine growth hormone have also been reported (Laron et al. 1964; Sonenberg et al. 1968). Murthy and McGarry (1971), studying antibodies induced in patients on long-term HGH therapy, found antibodies to HGH in three of 15 patients in low titre. Sera of 11 of 15 patients, however, bound [^{125}I]porcine growth hormone (PGH) and sera from eight bound PLR in titres ranging up to 1 : 320 000 and 51 200 respectively. The binding of PGH and PLR was shown to be due to a 7S IgG antibody. With the fluorescent antibody technique, none of the sera localized in cells of human pituitaries. All sera which bound [^{125}I]PGH and [^{125}I]PLR localized in somatotrophs and prolactin cells of porcine pituitaries (Marwaha and McGarry 1971) and evidence of separate antibody populations to each hormone was obtained.

Earlier reports that HGH and HPL display striking similarities in structure have already been mentioned (Catt, Moffat and Niall 1967; Friesen 1965). More recently Bewley and Li (1970), comparing the structure of ovine prolactin and HGH, have reported homologous amino acid sequences while Mills and co-workers (1970) have shown that ovine prolactin and porcine growth hormone show a high degree of identity of primary structure. Niall and co-workers (1971) have found that on a revised sequence for HGH, HPL and HGH as well as ovine prolactin display even more identity of amino acid sequence than was previously reported.

Thus, in animals and man and in various *in vitro* systems, a variety of related polypeptide hormone preparations, intact or degraded, from various species display overlapping biological characteristics. They also show much chemical and immunological identity, supporting the suggestion that they evolved from a parent molecule (Bewley and Li 1970; Niall et al. 1971). Minor differences in amino acid sequence, altering tertiary structure, may mask or expose biological activity which may be unmasked or destroyed by degradation *in vitro* or *in vivo*.

SUMMARY

The results presented here extend and confirm our previous report that an ovine prolactin preparation (PLR) mimics many of the metabolic effects of HGH in man. These metabolic effects were not evident when ovine prolactin prepared by different methods was used. Human placental lactogen, given to patients on balance study, produces some of the effects seen when HGH is administered. These findings and recent reports of striking homologies, chemical and immunological, shared by HGH, HPL and non-primate prolactin and growth hormone preparations, support the concept that these polypeptide hormones evolved from a parent molecule.

Acknowledgements

Ovine prolactin PLR was provided through the courtesy of Merck, Sharpe, and Dohme, Research Laboratories, Westpoint, Pennsylvania, and Rahway, N.J. Ovine prolactin PLC was provided through the courtesy of K. A. Ferguson and A. L. C. Wallace, Ian Clunies Ross Animal Research Laboratory, Paramatta, N.S.W., Australia. Porcine prolactin preparations were provided by Upjohn, Kalamazoo, Michigan.

We are indebted to the Endocrine Study Section of the United States Public Health Service for supplying the NIH ovine prolactin. The human placental lactogen utilized in this study was Purified Placental Protein—Human, lot number 716049, supplied by Lederle Laboratories.

The human growth hormone used was supplied by the Medical Research Council of Canada and prepared by the method of Raben.

This work was supported by the Medical Research Council of Canada, grant no. MT-631, and National Institutes of Health grant no. 5 RO1 HD 00511.

REFERENCES

BARRETT, R. J., FRIESEN, H. and ASTWOOD, E. B. (1962) *J. Biol. Chem.* **237,** 432–439.
BECK, J. C. and DAUGHADAY, W. H. (1967) *J. Clin. Invest.* **46,** 103–110.
BECK, J. C., GONDA, A., HAMID, M. A., MORGEN, R. O., RUBINSTEIN, D. and McGARRY, E. E. (1964) *Metabolism* **13,** 1108–1134.
BECK, J. C., McGARRY, E. E., DAWSON, K., GONDA, A., HAMID, M. A. and RUBINSTEIN, D. (1965) In *Proceedings of the Second Congress of Endocrinology,* pp. 1242–1256. Amsterdam: Excerpta Medica Foundation.
BECK, J. C., McGARRY, E. E., DYRENFURTH, I., MORGEN, R. O., BIRD, E. D. and VENNING, E. H. (1960) *Metabolism* **9,** 699–737.
BECK, J. C., McGARRY, E. E., DYRENFURTH, I. and VENNING, E. H. (1957) *Science* **125,** 884–885.
BERGENSTALL, D. M. and LIPSETT, M. B. (1958) *J. Clin. Invest.* **37,** 877.
BEWLEY, T. A. and LI, C. H. (1970) *Science* **168,** 1361–1362.
BLIZZARD, R. M., DRASH, A. L., JENKINS, M. E., SPAULDING, J. S., GLICK, A., WELDON, V. V., POWELL, G. F. and RAITI, S. (1966) *J. Clin. Endocrinol. & Metab.* **26,** 852–858.
BUTT, W. R. (1962) *Ciba Fdn. Colloq. Endocrinol.* **14,** *Immunoassay of Hormones,* p. 103. London: Churchill.
CATT, K. J., MOFFAT, B. and NIALL, H. D. (1967) *Science* **157,** 321.
CHADWICK, A., FOLLEY, S. J. and GEMZELL, C. A. (1961) *Lancet* **2,** 241–243.
DAUGHADAY, W. H. and PARKER, M. L. (1965) *Am. Rev. Med.* **16,** 47–66.
EHRHARDT, K. (1936) *Muench. Med. Wochenschr.* **83,** 1163–1164.
ELSAIR, J., GERBEAUX, S., DARTOIS, A. M. and ROGER, P. (1964) *Rev. Fr. Etud. Clin. & Biol.* **91,** 287–306.
FERGUSON, K. A. and WALLACE, A. L. C. (1961) *Nature (Lond.)* **190,** 632–633.
FLORINI, J. R., TONELLI, G., BREUER, C. B., COPPOLA, J., RINGLER, I. and BELL, P. H. (1966) *Endocrinology* **79,** 692–708.
FRANTZ, A. G., KILLIAN, P. and HOLUB, D. A. (1969) *J. Clin. Invest.* **48,** 25a (abst. 81).
FRIESEN, H. (1965) *Endocrinology* **76,** 369–381.

FUKUSHIMA, M. (1961) *Tokoku J. Exp. Med.* **74**, 161–174.
GRUMBACH, M. M., KAPLAN, S. L., ABRAMS, C. L., ELL, J. J. and CONTE, F. A. (1966) *J. Clin. Endocrinol. & Metab.* **26**, 478–482.
GRUMBACH, M. M., KAPLAN, S. L., SCIARRA, J. J. and BURR, I. M. (1968) *Ann. N.Y. Acad. Sci.* **148** (Art. 2), 501–531.
HAMID, M. A., RUBINSTEIN, D., FERGUSON, K. A. and BECK, J. C. (1965) *Biochim. & Biophys. Acta* **100**, 179–192.
HENNEMAN, P. H., FORBES, A. P., MOLDAWER, M., DEMPSEY, E. F. and CARROLL, E. L. (1960) *J. Clin. Invest.* **39**, 1223–1238.
HIGASHI, K. (1961) *Endocrinol. Jap.* **8**, 288–296.
ITO, Y. and HIGASHI, K. (1961) *Endocrinol. Jap.* **8**, 279–287.
JOSIMOVICH, J. B. (1965) *Endocrinology* **78**, 707–714.
JOSIMOVICH, J. B. and MACLAREN, J. A. (1962) *Endocrinology* **71**, 209–220.
JOSIMOVICH, J. B. and MINTZ, D. H. (1968) *Ann. N.Y. Acad. Sci.* **148** (Art. 2), 488–500.
KAPLAN, S. L. and GRUMBACH, M. M. (1964) *J. Clin. Endocrinol. & Metab.* **24**, 80–100.
KAPLAN, S. L., GURPIDE, E., SCIARRA, J. J. and GRUMBACH, M. M. (1968) *J. Clin. Endocrinol. & Metab.* **28**, 1450–1460.
KNOBIL, E. and GREEP, R. O. (1959) *Recent Prog. Horm. Res.* **15**, 1–69.
KNOBIL, E. and HOTCHKISS, J. (1964) *A. Rev. Physiol.* **26**, 47–74.
KOWARSKI, A., THOMPSON, R. G. and BLIZZARD, R. M. (1970) *Program of the Endocrine Society, 52nd Meeting*, p. 114.
KUROSAKI, M. (1961) *Tohoku J. Exp. Med.* **75**, 122–136.
LARON, Z., YED-LEKACH, A., ASSA, S. and KOWADLOSILBERGELD, A. (1964) *Endocrinology* **74**, 532–537.
LI, C. H. (1962) *Ciba Fdn. Colloq. Endocrinol. 14, Immunoassay of Hormones*, p. 364. London: Churchill.
LUFT, R., IKKOS, D., GEMZELL, C. A. and OLIVECRONA, H. (1958) *Lancet* **1**, 721–722.
LYONS, W. R., LI, C. H. and JOHNSON, R. E. (1961) *Program of the Endocrine Society, 43rd Meeting* (abst.).
MARWAHA, R. and MCGARRY, E. E. (1971) Submitted for publication.
MCCALISTER, M. and WELBOURN, R. B. (1962) *Br. Med. J.* **1**, 1669–1670.
MCGARRY, E. E. and BECK, J. C. (1962) *Lancet* **2**, 915–916.
MCGARRY, E. E. and BECK, J. C. (1969) *Ann. R. Coll. Physicians & Surgeons Can.* **2**, 20 (abst.).
MCGARRY, E. E. and BECK, J. C. (1971) In *Monograph on Human Growth Hormone*, ed. Stuart-Mason, A. London: Heinemann. In press.
MCGARRY, E. E., RUBINSTEIN, D. and BECK, J. C. (1968) *Ann. N.Y. Acad. Sci.* **148** (Art. 2), 559–569.
MEDICAL RESEARCH COUNCIL (OF GREAT BRITAIN) (1959) *Lancet* **1**, 7–12.
MELBY, J. C., DALE, S. L., WILSON, T. E. and NICHOLS, A. S. (1966) *Clin. Res.* **14**, 283 (abst.).
MILLS, J. B., HOWARD, S. C., SCAPA, S. and WILHELMI, A. E. (1970) *J. Biol. Chem.* **245**, 3407–3415.
MURTHY, G. and MCGARRY, E. E. (1971) *J. Clin. Endocrinol. & Metab.* **32**, 641–646.
NAJJAR, S. and BLIZZARD, R. M. (1966) *Pediat. Clin. N. Am.* **13**, 437–457.
NIALL, H. D., HOGAN, M. L., SAUER, R., ROSENBLUM, I. Y. and GREENWOOD, F. C. (1971) *Proc. Natl. Acad. Sci. (U.S.A.)* **68**, 866–869.
PEARSON, O. H., DOMINIGUEZ, J. M., GREENBERG, E., PAZIANOS, A. and RAY, B. S. (1960) *Trans. Ass. Am. Physicians* **73**, 217–226.
RABEN, M. S. (1959) *Recent Prog. Horm. Res.* **15**, 71–144.
RABEN, M. S. (1962) *New Engl. J. Med.* **266**, 31–35 & 82–86.
REISFELD, R. A., TONG, G. L., RICKES, E. L. and BRINK, N. G. (1961) *J. Am. Chem. Soc.* **83**, 3717–3719.

Samaan, N., Yen, S. C. C., Gonzalez, D. and Pearson, O. H. (1968) *J. Clin. Endocrinol. & Metab.* **28**, 485–491.
Shultz, R. B. and Blizzard, R. M. (1966) *J. Clin. Endocrinol. & Metab.* **26**, 921–924.
Sinkoff, M. W. and DeBodo, R. C. (1953) *Arch. Exp. Path. & Pharmakol.* **219**, 100–110.
Solomon, S. and Friesen, H. G. (1968) *A. Rev. Med.* **19**, 399–429.
Sonenberg, M., Kikutani, M., Free, C. A., Nadler, A. C. and Dellacha, J. M. (1968) *Ann. N.Y. Acad. Sci.* **148** (Art. 2), 532–558.
Wallace, A. L. C. and Ferguson, K. A. (1961) *J. Endocrinol.* **23**, 285–290.
Wilhelmi, A. E. (1961) *Can. J. Biochem.* **39**, 1659–1668.
Winegrad, A. I., Shaw, W. N., Lukens, F. D. W. and Stadie, W. C. (1959) *J. Biol. Chem.* **234**, 3111–3114.

DISCUSSION

Nicoll: Can you compare the potencies of the ovine prolactin (PLR) and human placental lactogen (HPL) preparations in their metabolic parameters?

Beck: We gave five doses of 750 mg of HPL, which would be equivalent to roughly 1 mg of HGH since this material was a 70 per cent pure preparation (it was the second HPL preparation of the Lederle Company). PLR was given in doses of 40 mg per day. The PLR was more potent than the HPL at these two particular doses; the degree of nitrogen retention, fall in BUN and hypercalciuria were all more marked on PLR than on 750 mg of HPL.

Nicoll: Does HPL cause real body growth?

Beck: We have not done this experiment. The only data that are available are on patients in whom Grumbach has reported minimal growth (Burr, Grumbach and Kaplan 1967).

Sherwood: Dr Blizzard also presented data on two hypopituitary dwarfs who failed to respond to HPL (Shultz and Blizzard 1966).

Beck: We have done one study on the effects of ovine prolactin on growth. The amount of ovine prolactin was limited. We thought we had chosen the most suitable experimental subject with hypopituitarism that we could find, and although he was suitable he also taught us something which came up earlier in discussion, on the importance of food and particularly nitrogen intake during the administration of growth-promoting agents (p. 322). Fig. 1 illustrates what happened. This patient received PLR for ten days preceded by an 18-day control period and with a 24-day post-control period on balance study. He was readmitted one month later and, using a similar protocol, received HGH for ten days. He was then given PLR for four days. The patient was then placed on 60 mg PLR a week and discharged. Because he failed to grow, the dose was increased to 120 mg per week, but he still did not grow. With the help of a social service worker it was pointed out that this youngster's socio-economic environment was such that he was receiving inadequate dietary intake at home. He was therefore brought into hospital and put on the

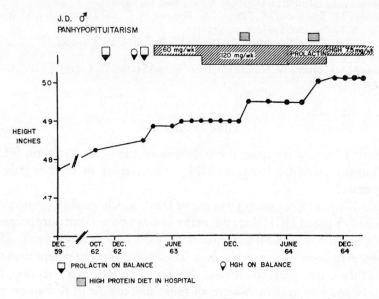

FIG. 1. (Beck). Study of effects of ovine prolactin on growth in a hypopituitary dwarf.

Clinical Research Centre balance diet once more; during this period his dietary intake including nitrogen intake was made the same as at the initial study in which a response occurred. There was a prompt resumption of linear growth. We then took him off the balance diet and let him eat as he pleased while still under observation in the ward, and his growth ceased. We introduced the same diet again which led to a prompt growth response. He was then discharged and allowed to go on to human growth hormone, because we were running out of the preparation of ovine prolactin. Again, on an inadequate diet at home, he failed to grow with HGH administration, as he had failed on PLR.

Wilhelmi: How does the Reisfeld preparation of ovine prolactin compare in potency and other biological characteristics with the NIH preparation?

Beck: This has been looked at by Dr K. Ferguson in only the crop sac assay, and they appear to have the same potency in that assay.

Nicoll: We have confirmed that.

Greenwood: How do you explain the three-week persistence of the effect of growth hormone on calcium excretion? I am interested in the question of whether these hormones are continuously acting as stimuli or whether growth hormone is acting as a programmer, only necessary to initiate a response. This seems to be an authentic case of a hormone programme—a longer-term effect than is consistent with its half-life in plasma and its time of survival in tissue.

Beck: We can't explain it. We believe the hypercalciuria to be on two bases: (1) there is an increase in renal plasma flow and glomerular filtration rate, and therefore an increased filtered load of calcium. (2) In addition, though the data are not definitive in man, and could only be proved in animals by micropuncture studies, it appears that there is also increased tubular rejection of calcium. We know that the persistence of the decreased urinary excretion of calcium long after the growth hormone injections cease is not associated with the maintenance of an elevated filtered load, since this disappears within four to seven days after the cessation of either HGH or PLR. The renal haemodynamics then return to those normally seen in the hypopituitary subject. So we have no explanation why the patients may have a very definite reduction in urine calcium persisting for long periods of time, and may also have long periods of positive calcium balance.

Wilhelmi: Does it bear any relation to the persistence of sulphation factor? Because this is an instance of what Dr Greenwood is talking about.

Beck: We only have limited data about this, so we have to say we don't know. In our hypopituitary subjects, we begin to see sulphation factor appear after a single injection of human growth hormone within the first 12–24 hours and it may persist for up to five days. We have never followed it beyond this.

Turkington: It has been shown that a degree of osteoporosis can develop in response to growth hormone, which represents endosteal resorption of bone. I wonder if this might contribute to the hypercalciuric effect of growth hormone and HPL.

Beck: I think there's no doubt that it does. Actually, the best radiographic examples of osteoporosis in textbooks of medicine are in acromegalic subjects. The possibility that you raise—that the hypocalciuria when you stop giving growth hormone or PLR, and the positive balance, are the result of the restoration of bone mass—is a very probable one, but we have not yet done the necessary dynamic studies to give a definitive answer.

Frantz: I think that the osteoporosis sometimes seen in acromegaly may result from the hypogonadism that can be associated with pituitary tumours. I don't think all acromegalics are osteoporotic.

Beck: It depends on the duration and intensity of the disease. Many of them have hypercalciuria; these are old observations that have been lying around in the medical journals for years and have not been remarked on (Scriver and Bryan 1935).

Sherwood: One has to be careful about two factors in osteoporosis: one is diet and the other is immobilization. Dr William Harris has shown that dogs with immobilization osteoporosis respond to bovine growth hormone with bone growth. This is potentially one of the important areas for a clinical trial of growth hormone when it is available (Harris and Heaney 1969).

Beck: Our subjects were not immobilized, and they were on a known calcium intake.

Greenwood: Were there any eugonadal females in this series? With all these lactogenic hormones being given, I find it aesthetically unpleasing that there wasn't a single target tissue in this series receptive to the lactogenic activity of these three hormones!

Beck: There were eugonadal females among the adults with bone marrow hypoplasia. However, they are not a fair sampling. Dr McGarry has noticed that some of the female hypopituitary dwarfs on long-term therapy with HGH (the Canadian MRC preparation) showed breast development.

Li: A most interesting observation is the stimulation of the bone marrow and the production of red blood cells. How many cases have you studied now on this?

Beck: There are only four patients in this study. Dr Friesen and Joanne Jepson in our department looked at this with HPL, and Joanne Jepson looked at it with other preparations of ovine prolactin in mice (Jepson and Friesen 1968).

Friesen: These studies were done in the polycythaemic mouse which is used as an assay for erythropoietin. Both prolactin and HPL stimulate erythropoietin production in the rat, and also cause increases in plasma volume. So these data are quite consistent with the findings in man.

Beck: We also have information which suggests that HGH and perhaps PLR stimulate production of renin, and that the influence on aldosterone may be via the renin-angiotensin mechanism.

Cotes: I don't know if Dr Friesen's finding (Jepson and Friesen 1967, 1968) of a stimulatory effect of HPL on erythropoiesis is a property of HPL or whether it might be attributable to the action of another constituent present in some, if not all, placental extracts. In my laboratory, using the appearance of radioactive iron in red cells of ex-hypoxic polycythaemic mice, we saw no erythropoietic effect after treatment with HPL alone in doses between 2·5 and 20 mg. When HPL was administered with erythropoietin, the effects were not significantly different from those seen with erythropoietin alone (Fig. 1).

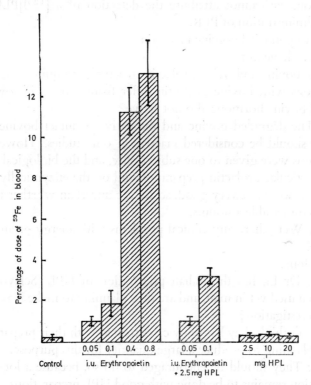

FIG. 1. (Cotes). Effects of erythropoietin and human placental lactogen on erythropoiesis in polycythaemic mice. The appearance of ^{59}Fe in the blood of ex-hypoxic polycythaemic female CBA mice 72 hours after the administration of an intravenous dose of ^{59}Fe citrate is shown, as a percentage of the dose given ±S.E.M. Erythropoietin from human urine (calibrated in international units by bioassay: Cotes and Bangham 1966; Cotes 1971) and human placental lactogen (Lederle Purified Placental Protein, Batch E-12 Lot 1359) were each administered intraperitoneally as two doses 72 and 48 hours before the ^{59}Fe.

Turkington: Did you have an opportunity to measure liver function, for example by excretion of bromsulphonphthalein?

Beck: We did, and all indicators remain unchanged.

Li: When you inject the prolactin preparation are there any indications of antibody formation?

Beck: No patients developed antibodies to prolactin that were detectable by haemagglutination or by ring test. More recently, however, we have shown that many patients treated only with HGH develop 7S IgG which binds porcine growth hormone (Murthy and McGarry 1971) and, in some instances, PLR (Murthy and McGarry, submitted for publication). Since all patients in the present report received HGH as well as a prolactin preparation, we cannot attribute the detection of a [^{125}I]PLR–binding IgG to administration of PLR.

Li: Have you tried porcine prolactin?

Beck: We have not.

Li: The bovine and ovine prolactin structures are quite close, yet your observations with bovine prolactin differ from those with ovine. Was degraded bovine hormone also not active?

Beck: The degraded bovine and the native or intact bovine prolactin probably should be considered inactive in our studies. However, these preparations were given to one subject only, and the biological assay data on this particular prolactin preparation and on the enzymically degraded preparation were not very good, so we are uncertain whether the patient received comparable amounts.

Frantz: Were there any clinical problems with allergy or anaphylactic reactions?

Beck: None.

Friesen: Dr Li, has the Italian preparation of HPL (Sclavo Institute, Siena) been used yet in man, and are they planning to make it available for clinical investigation?

Li: I don't think they have any cases treated with their preparation, but they are planning to produce a large quantity for this purpose.

Friesen: They should be encouraged to do so, because a lot of clinical investigation remains to be done with good HPL preparations.

Beck: It is of great importance that an adequate supply of HPL be made available for clinical study, because the information to date is really very limited.

Li: Our experience with the Lederle material, given to me by Dr M. M. Grumbach, when compared on the tibia assay for growth-promoting activity with the Siena preparation, is that the Siena one appears more active.

Frantz: Dr Grumbach said that when HPL was given in an acid vehicle, it seemed to be more potent. Have you any information bearing on this?

Beck: No. When one looks at these data, one really wonders whether one can categorically state that pH plays a role.

Li: Dr Grumbach has mentioned to me that at a pH of 3·5 the HCS (HPL) may become monomer.

Frantz: Have you found in your studies of its effect on the rat tibia that injecting it in acid media makes it more effective?

Li: We have not tried that.

REFERENCES

BURR, I., GRUMBACH, M. M. and KAPLAN, S. L. (1967) *Program of The Endocrine Society' 49th Meeting*, p. 39 (abst. 22).
COTES, P. M. (1971) In *Kidney Hormones*, pp. 243–267, ed. Fisher, J. W. London: Academic Press.
COTES, P. M. and BANGHAM, D. R. (1966) *Bull. Wld. Hlth. Org.* **35,** 751–761.
HARRIS, W. H. and HEANEY, R. P. (1969) *New Engl. J. Med.* **280,** 253–259, 303–311.
JEPSON, J. H. and FRIESEN, H. G. (1967) *Ann. Intern. Med.* **66,** 1042.
JEPSON, J. H. and FRIESEN, H. G. (1968) *Br. J. Haematol.* **15,** 465–471.
MURTHY, G. and MCGARRY, E. E. (1971) *J. Clin. Endocrinol. & Metab.* **32,** 641–646.
SHULTZ, R. B. and BLIZZARD, R. M. (1966) *J. Clin. Endocrinol. & Metab.* **26,** 921–924.
SCRIVER, W. and BRYAN, A. H. (1935) *J. Clin. Invest.* **14,** 212–219.

GENERAL DISCUSSION

LUTEOTROPIC ACTIVITY AND PROLACTIN

Short: On the question of gonadotropic activity, prolactin has a very important action on the corpus luteum, and Dr Denamur and I have shown that if you hypophysectomize sheep which have previously been hysterectomized, then prolactin is essential for luteum maintenance; you cannot maintain the corpus luteum with LH alone. However, LH has an important synergistic action, so that LH+prolactin is more effective than prolactin alone (Denamur and Short 1972). The two hormones together therefore seem to make up the luteotropic complex in the sheep.

I propose that we abandon the term 'the luteotropic hormone', especially when it is used synonymously with prolactin. The evidence from several species suggests that the corpus luteum requires more than one hormone for its maintenance.

Wilhelmi: I would second these suggestions.

Li: The luteotropic effect was first shown over 30 years ago in the rat by H. M. Evans, E. B. Astwood and their co-workers, but it has now been found that the most important factor in the luteotropic effect in the ovary is ICSH (or LH). So to some extent the main luteotropic hormone in the rat is ICSH, not prolactin; so I agree with Dr Short.

Frantz: Dr Short, I wonder if you think the name 'luteotropic' hormone, or even 'luteotropic complex', is necessary in man, in view of the fact that in hypopituitary humans, pregnancy can be successfully induced with a combination of FSH and human chorionic gonadotropin (HCG)?

Short: I think that an important component of the luteotropic stimulus acts before the corpus luteum is formed; Dr Denamur has shown that you can hypophysectomize sheep soon after ovulation and the corpus luteum will continue to secrete for several days (Denamur, Martinet and Short 1966), suggesting some long-term programming of the secretory activity of the cells which is initiated before ovulation takes place, as Channing (1970) has suggested.

Frantz: In individuals who are completely hypopituitary, since neither human menopausal gonadotropin nor HCG has luteotropic activity, as far as I am aware, there could not have been that priming by a pituitary hormone.

Short: Perhaps the gonadotropins that have been given to induce ovulation have themselves programmed the granulosa cells.

Friesen: It is important to recognize that most idiopathic hypopituitary patients have normal circulating levels of prolactin, so that if prolactin has a permissive role, one cannot exclude a possible luteotropic function for human prolactin in the clinical situation described by Dr Frantz. In fact Dr W. F. Daughaday has reported one patient with hypopituitarism with a prolactin level of 500 ng/ml (at the Second International Symposium on Growth Hormone, 1971).

Li: For a long time I thought that prolactin was the luteotropic agent, but now we have two lines of evidence that it is not. When the rat ovary was incubated *in vitro* with prolactin, we, as well as other workers, were not able to stimulate the production of progesterone. Secondly, one of my colleagues, Dr N. R. Moudgal (1969), injected a single dose of antiserum to ovine ICSH into rats on day 8 of pregnancy. Pregnancy was interrupted in almost all cases. When he took pregnant rats at eight days and hypophysectomized them, and then maintained with ICSH, they went to full term, but this could not be done with prolactin alone. So it is clear that prolactin is not luteotropic. But prolactin shows a minimal luteotropic property with ICSH, either permissive or synergistic. By itself prolactin has very little role in maintaining pregnancy, or in the production of progesterone *in vitro* by the isolated corpus luteum.

Short: I think we are both seeing the same fence from different sides! Firstly, we shouldn't expect a luteotropic hormone necessarily to stimulate the secretion of progesterone. It may just act by prolonging the life of the corpus luteum, and our evidence in sheep suggests that it acts in this way. The effect of an antiserum to LH terminating the life of corpora lutea in the rat is, I think, quite understandable. Certainly in the sheep you cannot maintain a fully functional corpus luteum with prolactin alone, in the absence of LH. And in pregnant rats, we have known for some time that the placenta secretes a prolactin-like hormone, so it is a difficult situation to analyse.

Meites: It should be said in defence of the luteotropic action of prolactin, Dr Li, that before day 8 of pregnancy in the rat, prolactin is essential. If you hypophysectomize a pregnant rat early in pregnancy, then pregnancy is terminated. We have shown (see Meites 1970) that if you inhibit prolactin secretion by implanting prolactin in the hypothalamus of rats early in pregnancy or pseudopregnancy, that terminates pregnancy or pseudopregnancy. In addition, we should be aware that prolactin may have a luteolytic as well as a luteotropic action. It has been pointed out by Malven (1965) and our laboratory (Piacsek and Meites 1967) in hypophysectomized rats with corpora lutea that prolactin can induce the disappearance of the corpora lutea. As I shall show, the major action of prolactin during the oestrous cycle of the rat appears to be to induce luteolysis of the previous crop of corpora lutea (see p. 394).

Spellacy: In support of Dr Frantz's comment, Vande Wiele and co-workers (1970) have presented a case of a hypophysectomized woman whose corpora lutea function was normal with only FSH and LH replacement therapy, demonstrating that prolactin is probably unnecessary.

Nicoll: Dr Short suggested earlier that the persistence of luteal function after hypophysectomy in sheep indicates that prolactin and perhaps LH have programmed the tissue to function for some time. Dr Turkington, have you tried adding prolactin to your *in vitro* mammary gland system and then removing it, to determine how long the biochemical effects persist after taking away the hormonal stimulus?

Turkington: We haven't done that, partly because the durations of these responses differ very much. For example, some RNA's might only last half an hour, whereas others, such as the transfer RNA's, might persist for several days. I'm not sure that one could tell from the decay of the induced response when the hormone was acting, or stopped acting.

Nicoll: We need to know the rate of change of these various parameters, particularly the final product (the rate of casein synthesis), in order to know how long it takes for the effect of prolactin to wear off after the hormonal stimulus is removed. You have a definite lag period in your induction of casein synthesis of about four hours. There is evidence from the toad bladder that you can expose the tissue to aldosterone and remove it, and the biochemical effects persist over an extended period in the absence of the hormone.

Greenwood: I asked Dr Beck this six years ago and the answer was that if you inject someone with growth hormone and then remove the hormone after say 24 hours, it's as if the patient had never received growth hormone.

Beck: I don't think I've ever said that! It depends which parameter you look at. If you look at calcium metabolism the effects of injected HGH seem to persist for long periods of time after HGH administration ceases, whereas the BUN for example promptly returns to the preinjection control value.

Greenwood: Isn't Dr Nicoll asking whether a short exposure of a tissue to a hormone can programme the tissue and hence produce effects long after the hormone is presumed to have gone—that is, some effect akin to differentiation?

Turkington: We studied the induction of the cyclic AMP-activated protein kinase, which is induced within minutes. When we induce it and then remove prolactin, the enzyme decays very slowly, over a period of hours. But I don't see how that helps us to know when the hormones stop acting.

Nicoll: On the basis of that one parameter (protein kinase), the latency of the effect is rather short. But for the synthesis of casein, which is one of the final steps in the sequence of events, the latency is quite long, over a period of several hours.

Turkington: That is more a function of how the cell manages its metabolism, rather than the duration of a discrete stimulus.

Nicoll: Yes, but it all ultimately depends on the initial stimulus.

A LUTEOLYTIC FUNCTION FOR PROLACTIN

Meites: We have evidence to suggest that prolactin has a luteolytic function during the oestrous cycle. Our assays of the three hormones, prolactin, FSH and LH during the oestrous cycle of the rat show that all three hormones reach a peak level on the afternoon of pro-oestrus, in agreement with the results of Gay, Midgley and Niswender (1970) and others. FSH and LH presumably have a function in stimulating ovulation and the formation of the corpus luteum. What is the function of prolactin? Our evidence is that its function at this stage of the cycle is to induce luteolysis of the previous crop of corpora lutea.

In our control rats, serum levels of prolactin on the afternoon of pro-oestrus rise steadily, reach a peak at about 5 p.m. and remain high until about 7 or 8 p.m. and then go down gradually. They are still higher on the day of oestrus than during the dioestrous phase. A single dose of ergocornine injected at 1·30 p.m. on the day of pro-oestrus completely prevents the rise in serum prolactin (Wuttke, Cassell and Meites 1971). In earlier experiments with Dr Nagasawa (Nagasawa and Meites 1970) we found that long-term injection of ergocornine did not interfere with the regularity of the cycle, but when we killed the rats they had very large ovaries full of corpora lutea. So we did a series of experiments in which we injected the rats with ergocornine for three days during one cycle, starting on the last day of dioestrus, and killed them on the next first day of dioestrus. Other rats were similarly injected with ergocornine for three days, but on the afternoon of pro-oestrus we also gave a single injection of 1 mg of NIH ovine prolactin (28 i.u. per mg). The ovaries of the rats given ergocornine alone were heavier than the ovaries of the controls; in the rats given ergocornine and one injection of prolactin, the ovaries weighed the same as in saline-injected controls. The average number of corpora lutea showed a significant increase in rats injected with ergocornine alone, and a normal number of corpora lutea in rats given ergocornine together with a single injection of prolactin.

In another experiment we injected ergocornine daily for three oestrous cycles starting on the last day of dioestrus, and killed the rats on the first day of the next dioestrus. Other rats were given ergocornine every day during the three cycles but in addition received injections of NIH ovine prolactin (1 mg) on the afternoon of pro-oestrus and on the day of oestrus; thus these rats received a total of six injections of 1 mg prolactin each during the three cycles. The ergocornine-treated rats had heavier ovaries than those treated with ergocornine for one cycle, and a significantly larger number of corpora lutea. Prolactin brought the ovaries essentially back to normal and prevented the increase in the number of corpora lutea. I believe these experiments show that prolactin has a luteolytic function during the oestrous cycle.

It has been known for many years that if a rat has corpora lutea at the time of hypophysectomy, the corpora lutea will persist although they are non-functional. If one injects prolactin (Malven 1965) or transplants a pituitary (Piacsek and Meites 1967) into these hypophysectomized animals, the corpora lutea disappear. Presumably this is what prolactin does in the normal cycling rat to the old corpora from the previous cycle, and in this way prevents a gradual build-up of luteal tissue. Rats ovulate up to 15 or more follicles at one time and if you continue to give ergocornine for many cycles, an enormous number of corpora lutea appear in the ovaries. One doesn't know what this would eventually do to ovarian function, but apparently prolactin prevents this from occurring.

Short: These data certainly look very convincing. But the rise in prolactin secretion that one gets in sheep during pro-oestrus occurs after the corpus luteum has regressed.

Friesen: Mathies and Lyons (1971) have shown the same thing with rat placental lactogen in the rat; extracts of placentas from the rat caused luteolysis of corpora lutea when given in large amounts.

Greenwood: In the rat do you find prolactin throughout the whole of the cycle as we did in the sheep and goat, with a surging as the LH peak comes down?

Meites: Yes. In the rat we find a rise (from a basal level during dioestrus) on the morning of pro-oestrus; it reaches a peak on the afternoon of pro-oestrus and then declines, but it is still higher on the day of oestrus than on the days of dioestrus (Amenomori, Chen and Meites 1970).

Bryant: Does the LH rise before prolactin?

Meites: On the contrary, prolactin starts to rise before LH (Wuttke and Meites 1970). The rise in LH in rats occurs only during a period of about two hours, between 5 and 7 p.m., and if you miss it then you see no rise during the rest of the cycle.

Sherwood: From the evidence you presented, Dr Meites, prolactin inhibits its own secretion but stimulates LH and FSH. The so-called positive feedback that triggers the LH surge could be mediated by prolactin. Is it possible that there is a sequence oestrogen, prolactin, LH?

Meites: We have wondered about this, and it is true that prolactin increases before either LH or FSH in the rat, so that possibility cannot be ruled out.

Greenwood: It's not so in the sheep, I think. Oestrogen induces LH and then prolactin, provided initially the progesterone is low, so my reading is: high oestrogen, low progesterone, producing LH followed by prolactin release.

Denamur: What is the mechanism inducing pseudopregnancy in the rat?

Meites: Prolactin is needed for pseudopregnancy but so is LH. If you stimulate the uterine cervix on the morning of oestrus, you get pseudopregnancy and soon afterwards there is a rise in prolactin and LH (W. Wuttke, M. Gelato and J. Meites, unpublished). If we give ergocornine we can prevent the rise in prolactin but not LH, but we still get pseudopregnancy. We conclude that one can obtain pseudopregnancy in the presence of very low levels of prolactin provided that there is an adequate amount of LH, and that the two hormones are both involved in the luteotropic process. Prolactin during pseudopregnancy is very low (20 ng/ml serum), even lower than during dioestrus. There is no question that some prolactin is necessary, because hypophysectomy, or a prolactin implant in the median eminence to shut off prolactin, or a large dose of ergocornine, can terminate pseudopregnancy.

Denamur: Are the corpora lutea from the previous cycle completely regressed at the beginning of pseudopregnancy?

Meites: At the beginning of pseudopregnancy they have probably regressed, presumably as a result of the rise in prolactin on the afternoon of pro-oestrus.

Spellacy: Dr Meites, is there any evidence that prolactin has any effect on prostaglandins?

Meites: I know of none.

REFERENCE PREPARATIONS OF HUMAN PROLACTIN

Beck: It is becoming critical that a reference standard preparation of human prolactin for both the biological and immunological assay systems be developed. I think Dr Wilhelmi is the most senior statesman on reference standards!

Wilhelmi: I think that the strict rule should be obeyed that if you are assaying human prolactin, human material should be used as a reference standard, so that we need a suitable source of human prolactin. It seems to me that the bioassays being developed that are most useful are those like Dr Frantz's and Dr Turkington's. They are very sensitive assays, so one could almost use the same reference material for both the immunoassays and these more sensitive biological assays. One possibility would be to follow up Dr Greenwood's suggestion (p. 218) that we approach Dr G. Ross to make it known, since he is willing to distribute this material, that there is available a human plasma containing a significant amount of prolactin activity, which could be used as a reference material. It could then be assayed by both systems and a small meeting could try to agree on a unit. When enough human prolactin has been isolated, then another kind of reference material can be provided.

Cotes: I agree that like must be compared with like. A second important attribute of a standard is that it should be stable. If a standard is to be handled in more than a very small number of laboratories, it is unsatisfactory to have to send it as a frozen solution. Inevitably, some delays occur with air freight services. Thought should be given to the possibility of preparing a dried batch of serum; the stability of such a preparation can be tested more easily and it can be made available to a larger number of laboratories.

Turkington: We have had the opportunity to assay sera that have been stored for progressively longer periods of time (one to five years) and we found that with storage of serum, its biological activity falls off progressively. Dr Ross's material (plasma) has been frozen for a couple of years already, and there is a possibility that its immunoreactivity may be quite different from the remaining biological activity. But since there is the possibility that in the next year or two human prolactin will be available as a standard in a relatively pure form, it might be better to wait for that rather than to use serum that has been frozen. Alternatively, it would be feasible to prepare in large quantity a serum that has been freshly drawn, because such serum is not hard to come by. There are patients taking phenothiazines who have as much as 1000 ng/ml of prolactin activity or more in their serum, and it would be possible to get several litres of that over a short period of time.

Bryant: We have found that ovine prolactin in serum decreases in immunoactivity over two or three years, stored frozen at $-20°C$, without enzyme inhibitors.

Friesen: Dr Ross's plasma may be particularly bad as a standard if you find loss of biological activity with prolonged storage, because there may be a marked dissociation between the immunoassay and bioassay in a sample that has been stored for a long time. A fresh sample would be better as a standard.

Wilhelmi: I would agree with that. I didn't know how old Dr Ross's material was.

Friesen: We found that serum left out inadvertently in the lab. for four days showed very little difference in immunological activity upon re-assay, so over a short period of time the stability of human prolactin appears good.

Greenwood: Dr Bryant assayed some crucial human plasmas for prolactin in January 1970, and Dr Siler got the same results nine months later.

There are already about ten groups that would like a standard preparation now. Dr Cotes has a good point, but it will take time for such a serum to be pooled, lyophilized and distributed. I think probably Dr Ross's plasma would do in the interval before a pituitary preparation becomes available.

Dr Cotes, how soon could a standard plasma pool be made available?

Cotes: There are large numbers of subjects taking phenothiazine drugs from whom it would be easy to get plasma or serum. Since the present number of interested laboratories is small, a batch of several hundred ampoules of serum would probably meet the interim need.* We should appreciate the collaboration of people in the group to make preliminary checks of the suitability of the serum. At this stage, an interim yardstick with some likelihood of its being stable and which could eventually be calibrated in terms of pure preparations would have big advantages.

Beck: I think the advantages are very obvious. The problem then is how to aid Dr Cotes in securing an adequate supply of lyophilized samples.

Cotes: Tell us first whether it should be serum or plasma. Secondly, is there any necessity for enzyme inhibitors to be added, or should people who are running either tissue culture systems or immunoassay systems add what they need themselves?

Frantz: I think no additives would be better for the bioassays, and serum is preferable because it's easier to handle.

Wilhelmi: Dr Turkington, do you know whether the serum activity stands up to lyophilization?

Turkington: No, we haven't really studied that point.

Forsyth: It does in partially purified fractions; I don't know about whole serum. We have put human plasma through Sephadex G-100 columns,

---

* A batch of some 300 ampoules each containing the freeze-dried residue of 1 ml serum from subjects under treatment with phenothiazine has been prepared. Ampoules of this preparation (71-167) may be obtained on request to the Director, Division of Biological Standards, National Institute for Medical Research, Mill Hill, London NW7 1AA, England.

GENERAL DISCUSSION

lyophilized the fractions, and collected and assayed them in our *in vitro* bioassay (Forsyth and Myres 1971). We certainly recovered substantial activity, but because our assay is only semi-quantitative I can't estimate accurately losses on the column or in lyophilization.

Bryant: I made up pools of lyophilized sheep plasma and sent it to various people who wanted high and low ovine prolactin standards. I assayed it before and after lyophilization and there was no difference. They found the same values after postage to various parts of the world.

Beck: It will be critical to be certain that Dr Frantz's and Dr Turkington's biological assays are uninfluenced by lyophilization of plasma or serum. It will also be necessary for someone to measure the levels in serum from phenothiazine-treated subjects which Dr Cotes might acquire.

Frantz: I would be glad to do that.

Friesen: We would be anxious to do that as well.

Greenwood: We would like to do this, so you would have two bioassays at least and two radioimmunoassays.

Friesen: It would also be important to have a pituitary prolactin extract as an interim preparation even if it is relatively crude, because there clearly is a suggestion of some difference between serum prolactin and pituitary prolactin.

Sherwood: Do you have evidence of differences in the immunoreactivity of the crude pituitary extract and serum? I have not detected any.

Friesen: No; they are parallel in our immunoassay. These are differences in results between some of Dr Turkington's bioassay data and ours. They may be due to differences in the assay rather than the preparation. But I think the possibility of a difference between serum and pituitary preparations is a real one until it can be ruled out. We could certainly provide a fair amount of pituitary material relatively free of growth hormone, because the latter may also produce effects in the bioassay.

Turkington: I think that material would not be suitable for bioassay; it probably contains enzymes and other contaminants which might destroy the active materials during the incubation that's necessary.

Friesen: Have you ever re-assayed a preparation at the end of the assay? How much of the material disappears during incubation?

Turkington: I've not done that experiment; I probably ought to.

Greenwood: We did that with Dr Prop some years ago. Sheep prolactin was added to mouse mammary gland cultures, and after five days of incubation there was no loss of prolactin as measured by radioimmunoassay, so there was no apparent uptake of the hormone by the tissue despite a histological response! In fact in a few instances I think we found more 'prolactin' at the end than we started with.

Beck: Certainly if one looks at the membrane receptor studies on insulin, it seems that this hormone can exert its biological action and then be removed in the same amount as was added.

HUMAN PLACENTAL LACTOGEN: NOMENCLATURE

Sherwood: I think it important that we come to some decision on the terminology of the placental hormone known both as human placental lactogen and also, since 1967, as human chorionic somatomammotropin (see Li *et al.* 1968). The editors of the endocrinological journals would also like us to decide. I feel strongly that we should keep the original name, human placental lactogen, for two reasons, one historical and the other physiological. John Josimovich would be the last person to insist on keeping his original name (Josimovich and MacLaren 1962), but we still don't know for certain the major role of this hormone in pregnancy. We know that it is a potent lactogenic hormone but we don't know how important its growth-promoting activity is. There are very high concentrations of HPL in pregnancy, but from Dr John Beck's data, ovine prolactin has much more somatotropic activity than HPL per unit weight. Although chemically it is very similar to growth hormone, I am reluctant at this time to call it a 'somatotropin'. If new functions are found or the function redefined, the name might be changed again. We should therefore keep the original name until more data are available.

Li: I was really the principal father of 'human chorionic somatomammotropin', in Siena in 1967, and I don't want necessarily to defend it, but to clarify why the name was chosen. Of course hormone functions change all the time as our knowledge becomes wider, and we do not know what HCS does because it's a new hormone, only purified (to some extent) in 1962. I proposed this name because 'human chorionic gonadotropin' (HCG) had already been accepted as a name for that human placental hormone and it seemed consistent to have a series of terms for other placental hormones, as they are found: we can expect human chorionic melanotropin (HCM), human chorionic thyrotropin (HCT) and human chorionic lipotropin (HCL). The structure of this hormone is very close to human growth hormone (HGH); 160 of the 190 amino acids are the same. We don't know what it does yet, but chemically it is very close to growth hormone, and the name reflects this.

Greenwood: I'm standing in for Dr Grumbach here! Dr Friesen has shown that human prolactin rises in pregnancy and so we do not need to postulate a lactogenic function for HPL. Originally I liked Dr Josimovich's term; I then preferred Dr Friesen's 'human placental protein'.

We have at least advanced to the stage where we could call it 'human placental hormone'. Until it is really pinned down I would prefer to go by history and priority and call it HPL.

Friesen: I have always been non-committal about the terminology, except that I do like Josimovich's simple term.

Frantz: I agree. And it's much easier to use!

Turkington: Dr Josimovich should certainly be given the credit; he discovered the hormone to a certain extent and named it in accordance with the knowledge available to him then. On the other hand, we have to recognize that we now have these two sets of hormones, pituitary hormones and placental, or chorionic hormones, and some consistency in naming the placental hormones would be in order. In that sense, the term 'chorionic' in the name seems appropriate. It seems to me that 'chorionic prolactin' would designate a major activity of the hormone that has been clearly demonstrated.

Wilhelmi: This discussion creates a problem for me, because I am distributing this substance. One technical problem was how to get 'human chorionic somatomammotropin' on a small label! As a (temporary) solution to the problem, the label now reads 'human placental lactogen', but this could be altered. However, the NIH material will go now to many people, a usage will be created, and unless a decision is taken now, at the start of distribution, that can be acted upon and will have a broad influence, usage will determine the name regardless of the editors of *Endocrinology* or any of us. So I seek your advice.

Li: I think HCS has been extensively used in Europe. HPL is also used widely among many workers. Lederle has adopted the term Purified Placental Protein (Human).

Greenwood: I agree with you that it's chorionic, we know it's human, but we do not know whether it's a lactogenic or growth hormone or has some other function.

Spellacy: It seems clear from the earlier discussions that you can't talk about pituitary prolactin as a growth factor since it doesn't produce growth, and the same logic can be applied to the placental hormone. The literature is very confusing now with four or five names being given to the placental hormone and it would be logical and simpler to go back to an old and appropriate term rather than create a new one. I believe that 'human placental lactogen' should be the choice.

Beck: This is in essence what Dr Wilhelmi through his distribution system is doing and I think it would be useful for editors as well as for many others that we advise him to do this, because his influence with respect to labelling on what people do in the next ten years is probably greater than anyone else's.

Forsyth: I have a prejudice in favour of human placental lactogen, but I can see a problem over contraction, because human pituitary prolactin may also be abbreviated to HPL.

Li: The chief problem remains: what is the function of this hormone; what is a lactogen doing to the foetus, in the placenta? The name human chorionic somatomammotropin gives an indication that the hormone increases as the size of the placenta increases, which suggest some connexion with the growth of the foetus.

Meites: Personally, I would be in favour of giving Josimovich the priority and calling it 'human placental lactogen', since he did the initial work and most of the activity described is lactogenic.

Beck: We appear to have a general consensus, except for C. H. Li abstaining, that HPL be retained until the physiological role of this hormone is determined.

REFERENCES

AMENOMORI, Y., CHEN, C. L. and MEITES, J. (1970) *Endocrinology* **86,** 506–510.
CHANNING, C. P. (1970) *Recent Prog. Horm. Res.* **26,** 589–622.
DENAMUR, R., MARTINET, J. and SHORT, R. V. (1966) *Acta Endocrinol. (Copenh.)* **52,** 72–90.
DENAMUR, R. and SHORT, R. V. (1972) Proceedings N.I.H. Conference, Washington, Sept. 1970. In press.
FORSYTH, I. A. and MYRES, R. P. (1971) *J. Endocrinol.* **51,** 157–168.
GAY, V. L., MIDGLEY, A. R., JR and NISWENDER, G. D. (1970) *Fed. Proc. Fed. Am. Soc. Exp. Biol.* **29,** 1880–1887.
JOSIMOVICH, J. and MACLAREN, J. A. (1962) *Endocrinology* **71,** 209–220.
LI, C. H., GRUMBACH, M. M., KAPLAN, S. L., JOSIMOVICH, J. B., FRIESEN, H. and CATT, K. J. (1968) *Experientia* **24,** 1288.
MALVEN, P. V. (1965) *Anat. Rec.* **151,** 381.
MATHIES, D. L. and LYONS, W. R. (1971) *Proc. Soc. Exp. Biol. & Med.* **136,** 520–523.
MEITES, J. (1970) In *Hypophysiotropic Hormones of the Hypothalamus: Assay and Chemistry,* pp. 261–281, ed. Meites, J. Baltimore: Williams and Wilkins.
MOUDGAL, N. R. (1969) *Nature (Lond.)* **222,** 286–287.
NAGASAWA, H. and MEITES, J. (1970) *Proc. Soc. Exp. Biol. & Med.* **135,** 469–472.
PIACSEK, B. E. and MEITES, J. (1967) *Neuroendocrinology* **2,** 129–137.
VANDE WIELE, R. L., BOGUMIL, J., DYRENFURTH, I., FERIN, M., JEWELEWICZ, R., WARREN, M., RIZKALLAH, T. and MIKHAIL, G. (1970) *Recent Prog. Horm. Res.* **26,** 63–95.
WUTTKE, W., CASSELL, E. and MEITES, J. (1971) *Endocrinology* **88,** 737–741.
WUTTKE, W. and MEITES, J. (1970) *Proc. Soc. Exp. Biol. & Med.* **135,** 648–652.

CONCLUSIONS

J. C. BECK

In my introduction I pointed out that advances in technology and the advent of a new generation of creative younger scientists, building upon information acquired by their elders, have set the stage for a rapid advance in the area of the lactogenic hormones. I think that this meeting has demonstrated this phenomenon to an unusual degree. It is difficult and perhaps unnecessary to summarize the contents of the symposium; perhaps more appropriate would be the delineation of certain areas which deserve special attention in the future. During the discussion of the chemistry of the lactogenic hormones it became clear that a re-evaluation was needed of the methods of bioassay for growth-promoting and lactogenic activities in routine use in laboratories concerned with the chemical aspects of the problem. I fully recognize the problem of the small amounts of material available, but I also feel that there is a danger of being misled if only the tibia test and the pigeon crop sac are used to evaluate the biological activity of the hormone fragments obtained by cleavage or by subsequent synthesis, and I think this is an area where help is critically needed.

Before moving on to another focal area, I should also like to urge Dr Li, Dr Sherwood and Dr Niall to examine together in detail their respective proposals for the structure of the lactogenic hormones. It is also clear that information on the three-dimensional structure of these molecules is critically needed and I hope that crystals of these hormones may soon be available for study by colleagues in X-ray crystallography. I suspect that those of you with chemical capabilities will rapidly begin, or more probably have already begun, the synthesis of fragments of the parent molecules. It would appear from the evidence presented in this symposium that these functions are probably located in different portions of the molecule with only slight or no overlap. The synthesis of a fragment with the biological activity of growth hormone would be an exciting and important achievement.

Another area which has required attention and was discussed (pp. 396–400) is the need for a series of reference standards for both the bioassay and the immunoassay of the lactogenic hormones. This becomes crucial not only from the point of view of the obvious contribution it would make to the research aspects of this problem, but also when these findings are brought into general clinical use. It is at the clinical level that there is enormous pressure, from the point of view of HPL and a reference preparation. In the same vein, I believe that it is important that those of us concerned with human pituitary collection and extraction programmes should modify our methods so that human prolactin, which undoubtedly seems to be here to stay, can be made available for future chemical and biological characterization.

It also seems appropriate to mention for further discussion the question of whether the nomenclature currently in use for the lactogenic hormones is the ideal one or whether revision is justifiable, and again our discussion may be able to influence this point. I personally have had misgivings about the designation 'human chorionic somatomammotropin' (HCS), and our general consensus in favour of human placental lactogen (HPL) may help to resolve the situation.

It also became clear during the meeting that knowledge about the control mechanisms for the synthesis and release of the lactogenic hormones is very primitive indeed, perhaps particularly with respect to the placental hormone, where at the moment there seem to be no appropriate data. The physiological role of HPL and prolactin in man requires much clarification, and this obviously remains a fertile area for future investigation.

I would be remiss if I didn't express my concern about the use of normal male plasma or serum in the new and very elegant bioassay systems for prolactin. With the information that such serum has appreciable prolactin content, as measured by immunoassay, one wonders about the wisdom of this manoeuvre in further attempts to increase the sensitivity of the assays.

In conclusion, I leave this symposium with great optimism. I had hoped for a good meeting, but had constantly been plagued by the question of its timing. However, the information presented has assured me that it was held at a highly appropriate time. The lactogenic hormones are clearly the new 'thing' in endocrinology, and with the many new investigators undoubtedly joining us in the race, the opportunities for small interdisciplinary meetings such as this rapidly fade on the horizon. I believe this opportunity given by the Foundation to meet and to communicate with each other will serve as a source of great strength in the future development of knowledge in this important area of endocrinology.

INDEX OF CONTRIBUTORS*

Page numbers in bold type indicate papers; other page numbers are contributions to the discussions.

Apostolakis, M. . . . **349**, 357
Beck, J. C. **1**, 24, 25, 49, 77, 78, 104, 128, 129, 134, 187, 194, 219, 220, 238, 239, 322, 355, 356, 357, **361**, 383, 384, 385, 386, 387, 388, 396, 398, 399, 400, 401, 402, **403**
Beck, J. Swanson 103, 237, 277, 278, **287**, 294, 295, 296
Belanger, C. **83**
Bern, H. A. **299**
Bryant, Gillian D. 25, 48, 193, 196, **197**, 217, 218, 220, 279, 395, 397, 399
Cotes, P. Mary 47, 49, 236, 286, 355, 357, 386, 397, 398
Cowie, A. T. 50, 133, 187, 189, 191, 319, 323, 342, 355, 357
Denamur, R. 128, 132, 190, 317, 321, 342, 345, 346, 396
Forsyth, Isabel A. 46, 48, 49, 106, 133, **151**, 184, 185, 188, 195, 220, 221, 319, 398, 402
Frantz, A. G. 24, 49, 105, 106, 108, 129, **137**, 184, 185, 186, 187, 191, 194, 195, 196, 217, 220, 221, 222, 236, 237, 239, 278, 281, 318, 341, 357, 385, 388, 391, 398, 399, 401
Friesen, H. 24, 25, 48, 50, 80, 81, **83**, 103, 104, 105, 106, 107, 108, 109, 110, 129, 130, 134, 185, 188, 194, 196, 218, 220, 222, 235, 237, 239, 280, 281, 282, 283, 285, 294, 295, 317, 342, 355, 356, 357, 386, 388, 392, 395, 397, 398, 399, 401
Greenwood, F. C. 25, 45, 47, 49, 50, 80, 82, 105, 108, 109, 128, 129, 133, 185, 191, 192, 194, 195, **197**, **207**, 217, 218, 219, 220, 221, 236, 237, 238, 239, 278, 281, 282, 285, 286, 295, 296, 317, 318, 323, 324, 339, 342, 345, 346, 347, 355, 356, 357, 384, 386, 393, 395, 396, 398, 399, 400, 401
Guyda, H. **83**
Handwerger, S. **27**
Herlant, M. . . 196, 281, 296, 319
Hwang, P. **83**
Kapetanakis, S. **349**
Kleinberg, D. L. **137**
Lazos, G. **349**
Lehmeyer, Joyce E.. . . . **53**

Li, C. H. **7**, 23, 24, 25, 46, 49, 50, 77, 105, 195, 237, 283, 285, 322, 342, 343, 386, 387, 388, 391, 392, 400, 401, 402
McGarry, E. E. **361**
McLaurin, W. D. **27**
MacLeod, R. M. **53**, 76, 77, 78, 79, 80, 81, 109, 130, 186, 191, 237, 323, 340, 342, 344, 347, 355, 356
Madena-Pyrgaki, A. . . . **349**
Meites, J. 48, 78, 80, 81, 104, 110, 132, 185, 187, 188, 190, 191, 192, 220, 221, 222, 238, 283, 294, 320, 321, 322, 323, **325**, 338, 339, 340, 341, 342, 343, 344, 345, 346, 347, 356, 357, 392, 394, 395, 396, 402
Morgenstern, L. L. **207**
Nicoll, C. S. 23, 24, 76, 77, 79, 81, 108, 131, 188, 218, 220, 237, **257**, 279, 281, 282, 283, 284, 285, 286, **299**, 317, 318, 319, 320, 321, 322, 323, 324, 343, 344, 345, 346, 383, 384, 393, 394
Noel, G. L. **137**
Pasteels, J. L. 78, 80, 106, 131, 184, 219, **241**, **269**, 277, 278, 279, 280, 281, 284, 295, 339, 340, 342
Prop, F. J. A. 130, 131, 184, 185, 189, 192, 193
Sherwood, L. M. 22, 24, **27**, 45, 46, 48, 49, 76, 78, 105, 110, 127, 128, 132, 191, 194, 218, 219, 220, 221, 236, 239, 279, 283, 285, 295, 296, 317, 338, 347, 355, 356, 383, 386, 396, 399, 400
Short, R. V. 131, 319, 357, 391, 392, 395
Siler, Theresa M. **207**
Spellacy, W. N. 79, 103, 194, 219, **223**, 236, 237, 238, 239, 318, 354, 355, 393, 396, 401
Turkington, R. W. 25, 45, 76, 77, 81, 106, 107, 109, **111**, 127, 128, 129, 130, 132, 133, 134, **169**, 184, 185, 186, 187, 188, 189, 190, 191, 192, 193, 194, 217, 218, 219, 236, 237, 280, 281, 283, 286, 295, 296, 318, 319, 355, 385, 387, 391, 393, 394, 397, 398, 399, 401
Wilhelmi, A. E. 23, 25, 50, 109, 110, 192, 218, 236, 283, 295, 296, 321, 322, 357, 384, 385, 397, 398, 401

*Indexes prepared by William Hill

INDEX OF SUBJECTS

Abortion, human placental lactogen levels, 230
Acetylcholine, effect on prolactin secretion, 328
Achondroplasia, 370
Acromegaly,
   lactogenic activity, 155
   prolactin secretion, 92, 98, 145, 147, 258
Adenyl cyclase in prolactin release, 76
Adrenaline,
   effect on pituitary structure, 61
   effect on prolactin secretion, 62, 81, 328
ACTH, release, 210, 216
Aldosterone,
   excretion, effect of human placental lactogen, 375
   secretion, effect of prolactin, 379
Alveolar cells, effect of prolactin, 187–189
Amino acid sequences,
   human growth hormone, 7, 18, 19, 20, 380
      compared with human placental lactogen, 48
   human placental lactogen, 15, 16, 18, 19, 20, 33, 34, 35, 37, 40, 45
      compared with growth hormone, 39, 40
   monkey placental lactogen, 48
   ovine prolactin, 12, 13, 18, 19, 20
   prolactin, 215, 380
Amino acid substitutions in human placental lactogen, 41
Amitriptyline, effect on prolactin levels, 178, 349
Amphibians, prolactin action in, 302, 304, 308, 317
Androgen, synergism with prolactin, 308
Asthma, 370

Birds, prolactin action in, 302, 306, 308
Blood,
   lactogenic activity, detection, 151–167
   prolactin activity, 274
Bone marrow,
   effect of growth hormone, 386
   effect of prolactin, 369

Brain lesions, effect on prolactin and growth hormone synthesis, 68, 73
Brain tumours, prolactin levels, 141
Breast,
   as target organ for prolactin, 194, 219
   effect of growth hormone, 112
   growth in neonates, 220
   prolactin action on, 124–126
      growth, 187–193
   steroid action on, 133
   stimulation,
      by placenta, 48
      prolactin secretion and, 271, 275
   tissue from mid-pregnant mouse as assay of prolactin, 137
Breast cancer, 317
   inhibition, 74
   prolactin activity, 110, 185, 195
      following pituitary stalk section, 175
   tranquillizers and, 356

Cadaverine, 62
Calcium,
   effect on prolactin secretion, 55–56
   excretion,
      effect of growth hormone, 384
      effect of HGH and prolactin, 379
   metabolism,
      effect of growth hormone, 393
      effect of human placental lactogen, 374
      effect of prolactin, 364, 366, 370, 372
Carbohydrate metabolism,
   effect of bovine growth hormone, 378
   effect of human placental lactogen, 375
   effect of prolactin, 368, 372, 373, 378
Casein,
   synthesis, 115, 116, 123
      effect of prolactin, 169, 393
      in Forbes-Albright syndrome, 173
Catecholamines,
   effect on lactation, 342
   effect on prolactin secretion, 58–64, 73, 78, 331–334, 346
   in hypothalamus, 58, 65, 73, 78–79, 80–81, 340
Cell membranes,
   growth hormone binding to, 128, 129

INDEX OF SUBJECTS

Cell membranes—*continued*
  human placental lactogen binding to, 127, 128, 129
  prolactin-binding sites, 112–113, 118, 124, 127, 128
  prolactin stimulation of, 114
Chemistry of lactogenic hormones, 7–26
Chickens, broodiness in, 319
Chlorpromazine, 349
  effect on pituitary hormones, 346
  effect on prolactin secretion, 66, 143, 187, 330, 331
Chlorprothixine, 349
Chromatography, affinity, 94
Chymotrypsin,
  cleavage of human placental lactogen, 28, 40
  peptides from, 31, 37
Co-lactogens, 3
Colchicine, effect on prolactin secretion, 279–280
Corpus luteum, effect of prolactin, 391–396
Cortisol, effect on prolactin activity, 185
Cyanogen bromide cleavage,
  of human placental lactogen, 29, 40, 45
  peptides from, 36
Cyclic AMP, activating protein kinase, 116, 130
Cyclic AMP-binding protein, 116, 130
Cystine residues in lactogenic hormones, 23

Deoxyribonucleic acid,
  synthesis,
    during lactation, 192
    effect of prolactin, 190
Deoxyribonucleic acid-RNA hybridization, 120
Diabetes, 88
  hypophysectomy, effect of prolactin, 368
  in pregnancy, human placental lactogen levels, 230
  retinopathy in,
    prolactin levels, 175, 186
Dibutyryl cyclic AMP,
  and mechanism of action of prolactin, 130
  prolactin synthesis and, 76
Digitalis, 357
Dimethylnaphthylene sulphonyl chloride, 31
Disulphide bridges, 38
  disruption, 28
  in human growth hormone, 10, 22
  in human placental lactogen, 15, 22

Disulphide bridges—*continued*
  in prolactin, 12, 23
Dopa, effect on prolactin secretion, 342, 347
Dopamine, effect on prolactin secretion, 60, 65, 73, 79, 107, 178, 329, 331, 332, 334, 347

Eating, effect on human placental lactogen levels, 226
Eclampsia, 88, 318
Edman degradation,
  of human placental lactogen, 31, 46
  peptides from, 37
Electron microscopy of pituitary cells, 61, 248–252
Ergot derivatives,
  effect on prolactin secretion, 68–72, 74, 80, 251, 320, 335, 394
Erythropoiesis,
  effect of prolactin and growth hormone, 386
Evolution,
  of prolactin, 299 *et seq.*, 318, 320
  placental lactogens in, 48

Fasting, effect on human placental lactogen synthesis, 89
Fish, prolactin action in, 304, 320
Fluphenazine, 349
  effect on prolactin levels, 178
Foetus,
  anencephalic, human placental lactogen in mothers, 237
  malnutrition, 88
  pituitary differentiation in, 216
  prolactin levels, 99, 219, 281
Folley, Sydney John, 1–5
Follicle-stimulating hormone,
  assay, 208
  release of, 210, 216, 334
Forbes–Albright syndrome,
  prolactin secretion in, 173, 176

Galactorrhoea,
  complicating drug therapy, 178, 186, 350
  lactogenic activity in blood, 155
  α_{2x} milk protein in, 352
  prolactin activity, 92, 98, 108, 109, 141, 147, 148, 161, 178, 185, 201–202, 349–359
  with normal menses, 143
Gastrin, pentapeptide, 45

Glucose,
  and human placental lactogen synthesis, 103, 226-228
  effect on prolactin release, 105, 203
Goats, pregnant and lactating, plasma assay, 155-161
Gonadal dysgenesis, 370
Growth,
  action of prolactin, 21, 23, 24, 306-308, 321, 383
  effect of prolactin and HPL in hypopituitary dwarf, 383
Growth hormone, 7, 322
  activity, 46-47, 131
  amino acid sequences, compared with placental lactogen, 18-21, 39, 40
  assay, 144
  effect on calcium metabolism, 384-385
  inhibition of action, 322
  in prepubertal hypopituitarism, 363
  lactogenic activity, intrinsic, 10, 144-145, 151 et seq.
  neutralization with anti-HGH serum, 271
  osteoporosis and, 385
  precursor, 80
  release, effect of ionic environment, 56
  secretion, 343
    by pituitary tumours, 70
    effect of suckling, 284
    in lactation, 147-148, 283
    stimulation, 205
  separation from prolactin, 94, 361
  stimulating mammary epithelial cells, 112
  structure, 7-10, 311, 323
  synthesis, site of, 101, 295
Growth hormone, bovine, 3
  activity compared with prolactin, 24
  effect on carbohydrate metabolism, 378
  effect on rabbit mammary gland, 195
  site of synthesis, 295
Growth hormone, human,
  activity, 49
    compared with human placental lactogen, 23
  amino acid sequences, 7, 8, 9, 18, 19, 20, 380
    compared with human placental lactogen, 48
  antigenic sites, difference from human placental lactogen, 15
  antisera, use in prolactin bioassay, 146-147
  biological activities, 11

Growth hormone, human—*continued*
  chemistry of, 7-10
  circular dichroism spectra, 17
  competition for binding with prolactin, 127
  destruction, 271, 285
  disulphide bridges, 22
  effect on blood urea nitrogen, 393
  effect on bone marrow, 386
  effect on calcium metabolism, 393
  effect on cell growth, 23, 191
  effect on liver, 194
  effect on mammary growth, 187, 188
  effect on renin production, 386
  immunofluorescence studies of human pituitary, 292, 294
  induction of casein synthesis, 171
  lactogenic activity, 10, 48-19, 105, 140, 153-155
  levels, 140
    effect of steroids, 237
    following hypophysectomy, 201
    in lactation, 153, 155
  oestrogen-potentiated release, 105
  physicochemical properties, 7
  plasma assay, 153-155
  prolactin activity, 1, 106, 140, 146, 153, 161
  radioimmunoassay, 46, 153
  release,
    from pituitary tumour, 212
    from tissue culture, 209
  secretion,
    following hypoglycaemia, 176
    in pregnancy, 294
  structure, 7-10, 148
    compared with growth hormone and human placental lactogen, 15, 18-19, 20-21
    tertiary, 25
  synthesis, 10, 90
Growth hormone, monkey,
  electrophoresis, 266, 281
  identification, 259
  immunoadsorption, 95
  secretion, 257-268, 284
    amounts, 260, 262, 264
    during lactation, 263, 264
    in infants, 265
    in organ culture, 261
    rates of, 262
  synthesis, 90-94
  prolactin activity, 257

INDEX OF SUBJECTS

Growth hormone, primate, 83
Growth hormone, porcine, 380, 387
Growth hormone, ovine,
    activity compared with prolactin, 24
Guanethidine, effect on prolactin production, 65
Guanidine, human placental lactogen synthesis and, 88
Gynaecomastia, prolactin levels in, 179–181, 187

Haloperidol, 349
Histones, phosphorylation of, 115
Human chorionic somatomammotropin (HCS); see *Human placental lactogen (HPL)*
Human growth hormone,
    see *Growth Hormone, human*
Human placental lactogen (HPL),
    activity, 14, 131, 361–389
        compared with growth hormone, 23
        tertiary structure and, 42
    amino acid sequence, 15, 16, 18, 19, 20, 33, 34, 35, 37, 40, 45
        compared with growth hormone, 40, 48
    amino acid substitutions in, 41
    antigenic sites compared with growth hormone, 15
    as index of placental function, 234, 236
    binding to cell membrane, 127, 128, 129
    bioassay, 14–15, 181–182
    biological role in pregnancy, 238
    chemistry, 12–17
    circular dichroism spectra, 17
    cleavage,
        by Edman degradation, 31, 37, 46
        peptides from, 31
        with chymotrypsin, 28, 37, 40
        with cyanogen bromide, 29, 36, 40, 45
        with pepsin, 29, 37, 40
        with trypsin, 40
    competition with prolactin, 127
    dimers, 49–50
    disulphide bridges, 22
    eclampsia and, 318
        aldosterone excretion, 375
    effects,
        blood urea nitrogen, 374, 376
        calcium balance, 374
        carbohydrate tolerance, 375

Human placental lactogen—*continued*
    effects—*continued*
        growth, 24, 49, 383
        phosphorus levels, 375
        sodium balance, 375
        with growth hormone, 374
    function, 4, 27–51, 400, 402
    immunoassay, 223–239
    immunological activity, 46
    induction of proteins by, 171
    immunofluorescence studies, 292, 294
    in pregnancy, 223–239
    levels,
        effect of eating, 226
        fluctuations in, 239
        following abortion, 230
        hypoglycaemia and, 227, 228
        in abnormal pregnancy, 230–234
        in diabetic pregnancy, 230
        in hypoglycaemia, 238
        in normal pregnancy, 229–230
        in placental tumours, 230
        in Rh factor sensitization, 230
        in sleep, 239
        ovarian steroids and, 237
        regulation, 226–229
        relation to placental–foetal mass, 228
    molecular weight, 14, 24
    nomenclature, 5–6, 25–26, 400–402, 404
    potency, 383
    purification, 27
    radioimmunoassay, 47, 223–225
    reduction and alkylation, 42, 46
    relation to prolactin, 18–21, 43
    release, 216
    structure, 27–51, 400
        activity and, 45
        compared with growth hormone and prolactin, 15, 18–19, 20–21
    synthesis, 83–89
        correlation with placental weight, 88
        effect of fasting, 89
        effect of glucose, 103
        effect of ovarian steroids, 237
        in vitro, 84
        rate, 86, 100
        regulation, 88, 100
        sites of, 86, 296
    tertiary structure, activity and, 42
    trypsin cleavage, 28
Human prolactin, see *under Prolactin*
5-Hydroxytryptamine, 62

INDEX OF SUBJECTS

Hyperglycaemia,
  and regulation of human placental lactogen, 228
  effect of prolactin, 379
Hypertension, 88
Hypoglycaemia,
  effect on prolactin secretion, 145
  growth hormone secretion, 176
  human placental lactogen levels, 227, 238
Hypophysectomized rats, mammogenic stimulation, 271
Hypophysectomy,
  effects in pregnancy, 392
  growth hormone levels, 201
  serum prolactin levels following, 107
Hypopituitarism,
  prepubertal, effect of prolactin, 362, 363
  prolactin activity in, 99, 391, 392
Hypothalamic prolactin-inhibiting factor, 58, *see under Prolactin-inhibiting factor*
Hypothalamus,
  catecholamine levels in, 58, 65, 73, 78–79, 80–81, 340
  controlling pituitary hormone release, 215
  lesions,
    effect on prolactin production, 67, 73, 325
  role in growth hormone synthesis, 101, 284
  role in prolactin secretion, 73, 78, 81, 101, 108, 174, 176, 269, 284, 325–347
  tumours of, prolactin secretion, 176–178

Imipramine, 349
  effect on prolactin levels, 178
Immunofluorescence, 277
  absorption experiments, 289
  cytological localization of staining, 290–291
  inhibition, 289
  nature of serological factor, 289
  of prolactin cells, 246–248
  pituitary studies in pregnancy, 287–297
  selection of antisera for staining, 288
  serological specificity, 287–288
  state of antigen in tissue sections, 279, 290
Insulin,
  effect on prolactin secretion, 205
  stimulation of mammary gland stem cells, 111, 187–188
Insulin tolerance tests, 144
Integument, prolactin action on, 308

Krebs-Ringer buffers, pituitary gland incubation with, 54–55

Lactalbumin,
  effect of oestrogens on activity, 184
  induction, 170
  inhibition of synthesis, 128
Lactation, *see also Suckling*
  DNA synthesis during, 192
  effect of catecholamines, 342
  effect of prolactin, 130, 284
  growth hormone levels in, 155, 263, 283, 284
    in monkey, 264
  initiation, 133
  interaction of steroid and prolactin, 132
  oral contraceptives and, 354
  plasma hormones in, 153
  pregnancy and, 132
  prolactin activity during, 141, 144, 173, 199, 355
    in monkey, 263, 264
Lactogenic activity,
  bioassay, 152
  in acromegaly, 155
  in galactorrhoea, 155
  in pregnant and lactating goats, 155
Lactogenic response, specificity of, 152
Lactose synthetase system, effect of prolactin, 116, 169
Largactil, *see Chlorpromazine*
Levopromazine, 349
Liver,
  effect of growth hormone on, 194
  effect of prolactin on, 194
  sulphation factor production, 194
Luteinizing hormone,
  assay, 208
  release, 210, 216
  secretion, 334
Luteotropic activity, prolactin and, 391–396
Luteotropic hormone, *see Human placental lactogen*
Lysosomes, in pituitary tissue culture, 279

Magnesium ion, effects on prolactin synthesis, 55–56
Mammary gland organ culture, 111
  bioassay method, 152
    of human growth hormone, 153
  co-culture with foetal and placental goat tissue, 160
  detecting lactogenic activity, 151–167

INDEX OF SUBJECTS

Mammatrophs (prolactin cells), structure, 61, 249–252
Menopause, prolactin levels, 110
Menstrual cycle,
  effect of tranquillizers on, 357, 358
  serum prolactin in, 99
Methionine residues in human placental lactogen, 40, 42
Methyldopa, effect on prolactin secretion, 331
Methyl tyrosine, effect on prolactin secretion, 331, 344
Mid-pregnant mouse breast tissue, as bioassay, 137
Milk proteins,
  effect of progesterone on production, 184
  induction, 114–116
    measurement of prolactin activity by, 169–196
  in galactorrhoea, 352
  synthesis, 112, 124, 128
    action of prolactin, 121
Milk secretion, role of prolactin, 312
Monkey placental lactogen, 50
  amino acid composition, 48
Mouse mammary gland bioassay, 169–173

Nitrogen in urine,
  effect of human placental lactogen, 376
  effect of prolactin, 364 et seq.
Nitro-prolactin, 12, 25
Nomenclature, 5–6, 25–26, 400–402, 404
Noradrenaline,
  effect on pituitary structure, 61
  effect on prolactin synthesis, 62, 63, 64, 65, 80, 81

Oestradiol, effect on prolactin synthesis, 65
Oestrogens,
  causing mammary hyperplasia, 186
  effect on prolactin activity, 59, 184, 188, 335, 339
  metabolism, 104
Oral contraceptives, lactation and, 354
Osmoregulation, action of prolactin in, 304–306, 311, 319
Osteoporosis, 385
Ovariectomy, effect on prolactin synthesis, 58
Ovine prolactin, see under Prolactin
Oxytocin,
  effect on prolactin release, 220, 329
  estimation, 4

Pepsin, cleavage of human placental lactogen, 29, 37, 40
Peptides,
  from cleavage of human placental lactogen, 31, 40
  tryptophan-containing, 37
Perphenazine, 349
  effect on prolactin synthesis, 65, 66, 68, 328
pH, role in prolactin activity, 388
Phenothiazine drugs, effect on prolactin secretion, 143, 178, 180, 186, 200, 217
Phosphorus,
  effect of human placental lactogen, 375
  effect of prolactin, 366
  excretion,
    effect of growth hormone, 378
Pigeon local crop assay, 11, 271
  criticism of, 274
Pituitary,
  cells,
    counting, 245
    differential staining, 241–246
    electron microscopy, 248–252
    Herlant's tetrachrome staining, 242, 244
    immunofluorescence, 246–248
    methasol blue-PAS-orange G staining, 243
    nomenclature, 241
    prolactin-secreting, 61, 280, 286
    staining, 277 et seq.
  differentiation in foetus, 216
  effect of noradrenaline on fine structure, 61
  hormones,
    degradation, 109
    control of release, 215
  immunofluorescence studies,
    absorption experiments, 289
    inhibition, 289
    in pregnancy, 287–297
    nature of serological factor, 289
    selection of antisera, 288
    serological specificity, 287–288
    state of antigen, 290
    with antisera to HGH and HPL, 292–294
  incubation,
    in Krebs-Ringer buffers, 54–55
    with puromycin, 54
    ionic environment affecting prolactin production, 54

Pituitary—*continued*
  morphology, 61, 270
  prolactin content, 97, 109
  protein storage and synthesis, 266, 286
  rhesus monkey,
    growth hormone and prolactin in, 258 *et seq.*
  tissue culture experiments, 284
    lysosomes in, 279
    prolactin release from, 207–217
  transplantation,
    prolactin levels following, 339, 341, 342
  stalk section,
    effect on prolactin synthesis, 81, 174–176, 185
  tissue culture, 269–286
  tumours,
    effect of ergot derivatives on, 70
    growth-hormone secreting, 70
    hormone release from, 212
    MtTW5, 57
    prolactin activity in, 57, 71, 92, 147, 161, 175–178, 202
    StW5 and 7315a, 70
Placenta,
  abnormal function, 88
  function, 238
    human placental lactogen as index of, 236
  human placental lactogen in, 103
  incubation of tissue, 84
  lactogenic activity in, 12, *see also under Human placental lactogen*
  mammary-stimulating activity, 48
  release of proteins, 85
  tumours of,
    human placental lactogen levels in, 230
  weight, relation to human placental lactogen levels, 88, 228
Placental-foetal mass, relation to human placental lactogen levels, 228
Placental lactogens, in evolution, 48
Placental lactogen, human *see Human placental lactogen*
Placental lactogen, goat, 48, 163
Placental lactogen, monkey (MPL), *see Monkey placental lactogen*
Placental lactogen, ovine, 1, 48, 50
Placental lactogen, rat,
  effect on corpus luteum, 395
Polyoestradiol phosphate, effect on prolactin production, 58

Potassium,
  effect on prolactin synthesis, 55, 56
  excretion,
    effect of growth hormone, 378
    effect of prolactin, 367, 368
Pregnancy,
  action of prolactin in, 99, 104, 132, 221
  effect of hypophysectomy, 392
  human placental lactogen in, 223–239
  osmoregulation in,
    role of prolactin, 306
  pituitary immunofluorescence studies in, 287–297
  prostaglandin levels in, 226
Progesterone,
  effect on prolactin activity, 184, 328
  galactorrhoea and, 352
  inhibiting lactalbumin synthesis, 128
Prolactin, 3
  action of, 77, 131, 133, 299–324
    aldosterone secretion, 379
    anti-gonadal, 131, 319
    blood urea and urine nitrogen, 364 *et seq.*, 379
    bone marrow hypoplasia, 369
    broodiness in chickens, 319
    calcium balance, 364, 366, 370, 372
    carbohydrate metabolism, 368, 372, 373, 378
    casein synthesis, 393
    cell growth, 191
    common factor, 299, 311–312
    growth, 194, 306–308, 321, 322
    in amphibians, 302, 304, 308
    in birds, 302, 306, 308
    in fish, 304, 320
    in hypophysectomized diabetic patients, 368
    in hypopituitarism, 391, 392
    in pituitary tumours, 175
    in reptiles, 302, 306, 308
    in teleosts, 302, 304, 308, 312, 317, 320
    integument, 308
    inulin clearance, 378
    lactogenic, 12, 21, 49, 111 *et seq.*, 151 *et seq.*, 306
    lactose synthetase system, 116
    maternal behaviour, 319–320
    mammary growth, 187–193
    mechanism, 111, 118
    milk secretion, 312
    osmoregulation, 304–306, 311, 319

INDEX OF SUBJECTS

Prolactin—*continued*
  action of—*continued*
    phosphorus excretion and balance, 366–367
    polysomes, 122
    potassium excretion, 367, 368
    pregnancy, 306
    primary and secondary, 317
    puberty, 334
    renin, 386
    reproduction, 302–304
    ribosomes, 128
    sexual activity, 319
    sodium metabolism, 306, 317, 367, 368
    with steroids, 132
    amino acid, sequence, 215
    antibodies to, 277–278, 387
  assay, 162
  binding to cell membrane, 112–113, 127
  breast as target organ, 194, 219
  cells secreting, 61, 241 *et seq.*, 280, 286
    distinctive features, 249
    identification, 244, 249
    counting, 245
    differential staining, 241–246
    functional study, 250–252
    release of hormone, 250
    reliability and specificity, 243
    segregation, 250
  cell-membrane binding, 124, 128
  comparative endocrinology, 299, 300, 301
  comparison of preparations, 371
  competition with growth hormone, 127
  correlation between pituitary and plasma levels, 222
  definition, 321
  effect of one species on another, 321
  effect on cell division, 188
  electrophoresis, 285
  heterogeneity of, 76
  hyperglycaemia and, 145, 379
  immunoassay, 278
  immunofluorescence, 277
  induction of milk proteins by, 114–116
  in evolution, 299 *et seq.*, 318, 323–324
  levels,
    during lactation, 130, 141, 144, 355
    effect of phenothiazines, 178, 180, 217
    following pituitary transplantation, 339, 341, 342
    in foetus, 99, 219
    in galactorrhoea, 108, 141, 161
    in intracranial tumours, 141

Prolactin—*continued*
  levels—*continued*
    in pituitary tumours, 161, 202
    in pregnancy, 221
  luteolytic function, 394–396
  luteotropic activity, 391–394
  measurement, 91, 352
    by milk protein induction, 169–196
  mechanism of action, 111–126
  nitrophenylation of, 25
  pigeon crop gland assay, 12, 271
    criticism, 274, 403
  precursor, 80
  radioimmunoassay, chromatoelectrophoretic and incubation damage, 198
  receptors, 194, 219, 318, 323
  receptor-like activity in liver, 194
  receptor-site binding, 118
  release, 335–336
    cellular aspects, 250
    effect of oxytocin, 329
    regulation, 53–75
    role of adenyl cyclase, 76
  Ribonucleic acid synthesis stimulated by, 114, 119
  secretion, *see under Prolactin, release and Prolactin, synthesis*
  separation from human growth hormone, 94, 361
  species differences, 109
  stimulating rabbit mammary gland, 153
  stimulating ribosomes, 122
  storage, 251, 252
  structure, 47, 311, 323
  synergism with steroids, 308–310
  synthesis, 3, 4
    cellular aspects, 250
    destruction of excess material, 251
    effect of adrenaline, 62, 81
    effect of calcium and magnesium, 55, 56, 78
    effect of catecholamines, 58–64, 73, 78, 331–334, 346
    effect of chlorpromazine, 66, 331
    effect of dopa and dopamine, 60, 65, 73, 79, 81, 329, 331, 332, 334, 342, 347
    effects of psychotropic drugs, 65–68, 143, 178–179, 186, 349–358
    effect of ergot derivatives, 68–72, 74, 80, 320, 335, 394
    effect of guanethidine, 65
    effect of hypothalamic lesions, 67, 73, 325

Prolactin—*continued*
  synthesis—*continued*
    effect of Krebs-Ringer buffers, 55
    effect of methyltyrosine, 331, 344
    effect of noradrenaline, 62, 63, 64, 65, 80, 81
    effect of oestrogens, 59, 335
    effect of ovariectomy and adrenalectomy, 58
    effect of perphenazine, 65, 66, 68
    effect of pituitary stalk section, 81, 186
    effect of pituitary tumours, 57, 71
    effect of polyoestradiol phosphate, 58
    effect of reserpine, 65, 68, 331, 333, 340, 345
    effect of steroids, 339
    effect of stress, 79, 107, 131
    effect of suckling, 328
    effect of thyroid hormones, 335
    effect on potassium, 55, 56
    feedback, 334, 339, 396
    in galactorrhoea, 92
    in pregnancy, 104
    inhibition, 251, 326–329, 334
    ionic environment affecting, 54
    morphology, 241–255
    regulation, 53–75
    role of hypothalamus, 78, 79, 81, 101, 174, 325–347
    species differences, 104, 106, 342
  target tissue for, 129
  tissue culture studies, 284
    lysosomes in, 279
  transcription and, 119–124
Prolactin, birds, inhibition of secretion, 346
Prolactin, bovine, 362
  activity, 377
    compared with growth hormone, 24
    growth-promoting, 23
  structure, 388
    compared with ovine, 23
Prolactin, goat, 194–195, 318
  cross-reactions, 195
  effect of suckling on levels, 156
  levels,
    in multiparous animals, 158
    in primiparous animals, 157
Prolactin, human,
  activity,
    and breast cancer, 195, 356
    in galactorrhoea, 172, 349–359
    in lactation, 173

Prolactin, human—*continued*
  activity—*continued*
    in mammary growth, 187, 188
    on alveolar cell growth, 189
    on casein induction, 169
    on DNA synthesis, 190
    on lactose synthetase, 169
  antigens, 216
  antisera, 145–149
  assay, 137–140, 397
    specificity, 139
    variation in, 172
    with mid-pregnant mouse breast tissue, 137–140
  bulk preparation, 207, 219
  cells secreting, 280
  compared with ovine, 108, 148
  content of pituitary, 97, 109
  deamidation, 282–283
  during menstrual cycle, 99
  electrophoresis, 283
    compared with monkey, 282
  electrophoretic mobility characteristics, 266
  half-life, 177
  identification of molecule, 173
  in hypopituitarism, 99
  in plasma, correlation of radioimmunological and biological assay, 203
  isolation and radioiodination of, 209, 212, 257
  levels of, 140, 280
    at menopause, 110
    during pregnancy, 99
    effect of chlorpromazine, 143, 187
    effect of dopamine, 178
    effect of drugs on, 143, 349
    effect of hypertonic saline, 220
    effect of phenothiazines, 143, 186, 200, 349
    effect of pituitary stalk section, 185
    effect of reserpine, 178, 180
    following hypophysectomy, 107
    in breast cancer, 110, 185
    in diabetic retinopathy, 186
    in galactorrhoea, 109, 185, 201–202, 352–353
    in gynaecomastia, 179–181
    in lactation, 199
    in neonate, 219
    in subjects on glucose, 203
  measurement, 358
  molecular weight, 174

Prolactin, human—*continued*
  neutralization studies with antisera, 145–149
  plasma measurements, 197–206, 358
  purification, 272–273
  radioimmunoassay, 96–100, 149, 197–206, 208
    correlation with biological assay, 203
    specificity, 197
  reference preparations, 218, 396–400, 403
  relation to human placental lactogen, 43
  release,
    concentration, 210
    effect of glucose, 105
    effect of oxytocin, 220
    from pituitary tissue cultures, 207–217
    from pituitary tumour, 212
  secretion *see under release and synthesis*
  serum content, 106
  serum measurements, 98, 99
  sex differences in content, 110
  synthesis, 90
    effect of colchicine and vinblastine, 279–280
    effect of dopamine, 107
    effect of pituitary stalk section, 174–176
    effect of stress, 218
    effect of tranquillizing drugs, 143, 175
    evidence of, 271–272
    foetal, 281
    in acromegaly, 92, 98, 145, 147, 258
    in Forbes-Albright syndrome, 176
    in galactorrhoea, 98, 147
    in hypothalamic tumours, 176–178
    in lactation, 148
    in pituitary tumours, 92, 147, 176–178
    mechanism, 279
    morphology, 270
    rate, 163
    role of hypothalamus, 108, 176, 269
    studied by *in vitro* bioassay, 137–150
  tissue culture experiments, 269–286
    compared with blood activity, 274
    long-term, 270–275
Prolactin, monkey, 105
  content of pituitary, 97
  deamidated, 262
  electrophoresis, 282, 283
  electrophoretic mobility characteristics, 265–266
  half-life, 109
  immunoadsorption, 95
  radioimmunoassay, 96–100

Prolactin, monkey—*continued*
  secretion, 257–268, 284
    amounts, 262, 264
    during lactation, 260, 264
    in infants, 263, 265
    in organ culture, 261
  serum measurement, 98, 99
Prolactin, non-primate, biological effects, 361–389
Prolactin, ovine, 42, 92, 320
  activity, 49
    in prepubertal hypopituitarism, 362, 363
    compared with growth hormone, 24
    on blood urea, 308
    on growth, 23, 323, 383
  amino acid sequences, 12, 13, 18, 19, 20
  antibodies to, 247, 248, 277–278
  assay, 399
  circular dichroism spectra, 17
  compared with human, 108, 148
  deamidation, 283
  effect of oestrogens on bioassay, 184
  immunological activity, 25, 397
  insulin-like activity, 378
  levels,
    in foetus and neonate, 220
    in pregnancy, 221
  luteolytic effect, 394
  neutralization with antisera, 147
  physicochemical properties, 10–12
  potency, 383
    comparisons, 383, 384
  reaction to anti-HPL antiserum, 236
  structure, 380, 389
    compared with bovine, 23
    compared with growth hormone and human placental lactogen, 18–19, 20–21
  sulphation factor assay, 194
Prolactin, porcine, disulphide groups, 23
Prolactin, primate, 83
  isolation of, 163
  purification, 94–96
  synthesis, 90
    in vitro, 257–268
Prolactin, rabbit, levels in pregnancy, 221
Prolactin, rat,
  effect of tranquillizers on, 356
  effect on cell division, 188
  levels, 221–222
  regulation of, 53 *et seq.*, 325 *et seq.*
  release, 333, 346

Prolactin, rat—*continued*
  turnover, 344
Prolactin cells, structure, 61, 249–252
Prolactin-inhibiting factor, 58, 73, 78–79, 326–329, 332, 334, 335, 338, 339, 345
Prolactin-releasing factor, 81, 343, 346
Prolactin-stimulating factor, 345
Promazine, 349
Prostaglandins,
  and human placental lactogen levels, 226, 227
Prostatic carcinoma, 317
Protein synthesis,
  in pituitary, 266
  site of, 121
Protein phosphokinase, 132
  induction of, 116, 129–130, 393–394
Psychotropic drugs, galactorrhoea due to, prolactin activity, 142, 349–359
Puberty, onset, 334
Putrescine, 62

Rat tibia test, 11, 23–24, 272, 403
Renin, effect of HGH and prolactin, 386
Reproduction, action of prolactin, 302–304
Reptiles, prolactin action in, 302, 306, 308
Reserpine,
  effect on pituitary hormones, 356
  effect on prolactin levels, 65, 68, 178, 180, 328, 331, 333, 340, 345
Rhesus monkey, prolactin and growth hormone secretion, 257–268, 281
Rh sensitization, human placental lactogen levels, 230
Ribonucleic acid, prolactin stimulation, 114, 117, 119, 123, 124

Saline injection,
  effect on prolactin levels, 220
Serum glutamic-oxaloacetic transaminase, 318
Sexual activity, prolactin and, 319
Sleep, human placental lactogen levels in, 239
Sodium, balance, effect of human placental lactogen, 375
  excretion,
    effect of growth hormone, 378
    effect of prolactin, 306, 317, 367, 368
Species specificity, 4
Spermidine, 62
Spironolactone, 357

Spleno-hepatomegalia, 70
Steroids,
  effect on prolactin activity, 184–185, 328
  prolactin synergism with, 308–310
Stress, effect on prolactin synthesis, 79, 107, 131, 218
Suckling, 4, *see also Lactation*
  effect on growth hormone, 284
  effect on prolactin levels, 141, 199, 328
Sulphation factor, 194, 385

Teleosts, prolactin action in, 302, 304, 308, 312, 317, 320
Thiopropazate, 349
Thioridazine, 349
TSH,
  assay, 208
  release, 210, 216
Thyrotropin-releasing factor, 330
Thyroxine, prolactin release and, 335
Toxaemia of pregnancy, human placental lactogen levels, 230
Tranquillizers,
  breast cancer and, 356
  depressing milk yields, 355
  effect on menstrual cycle, 357, 358
  effect on prolactin secretion, 142, 143
Trifluoperazine, 349
Trypsin,
  cleavage of human placental lactogen, 28, 40, 88
  peptides from, 31
Tryptophan residue,
  in human growth hormone, 10
  in ovine prolactin, 10
MtTW5 Tumour, 57
7315a Tumour, 57
Tyrosine residues, in prolactin, 12

Urea levels in blood,
  effect of prolactin, 308, 364 *et seq.*, 379
  effects of human placental lactogen, 374, 376

Vaginal mucification, role of prolactin, 308
Vasopressin, 62
Vinblastine, effect on prolactin secretion, 279–280

Water and electrolyte metabolism, role of prolactin, 304–306, 311, 319